MEDICAL RADIOLOGY

Diagnostic Imaging and Radiation Oncology

Radiation Exposure and Occupational Risks

Contributors

G. Keller · J. F. Meissner · O. Messerschmidt · H. Mönig
H. Muth · C. Streffer · K.-R. Trott

Edited by

Eberhard Scherer, Christian Streffer,
and Klaus-Rüdiger Trott

Foreword by

Luther W. Brady, Martin W. Donner, Hans-Peter Heilmann,
and Friedrich Heuck

With 32 Figures

Springer-Verlag Berlin Heidelberg New York
London Paris Tokyo Hong Kong

Professor (em.) Dr. EBERHARD SCHERER
Universitätsklinikum der Gesamthochschule
Radiologisches Zentrum, Strahlenklinik und Poliklinik
Hufelandstraße 55
4300 Essen 1, FRG

Professor Dr. CHRISTIAN STREFFER
Universitätsklinikum der Gesamthochschule
Institut für Medizinische Strahlenphysik und Strahlenbiologie
Hufelandstraße 55
4300 Essen 1, FRG

Professor Dr. KLAUS-RÜDIGER TROTT
Department of Radiation Biology
The Medical College of St. Bartholomew's Hospital
Charterhouse Square
London EC1M 6BQ, UK

MEDICAL RADIOLOGY · Diagnostic Imaging and Radiation Oncology

Continuation of
Handbuch der medizinischen Radiologie
Encyclopedia of Medical Radiology

Library of Congress Cataloging-in-Publication Data. Radiation exposure and occupational risks / contributors, G. Keller
... [et al.]; edited by Eberhard Scherer, Christian Streffer, and Klaus-Rüdiger Trott; foreword by Luther W. Brady ... [et
al.]. p. cm. – (Medical radiology) Includes bibliographical references.

ISBN-13: 978-3-642-83804-0   e-ISBN-13: 978-3-642-83802-6
DOI: 10.1007/978-3-642-83802-6

1. Radiation - Health aspects. I. Keller, G. (Gert) II. Scherer, Eberhard. III. Streffer, Christian, 1934- . IV. Trott,
K.-R. (Klaus-Rüdiger). V. Series. [DNLM: 1. Occupational Diseases. 2. Radiation Injuries. 3. Risk.
WA 470 R1285] RA1231.R2R272 1989 616.9'897-dc20 DNLM/DLC

© Springer-Verlag Berlin Heidelberg 1990
Softcover reprint of the hardcover 1st edition 1990

2113/3130-543210 – Printed on acid-free paper

List of Contributors

Privatdozent Dr. Gert Keller
Institut für Biophysik der
Universität des Saarlandes
6650 Homburg/Saar, FRG

Professor Dr. Johannes F. Meissner
Mühloh 1
2062 Borstel, FRG

Professor Dr. Otfried Messerschmidt
Mortonstraße 13
8000 München 45, FRG

Professor Dr. Hans Mönig
Institut für Biophysik und Strahlenbiologie
der Universität Freiburg
Albertstraße 23
7800 Freiburg i. Brsg., FRG

Professor (em.) Dr. Hermann Muth
Am Gedünner 5
6650 Homburg/Saar, FRG

Professor Dr. Christian Streffer
Universitätsklinikum
der Gesamthochschule
Institut für Medizinische Strahlenphysik
und Strahlenbiologie
Hufelandstraße 55
4300 Essen 1, FRG

Professor Dr. Klaus-Rüdiger Trott
Department of Radiation Biology
The Medical College of St. Bartholomew's
Hospital
Charterhouse Square
London EC1M 6BQ, UK

Foreword

The aim of radiation protection standards is to make the radiation workplace as safe as is humanly possible. The gradual evolution over the last 20 years has been towards a more precise definition of the limits for occupational exposure. These have been created not only in terms of short-term effects but also more importantly in terms of long-term risks involving such problems as the potential for carcinogenesis and genetic change.

In the United States the National Committee for Radiation Protection has recommended that 5 rems (50 mSv) should remain as the maximum permissible dose equivalent for total body exposure. This would represent the sum of internal and external exposure and should be regarded as the upper limit allowed.

The community of radiation users is required to conduct its operations in such a manner that the absolute value of the individual's dose equivalent in rems does not exceed his age in years. There should be additional limits for tissues and organs based on short-term effects. Therefore, individual organs are limited to dose equivalents low enough to ensure that the dose threshold values are not exceeded.

From these basic considerations has emerged the concept of "as low as reasonably achievable" (ALARA). Dose limits in radiation protection are all subject to the concept of ALARA, although it is acknowledged that standard-setting committees would basically prefer recommendations that allow zero exposure of personnel. However, this is not only not reasonable but impossible. Exposure from natural sources is unavoidable, and some exposure from man-made devices is also unavoidable if humans are to realize the major benefits that can derive from the use of radiation and radioactive materials. Therefore, radiation exposure should be continually monitored and controlled with no unnecessary exposure allowed and with the best radiation protection guidelines constantly in force to keep it as low as reasonably achievable. Equipment and facilities should be designed so that exposure of personnel and the public is kept down to a minimum, not to a standard. No exposure whatsoever should be permitted without considering the benefits that may be derived from that exposure and the relative risks of alternative approaches.

SCHERER et al. present data natural radiation exposure, the occupational risks in handling radioactive substances, and the potential for occupationally induced radiation carcinogenesis. Within the volume are covered the medical aspects of radiation accidents and the potential benefits that might accrue as a consequence of radiation protection in mammals and in man.

These general background data are of major importance in recognising the entire problem with regards to radiation exposure and its potential for short-term and late effects as well as potential mechanisms by which radiation exposure can be controlled and limited to the greatest degree possible.

L. W. BRADY   H.-P. HEILMANN   F. HEUCK   M. W. DONNER
Philadelphia   Hamburg   Stuttgart   Baltimore

Preface

The discussion of radiation risks has become a very emotional subject during the past few decades, not only amongst the general public but also within the scientific community. In these discussions it is frequently evident that the participants are inadequately informed and that prejudices prevail. This book gives solid information in the highly important field of radiation exposure in man. Besides natural radiation exposure, special emphasis is placed on occupational radiation exposure and the related risks. Experts have written extensive reviews on the various fields. Some of the articles are based on a German edition published in 1985; these articles have been updated and other new articles have been added in order to supplement the volume.

Despite the increasing use of radionuclides and of ionizing radiation in science, medicine, and technology, radiation exposure in man still comes predominantly from natural sources. New data from measurements made in recent years give much better information about exposure from natural sources, especially from radon. These data are reported in the first chapter. Occupational exposure from working with radionuclides and ionizing radiation and knowledge about the related risks are reviewed in the following two chapters. Studies on possible carcinogenic effects on the workforce in nuclear installations are highlighted. Radiation accidents, the effects of high radiation doses, and medical treatment of the resulting damage are reviewed in a further chapter. For about four decades the effects of drugs which can reduce the radiation damage when administered before exposure have been studied, both with cells in vitro and in animals. During recent years initial results have been obtained from tumor patients who have been treated with ionizing radiation. These data are reviewed in the last chapter.

This volume provides extensive reviews on these topics, and includes many literature references. It will introduce physicians, scientists, and engineers to the field of radiation exposure and occupational risks, and will serve as a reference book for all specialists working in the field of radiology and radioprotection in the broad sense.

EBERHARD SCHERER
CHRISTIAN STREFFER
KLAUS-RÜDIGER TROTT

Contents

1 Natural Radiation Exposure

Gert Keller and Hermann Muth

CONTENTS

1.1 Introduction

Our knowledge of the different components of natural radiation exposure of man forms a vital scientific base for the assessment of the effects of additional non-natural radiation exposure, to which people are increasingly subject within the context of modern technical and scientific development. In general, every non-natural radiation exposure that adds to the natural radiation exposure is to be considered negative and undesirable as it entails an additional risk for man. Here the basic rule of radiation protection applies: "Keep the dose as low as you can."

Information about radiation doses from natural sources, and especially their natural range, is also considered an indispensable scientific standard when defining the dose limits for human protection (occupationally exposed people, segments of the population, the general population) against additional radiation exposure.

The more detailed the investigations and knowledge of natural radiation exposure became, and the more complex and varied the scientific and practical problems of protection grew, the greater was the need for a complex study of total natural radiation exposure. Distinction had to be made between "unmodified natural radiation exposure" and „modified natural radiation exposure" or "technologically enhanced natural radiation exposure." In the following study the individual components of these two groups of natural radiation exposure are considered and the most recent data, based on the international literature, summarized and evaluated.

1.2 Unmodified Exposure to Natural Radiation

Since the origin of life on earth all beings on our planet have been exposed to ionizing radiation. This unmodified radiation exposure from natural sources is highly important, as it forms the main part of the entire collective dose of the world population. It is characteristic for this radiation that all living beings are exposed to it and, although it varies regionally, the exposure has been relatively constant over long periods of time. The unmodified exposure to natural radiation is divided into external and internal exposure.

1.2.1 External Radiation Exposure

The external radiation exposure is due to cosmic rays from extraterrestrial sources and from terrestrial radiation emitted by radionuclides which are naturally present in the earth's crust, the water, and the air.

1.2.1.1 Cosmic Rays

The highly energetic radiation from the universe, hitting the earth, is known as primary cosmic radiation. If, on its way to the earth, this radiation interacts with atoms in the earth's atmosphere (spallation reactions), secondary particles and quantum radiation is formed, which is known as secondary cosmic radiation.

Primary Cosmic Radiation. The primary cosmic radiation consists mainly of highly energetic protons from interstellar space. About 10% of this radiation is accounted for by ^{4}He ions and in addition there are smaller amounts of heavier particles together with electrons, protons, and neutrinos. The broad energy spectrum of primary cosmic radiation ranges from 1 to 10^{14} MeV and has an energy peak at about 300 MeV. According to KOBZEV et al. (1975), protons of primary cosmic radiation with energies over 100 MeV are mainly of galactic origin. Primary cosmic radiation makes a direct contribution to the natural radiation exposure of man only during flights at high altitudes or space flights (see Sect. 1.3.1).

During periods of intense solar flares, the emission of protons and α-particles from the sun increases. However, because of their relatively low energies these sources do not contribute significantly to the exposure to cosmic radiation at the earth's crust.

Secondary Cosmic Radiation. During the complicated processes of energy dissipation to which the primary cosmic radiation is subjected when penetrating the earth's atmosphere, photons, particles, and radionuclides are produced, too, which can reach the surface of the earth (neutrons, protons, pions, kaons, and cosmogenic nuclides ^{3}H, ^{7}Be, ^{10}Be, ^{22}Na, and ^{24}Na). In view of the different biological effect of neutrons ("indirectly ionizing radiation") in comparison to the directly ionizing components of secondary cosmic radiation, as well as for dosimetric reasons, these two components are determined separately. Only in this way does an assessment of the total dose equivalent from cosmic radiation become possible.

According to MUTH (1974) a dose rate of $0.015 \ \mu Gy \cdot h^{-1}$ is absorbed per $ion \cdot cm^{-3} \cdot s^{-1}$, assuming a mean energy of 33.7 eV for the formation of ion pairs in air. 2.14 ion pairs $cm^{-3} \ s^{-1}$ are considered to be formed at sea level by the direct ionizing components of cosmic radiation. Without considering the shielding by buildings, this represents an effective annual dose equivalent in air of 0.28 mSv (according to UNSCEAR 1982). This component decreases with lower geographic latitudes and increases with the altitude above sea level (doubling every 1.5 km for the first few kilometers).

Initially there is about the same amount of fast protons and neutrons in the cascade of nucleons (SCHAEFER 1974); however, the mixture becomes ever more deficient in protons because of the increasing significance of energy loss by ionization in comparison with that caused by nuclear processes with decreasing proton energy. Thus at sea level practically only the neutron component is important. According to UNSCEAR (1982) a neutron flux density at sea level of 0.008 $cm^{-2} \ s^{-1}$ is presumed and a conversion factor for the neutron flux density to dose rate of $5 \cdot 10^{-8} \ Gy \cdot h^{-1} \cdot cm^2 \cdot s$ is used. If, in addition, a quality factor of 6 for neutrons of cosmic radiation at sea level is assumed, an annual effective dose equivalent of 0.021 mSv will result. This dose equivalent, determined by the neutron component of cosmic radiation, increases rapidly with altitude and reaches a maximum at a height of between 10 and 20 km.

1.2.1.2 Terrestrial Radiation

The terrestrial radiation is caused by the natural radionuclides which are contained in soils and rocks as well as in the hydrosphere and atmosphere. These natural radionuclides are either cosmogenic or primordial. The cosmogenic nuclides do not contribute significantly to the dose of terrestrial radiation at the surface of the earth.

Table 1.1. Average activity concentrations of potassium 40, uranium 238, and thorium 238 in soil and absorbed dose rate in air 1 m above the ground surface (UNSCEAR 1982)

Radionuclide or decay series	Dose rate per unit activity concentration in soil (10^{-10} $Gy \cdot h^{-1}$ per $Bq \cdot kg^{-1}$)	Average concentration in soil[a] ($Bq \cdot kg^{-1}$)	Absorbed dose rate in air[a] (10^{-8} $Gy \cdot h^{-1}$)
^{40}K	0.43	370 (100–700)	1.6 (0.4–3.0)
^{238}U	4.27	25 (10–50)	1.1 (0.4–2.1)
^{232}Th	6.62	25 (7–50)	1.7 (0.5–3.3)

[a] The typical range is given within parentheses

Sources of Terrestrial Radiation. Sources of terrestrial radiation are natural radionuclides of the uranium–radium and the thorium series which are found in the uppermost layers of the earth, in water, and in air (see Tables 1.12 and 1.13), as well as some radionuclides which are not part of these series, in particular ^{40}K. Only the γ-ray emitting nuclides are of any significance for the exposure from terrestrial background radiation (the contribution made by the β-rays is less than 1‰ of the total dose). The concentrations of ^{238}U, ^{232}Th, and ^{40}K in soil are of crucial importance for the external radiation exposure of the population. Table 1.1 (according to UNSCEAR 1982) shows the mean concentrations of these radionuclides in soil and the dose rate in air, assuming that all decay products are in radioactive equilibrium with their mother nuclides.

Radiation Exposure Outdoors. In a research project supported by the Ministry of the Interior in the Federal Republic of Germany, about 25000 measurements were made of the dose rates outdoors and about 30000 measurements of those indoors (see Sect. 1.3.2.2) (BUNDESMINISTER des INNERN 1978). Within this extensive study, the γ-ray dose rate outdoors due to terrestrial background radiation, excluding the component of cosmic radiation, was measured and evaluated by several scientific institutions.

An isodose map showing the distribution of the dose rates measured outside houses (1 $\mu R/h \cong 1 \cdot 10^{-8}$ $Gy \cdot h^{-1}$) is given in Fig. 1.1. The mean value of the dose rate outdoors in the Federal Republic of Germany is $6 \cdot 10^{-8}$ $Gy \cdot h^{-1}$, with a range for the different federal states of $4.2 \cdot 10^{-8}$ to $7.9 \cdot 10^{-8}$ $Gy \cdot h^{-1}$.

The global mean external terrestrial dose rate is about $4.5 \cdot 10^{-8}$ $Gy \cdot h^{-1}$, varying between countries from about 3.6 to $9.1 \cdot 10^{-8}$ $Gy \cdot h^{-1}$. According to the results of O'BRIEN (1978), a ratio of 0.7 can be taken for the relationship between mean body dose and dose in air.

Assuming an average time spent outdoors per day of about 5 h (factor = 0.2), the annual effective dose equivalent outdoors would be calculated at $5.5 \cdot 10^{-5}$ Sv ($4.5 \cdot 10^{-8}$ $Gy \cdot h^{-1} \cdot 0.7$ $Sv \cdot Gy^{-1} \cdot$ 8760 $h \cdot y^{-1} \cdot 0.2 = 5.5 \cdot 10^{-5}$ $Sv \cdot y^{-1}$).

The contributions of radon and thoron decay products in the air and of the terrestrial β-rays to external radiation exposure are negligible.

There are regions on earth where, in contrast to the "normal" terrestrial radiation found in most countries, higher concentrations of thorium and uranium exist in the uppermost layers and therefore relatively high dose rates from γ-radiation occur. Such areas with increased terrestrial radiation are a coastal area of the state of Kerala in south-west India (monazite sand), the Atlantic coast of the states of Espirito Santo and Rio de Janeiro in Brazil (monazite sand) and an abnormal geological fold from the coast to the state of Minas Gerais in Brazil (related to vulcanoes).

GOPAL-AYENGAR et al. (1970, 1972) performed a dosimetric study in a 55-km long coastal area in the state of Kerala, India, with a population of about 70000 people. The radiation exposure of 8513 people in 2374 households was measured to calculate the dose for the total population of this coastal area. According to this study, 16000 people receive an annual dose of over 0.005 Sv and of these, 4500 a dose of over 0.01 Sv and 470 a dose of over 0.02 Sv. Similar results have been published for the quoted regions in Brazil.

1.2.2 Internal Radiation Exposure

While in the case of cosmic rays and terrestrial background radiation, total body irradiation generally occurs, with internal radiation often only circumscribed areas of the body, in some instances of microscopic dimensions, are exposed to radiation. The radionuclides leading to internal exposure mainly enter the human body by ingestion with food and water. The naturally radioactive products responsible for the internal component of unmodified exposure to natural radiation are divided into two categories, the radionuclides produced by cosmic rays and the so-called primordial radionuclides.

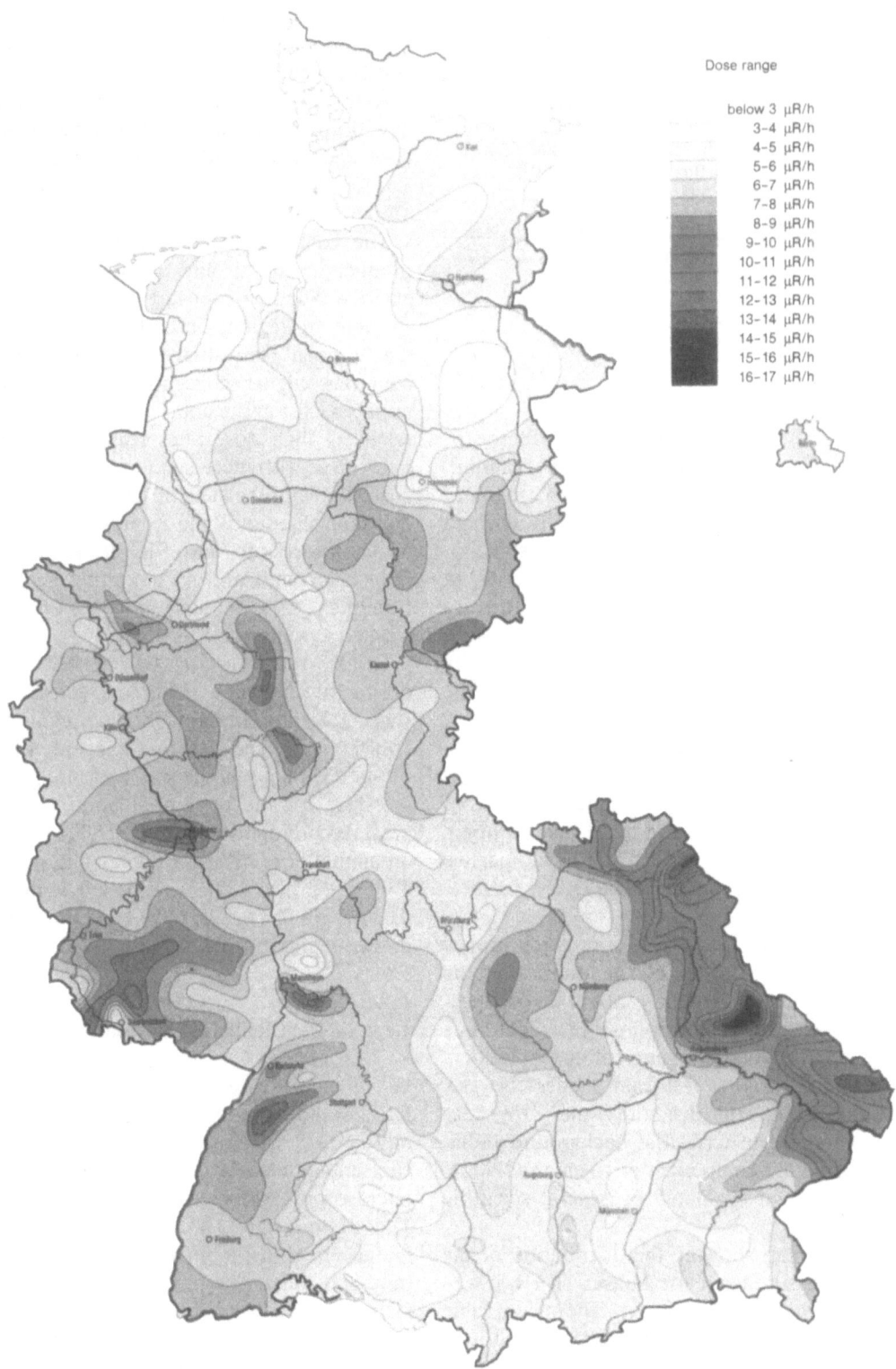

Fig. 1.1. Dose rates from outdoor terrestrial γ-irradiation (μR/h) in the Federal Republic of Germany

1.2.2.1 Cosmogenic Radionuclides

In total about 20 radionuclides are known to be produced by cosmic rays in the earth's atmosphere. Cosmogenic radionuclides contribute very little to natural radiation exposure: ^{14}C, ^{22}Na, ^{7}Be, and ^{3}H are the only components with some significance.

Tritium. A large amount of natural tritium is produced by interaction of cosmogenic neutrons with ^{14}N in the earth's atmosphere. Tritium is also formed directly by primary components of cosmic rays. According to studies by KAUFMAN and LIBBY (1954) performed before the bomb tests, the natural tritium concentration in freshwater was 200-900 $Bq \cdot m^{-3}$ and in seawater, about 100 $Bq \cdot m^{-3}$. The total inventory of ^{3}H in the world was estimated to be $1.3 \cdot 10^{18}$ Bq. Assuming that the tritium concentration in body tissues is about the same as in surface water, MUTH (1974) calculated a dose rate in soft tissues of $6 \cdot 10^{-9}$ to $25 \cdot 10^{-9}$ $Gy \cdot y^{-1}$ with a mean β-energy of 5.69 keV. The average dose rate in all tissues is 10^{-8} $Gy \cdot y^{-1}$.

Beryllium 7. The background concentration of ^{7}Be in air near the ground is, according to KOLB (1974), about 3 $mBq \cdot m^{-3}$ and in rainwater, according to AURAND et al. (1974), about 700 $Bq \cdot m^{-3}$. The main contribution of ^{7}Be to internal exposure is through ingestion of green vegetables, with a resulting body dose of less than 10^{-7} $Gy \cdot y^{-1}$.

Carbon 14. In the nineteenth century the specific activity of natural ^{14}C produced by cosmogenic neutrons in the upper atmosphere was, according to estimates of TELEGADA (1971), about 230 Bq per kg carbon. This results in an inventory of $140 \cdot 10^{15}$ Bq carbon 14 in the atmosphere. In our century the ^{14}C inventory has increased worldwide about 60-fold through the emission of carbondioxide by burning fossil fuels, although the specific activity has decreased.

Assuming a specific activity of natural ^{14}C of 230 Bq per kg carbon in the terrestrial biosphere and a mean concentration of carbon in the human body of 18%, an average dose for the total body of 10^{-5} $Gy \cdot y^{-1}$ can be calculated at a mean β-energy of 50 keV, with fluctuations between $5 \cdot 10^{-6}$ and $2.2 \cdot 10^{-5}$ $Gy \cdot y^{-1}$ for individual tissues or organs (Table 1.2).

Sodium 22. Although the concentration of ^{22}Na, at about $4 \cdot 10^{-7}$ $Bq \cdot m^{-3}$ in air near the ground, seems very low, its contribution to the body dose is, with

Table 1.2. Estimated annual absorbed doses from incorporated cosmogenic radionuclides (UNSCEAR 1982)

Tissue or organ	Annual dose (µGy)			
	^{3}H (β^-)	^{7}Be (γ)	^{14}C (β^-)	^{22}Na (β^+, γ)
Gonads	0.01	5.7	5.0	0.14
Lungs	0.01	-	5.7	0.12
Breast	0.01	-	-	0.13
Red bone marrow	0.01	1.2	24	0.22
Bone lining cells	0.01	-	22	0.27
Thyroid	0.01	-	5.9	0.12
Other tissues	0.01	-	13	-

about $2 \cdot 10^{-7}$ Gy, higher than that from ^{3}H and ^{7}Be. This is due to the special decay conditions and the metabolism of sodium 22. Based on information from UNSCEAR (1982), Table 1.2 shows the annual body doses through internal radiation for the cosmogenic radionuclides ^{3}H, ^{7}Be, ^{14}C, and ^{22}Na.

1.2.2.2 Primordial Radionuclides

In addition to cosmogenic radionuclides, primordial radionuclides also contribute to the unmodified exposure to natural radiation of man. The main sources of this radiation are radionuclides of the natural uranium and thorium series as well as ^{40}K and ^{87}Rb.

Potassium 40. The radionuclide ^{40}K is contained in the isotope mixture of the natural element potassium to 0.0119%. Owing to its great physiological relevance, several studies assessed the distribution of this element in the human organism and its dependency on age (OBERHAUSEN 1963).

The mean concentration for an adult is about 2 g potassium per kg body weight, which results in a mean activity concentration of about 60 $Bq \cdot kg^{-1}$ of ^{40}K in the organism. Information on the distribution of potassium and the resulting dose from γ- and β-radiation is given in Table 1.3.

Rubidium 87. In contrast to ^{40}K, few studies exist on the distribution of ^{87}Rb in the human organism. According to SPIERS (1968) the mean concentration of the element rubidium (the proportion of ^{87}Rb in total rubidium is 27.85%) in the body is 17 ppm and thus 10, 4.5, or 12 µg per g wet weight in bone, ovaries, or testes. Therefore a mean dose to the gonads of about $3 \cdot 10^{-6}$ $Gy \cdot y^{-1}$ can be estimated. Table 1.3 gives the distribution of rubidium in the body (INTERNATIONAL COMMISSION ON RADIOLOGICAL PROTECTION 1975) and the resulting dose from

Table 1.3. Average tissue concentrations in adult males and annual absorbed doses from ^{40}K and ^{87}Rb (UNSCEAR 1982)

Tissue or organ	Potassium			Rubidium		
	Mass concentration of element $(g \cdot kg^{-1})$	Activity concentration of ^{40}K $(Bq \cdot kg^{-1})$	Annual absorbed dose (μGy) β^-, γ	Mass concentration of element $(mg \cdot kg^{-1})$	Activity concentration of ^{87}Rb $(Bq \cdot kg^{-1})$	Annual absorbed dose (μGy) β^-
Gonads (testes)	2.1	64	180	20	18	10.0
Lungs	2.1	64	180	9.2	8.1	4.5
Red bone marrow	4.4	130	270	7.8	7.0	7.0
Bone lining cells			140			14.0
Thyroid	1.1	33	100	6.0	5.3	3.0
Other tissues	2.0	61	170	7.8	7.0	4.0

Table 1.4. Intakes in normal areas of ^{238}U, ^{232}Th, and their decay products (UNSCEAR 1982)

Source	Annual intake (Bq)	
	Inhalation	Ingestion
^{238}U series		
^{238}U	0.01	5
^{234}Th	0.01	5
^{234}Pa	0.01	5
^{234}U	0.01	5
^{230}Th	0.01	–
^{226}Ra	0.01	15
^{210}Pb	4	40
^{210}Po	0.8	40
^{232}Th series		
^{232}Th	0.01	–
^{228}Ra	0.01	15
^{228}Ac	0.01	15
^{228}Th	0.01	15

^{87}Rb through internal radiation for different tissues or organs according to UNSCEAR (1982).

Uranium and Thorium Decay Series. Out of the three natural radioactive decay series with the mother nuclides ^{238}U, ^{235}U, and ^{232}Th, the contribution of the ^{235}U series and their decay products to internal radiation exposure can be neglected. For the sake of clarity the two remaining series, ^{238}U and ^{232}Th (see Tables 1.12 and 1.13), are classified into smaller partial series or subgroups, whereby the activity of the decay products is defined by the activity of the precursor nuclide.

The subdivision was made as follows:
- Uranium (^{238}U $\rightarrow$ ^{234}U)
- Thorium (^{232}Th and ^{230}Th)
- Radium (^{226}Ra, ^{228}Ra $\rightarrow$ ^{224}Ra)
- Radon, thoron, and their short-lived decay products (^{222}Rn, ^{218}Po, ^{214}Pb, ^{214}Bi and ^{220}Rn, ^{216}Po, ^{212}Pb, ^{212}Bi)

- Long-lived decay products of radon (^{210}Pb, ^{210}Po)

The daily uptake of primordial radionuclides of the uranium or thorium series by humans through inhalation or ingestion in areas with "normal natural radioactivity" is shown in Table 1.4 according to UNSCEAR (1982).

Uranium. Assuming that ^{238}U is in radioactive equilibrium with ^{234}Th, ^{234}Pa, and ^{234}U, 1 kg uranium contains 12 MBq for each of the four radionuclides. The uranium concentration in the atmosphere is determined by the stirred-up dust particles of the soil. With a mean concentration of dust of 50 μg $\cdot$ m^{-3} and a mean ^{238}U activity concentration in soil of 25 Bq $\cdot$ kg^{-1}, UNSCEAR (1982) calculated an activity concentration in air near the ground of 1.2 μBq $\cdot$ m^{-3}. This results in an annual ^{238}U intake by inhalation of about 0.01 Bq (see Table 1.4).

About 15 mBq ^{238}U are ingested daily, of which very little is generally contributed by drinking water. Nevertheless, in some countries, like the USSR and Finland, various authors have reported uranium concentrations in water of about 10^5 Bq $\cdot$ m^{-3}. The mean ^{238}U concentration in humans is about 7 mBq $\cdot$ kg^{-1} for soft tissues and 150 mBq $\cdot$ kg^{-1} for ashed bone. The resulting annual dose for soft tissues is 4 $\cdot$ 10^{-7} Gy $\cdot$ y^{-1}, and for bone surface, 3 $\cdot$ 10^{-6} Gy $\cdot$ y^{-1} (see Table 1.5).

Thorium. The mean activity concentration of ^{232}Th in man is about 200 mBq $\cdot$ kg^{-1} for ashed bone, 20 mBq $\cdot$ kg^{-1} for the lung, and 2 mBq $\cdot$ kg^{-1} in other soft tissues. The annual dose is highest for the bone surface, at 2 $\cdot$ 10^{-6} Gy $\cdot$ y^{-1}. In terms of its distribution in the human organism, its occurrence, and its physical half-life in comparison to the human life span, ^{230}Th is similar to ^{232}Th; therefore it results in a similar annual dose for the individual tissues or organs (Tables 1.5 and 1.6).

Table 1.5. Estimated annual absorbed doses resulting from internal irradiation by radionuclides of the ^{238}U decay series (UNSCEAR 1982)

Tissue or organ	Dose (Gy·y^{-1})				
	^{238}U→^{234}U		^{230}Th	^{226}Ra→^{214}Po[a]	
	(α)	(β, γ)	(α)	(α)	(β, γ)
Gonads	$2\cdot10^{-7}$	$3\cdot10^{-8}$	$0.7\cdot10^{-8}$	$1.7\cdot10^{-7}$	$0.4\cdot10^{-8}$
Breast	$2\cdot10^{-7}$	$3\cdot10^{-8}$	$0.7\cdot10^{-8}$	$1.7\cdot10^{-7}$	$0.4\cdot10^{-8}$
Lungs	$2\cdot10^{-7}$	$3\cdot10^{-8}$	$4.7\cdot10^{-7}$	$1.7\cdot10^{-7}$	$0.4\cdot10^{-8}$
Red bone marrow	$5\cdot10^{-7}$	$2\cdot10^{-7}$	$5.6\cdot10^{-7}$	$4.8\cdot10^{-7}$	$0.8\cdot10^{-7}$
Bone lining cells	$4\cdot10^{-6}$	$4\cdot10^{-7}$	$7.4\cdot10^{-6}$	$5.4\cdot10^{-6}$	$2.4\cdot10^{-7}$
Thyroid	$2\cdot10^{-7}$	$3\cdot10^{-8}$	$0.7\cdot10^{-8}$	$1.7\cdot10^{-7}$	$0.4\cdot10^{-8}$
Other tissues	$2\cdot10^{-7}$	$3\cdot10^{-8}$	$0.7\cdot10^{-8}$	$1.7\cdot10^{-7}$	$0.4\cdot10^{-8}$

[a] Not containing the dose from inhalation of ^{222}Rn and its decay products

Table 1.6. Estimated annual absorbed doses resulting from internal irradiation by radionuclides of the ^{232}Th decay series (UNSCEAR 1982)

Tissue or organ	Dose (Gy·y^{-1})		
	^{232}Th	^{228}Ra→^{208}Tl[a]	
	(α)	(α)	(β, γ)
Gonads	$0.3\cdot10^{-8}$	$8.0\cdot10^{-8}$	$1.2\cdot10^{-8}$
Breast	$0.3\cdot10^{-8}$	$8.0\cdot10^{-8}$	$1.2\cdot10^{-8}$
Lungs	$4.0\cdot10^{-7}$	$2.4\cdot10^{-6}$	$4.5\cdot10^{-8}$
Red bone marrow	$1.7\cdot10^{-7}$	$3.5\cdot10^{-7}$	$6.9\cdot10^{-8}$
Bone lining cells	$2.0\cdot10^{-6}$	$4.4\cdot10^{-6}$	$1.9\cdot10^{-7}$
Thyroid	$0.3\cdot10^{-8}$	$8.0\cdot10^{-8}$	$1.2\cdot10^{-8}$
Other tissues	$0.3\cdot10^{-8}$	$8.0\cdot10^{-8}$	$1.2\cdot10^{-8}$

[a] Not containing the dose from inhalation of ^{220}Rn and its decay products

Radium. As for uranium and thorium, the intake of ^{226}Ra or ^{228}Ra by inhalation, at about 0.01 Bq·y^{-1}, is basically caused by resuspension of dust particles from the soil in air near the ground. With a mean activity intake of 15 Bq·y^{-1} by food for ^{226}Ra and for ^{228}Ra (see Table 1.4) the ingestion pathway plays the decisive role, while the contribution of drinking water is generally small. In populated regions with a high thorium or uranium concentration in the soil, such as Kerala (India) or Araxa-Tapira (Brazil), the daily radium intake may be increased by up to a factor of 100 in extreme cases, as compared to the mean in areas with normal natural radioactivity. In the organism radium reacts similarly to calcium. About 70%–90% of the incorporated radium is found in bone, while the remainder is nearly evenly distributed in soft tissues. In areas with "normal radioactivity" the mean ^{226}Ra concentration in human bones is about 170 mBq·kg^{-1}, varying between 70 and 700 mBq·kg^{-1}. In soft tissues the ^{226}Ra concentration is given as 2.7 mBq·kg^{-1} (UNSCEAR 1982). The mean activity concentration of ^{228}Ra is 90 mBq·kg^{-1} in ashed bone and 4 mBq·kg^{-1} in soft tissues.

When calculating the radiation dose from radium and its decay products, an average retention factor of 0.33 for ^{222}Rn or 1.0 for ^{220}Rn in bone and soft tissues and an even concentration of radium and its decay products over the entire mineral bone were assumed. The calculated annual doses for individual tissues or organs are given in Tables 1.5 and 1.6). In the mentioned areas in India and Brazil the mean dose rates are increased by a factor of 10 as compared to "normal areas" (MUTH 1974).

Radon, Thoron, and Its Short-Lived Decay Products. For radiation exposure by ^{222}Rn, ^{220}Rn, and their decay products, the inhalation of short-lived decay products is of significance. This problem is dealt with in more detail in Sect. 1.3.3.

According to UNSCEAR (1982) outdoors the mean equivalent equilibrium concentration of radon daughters is 2 Bq·m^{-3} worldwide and, therefore, about 10 times lower than indoors. Our own measurements in the Federal Republic of Germany (KELLER et al. 1982a) yielded a mean outdoor concentration of about 2.2 Bq·m^{-3}, while the indoor concentration was higher only by a factor of 4–5.

The resulting mean lung dose is calculated at about $1\cdot10^{-5}$ Gy·y^{-1} (varying between $2\cdot10^{-6}$ and $4\cdot10^{-5}$ Gy·y^{-1}), assuming 5 h spent outdoors per day.

The mean equivalent equilibrium concentration of short-lived thoron decay products in air is about 0.04 Bq·m^{-3} (UNSCEAR 1982) or 0.05 Bq·m^{-3} (KELLER et al. 1982b). The resulting lung dose is $1\cdot10^{-6}$ to $2\cdot10^{-6}$ Gy·y^{-1}. For other organs or tissues, the dose rates through inhalation of these radionuclides are insignificant.

Table 1.7. Estimated tissue concentration and annual absorbed dose due to ^{210}Pb, ^{210}Bi, and ^{210}Po in areas of normal and high dietary intake (UNSCEAR 1977)

Area and radionuclide	Annual tissue absorbed dose (Gy·y^{-1})				
	Gonads	Lungs	Bone lining cells	Red bone marrow	Thyroid, breast, remainder
Areas of normal dietary intake					
Nonsmokers					
^{210}Pb (β)	$6 \cdot 10^{-9}$	$6 \cdot 10^{-9}$	$8 \cdot 10^{-9}$	$5 \cdot 10^{-9}$	$6 \cdot 10^{-9}$
^{210}Bi (β)	$4 \cdot 10^{-7}$	$4 \cdot 10^{-7}$	$4 \cdot 10^{-6}$	$2 \cdot 10^{-6}$	$4 \cdot 10^{-7}$
^{210}Po (α)	$6 \cdot 10^{-6}$	$3 \cdot 10^{-6}$	$3 \cdot 10^{-5}$	$7 \cdot 10^{-6}$	$6 \cdot 10^{-6}$
Smokers					
^{210}Pb (β)	$8 \cdot 10^{-9}$	$8 \cdot 10^{-9}$	$1 \cdot 10^{-8}$	$6 \cdot 10^{-9}$	$8 \cdot 10^{-9}$
^{210}Bi (β)	$6 \cdot 10^{-7}$	$7 \cdot 10^{-7}$	$6 \cdot 10^{-6}$	$3 \cdot 10^{-6}$	$6 \cdot 10^{-7}$
^{210}Po (α)	$8 \cdot 10^{-6}$	$9 \cdot 10^{-6}$	$4 \cdot 10^{-5}$	$9 \cdot 10^{-6}$	$8 \cdot 10^{-6}$
Areas of high dietary intake					
Reindeer or caribou eaters					
^{210}Pb (β)	$1.4 \cdot 10^{-8}$	$1.4 \cdot 10^{-8}$	$1.9 \cdot 10^{-8}$	$1.3 \cdot 10^{-8}$	$1.4 \cdot 10^{-8}$
^{210}Bi (β)	$1 \cdot 10^{-6}$	$1 \cdot 10^{-6}$	$1 \cdot 10^{-5}$	$6 \cdot 10^{-6}$	$1 \cdot 10^{-6}$
^{210}Po (α)	$7 \cdot 10^{-5}$	$4 \cdot 10^{-5}$	$1 \cdot 10^{-4}$	$5 \cdot 10^{-5}$	$7 \cdot 10^{-5}$

Long-Life Decay Products of Radon. The main source of ^{210}Pb and ^{210}Po in the atmosphere is radon emission from the soil. The mean ^{210}Pb concentration in air near the ground is 0.5 mBq·m^{-3} and the resulting ratio of ^{210}Po to ^{210}Pb is 0.2 according to JACOBI (1979). Table 1.4 shows the annual intake of ^{210}Pb and ^{210}Po for nonsmokers. One cigarette contains about 20 mBq ^{210}Pb and 15 mBq ^{210}Po. About 10% of ^{210}Pb and 20% of ^{210}Po are inhaled when smoking a cigarette; therefore with an average consumption of 20 cigarettes per day, the daily intake for smokers increases to 40 mBy ^{210}Pb and 60 mBq ^{210}Po, which represents an increase by factors of 4 and 30 respectively, compared to nonsmokers (and considering inhalation only).

The mean daily intake of ^{210}Pb and ^{210}Po through ingestion is about 100 mBq. For some populations in the Arctic Circle, living mainly on reindeer and caribou, the daily intake can increase to more than ten times this mean.

As a so-called "bone seeker," lead is mainly deposited in bone. About 70% is found in the skeleton. The total activity of lead 210 is 15 Bq in skeleton and 6.4 Bq in soft tissues.

In contrast to lead, polonium is not a bone seeker and is therefore mainly found in soft tissues. The ^{210}Po in bones originates from the radioactive decay of the deposited ^{210}Pb.

According to UNSCEAR (1982) a ratio of about 0.8 can be assumed between the mean activity concentrations of ^{210}Pb and ^{210}Po in bone. The mean ^{210}Po concentration in ashed bone is 2.4 Bq·kg^{-1}. In soft tissues the total activity for ^{210}Po and ^{210}Pb is about equal.

The calculated radiation doses (Table 1.7) are mainly due to the high energy α-particles of ^{210}Po; the β-rays of ^{210}Pb and ^{210}Bi contribute only about 10% of the dose.

1.3 Technologically Modified Exposure to Natural Radiation

The unmodified exposure to natural radiation, dealt with in the preceding section, to which all life on our planet is exposed, has been modified in may ways by the technological and civilizing interventions of man. In a few cases the natural radiation exposure has decreased. For example, purifying drinking water (from surface water) reduces the natural concentration of radium. Nevertheless, in general the natural radiation exposure increases as a result of human interference with nature.

This increase in natural radiation exposure includes contributions from exposures to natural sources of radiation which would not exist without technology or civilization. For example, the naturally radioactive products in the coal stored deep in the earth would not have any influence on the radiation exposure of humans without coal mining for energy production. Only through mining, burning, and the resulting emission of fly ash through chimneys can the natural radionuclides contained in coal irradiate humans.

Further examples are the increased exposure to cosmic rays during flights at high altitudes, the increased terrestrial irradiation caused by using building materials with higher concentrations of naturally radioactive materials, the increased lung exposure through inhalation of radon and thoron decay products by living indoors, the radiation exposure through industrial or agricultural use of phosphates, and the radiation exposure from consumer goods and industrial products with a low concentration of natural radionuclides.

1.3.1 Radiation Exposure Through Cosmic Radiation

1.3.1.1 Flights at High Altitudes

The dose rate of cosmic radiation increases with height above sea level. In comparison the dependency of cosmic radiation on the geographic latitude and sun activity is low. Table 1.8 shows the dependency of the absorbed dose rate and the dose equivalent rate on the altitude according to O'BRIEN (1975). These dose rates are averaged over two geomagnetic latitudes (43° and 55°) and over two periods of sun activity, the minimum and the maximum.

According to data from the International Civil Aviation Organization (1978), the total passenger kilometers in civil air traffic worldwide in 1978 was about 934 billion. At a mean speed of about 600 km·h^{-1}, this results in about 1.5 billion passenger hours. The altitude of flights below sonic velocity can vary between 3 and 12 km; the mean height is about 8 km. This results (see Table 1.8) in

a mean dose rate from cosmic radiation of about 0.8 μGy·h^{-1} or a mean dose equivalent rate of about 1.3 μSv·h^{-1}. (When referring to the equivalent dose it has to be borne in mind that the proportion contributed by the individual radiation components, such as direct ionizing radiation, neutrons, or pion stars, varies considerably depending on the altitude.)

The collective dose equivalent of the world population from air traffic is therefore about 2000 man-Sv per year. For an intercontinental flight below sonic velocity the mean dose per person from cosmic radiation is about 0.03 mGy, the mean dose equivalent about 0.04 mSv.

For supersonic aircraft transport at altitudes between 16 and 20 km, the mean energy dose is probably about the same as for flights below sonic velocity. In fact the dose rate is higher, but the duration of the flight is considerably shorter. However, during times of maximal sun activity the dose rate per person using supersonic aircraft transport at very high altitudes can increase above the mean of about 0.03 mGy per flight.

1.3.1.2 Space Travel

The radiation exposure for astronauts during space travel is considerably higher than that for airline passengers. While staying in space astronauts receive the primary cosmic radiation particles of the galactic radiation, the radiation from intense solar flares, and the intensive radiation in the two van Allen radiation belts around the earth. [According to measurements by SAVUN et al. (1973) the maximum dose rate in the inner belt is about 0.22 Gy·h^{-1} and that in the outer belt, about 0.05 Gy·h^{-1}.] The different results obtained by CURTIS (1974), ENGLISH et al. (1975), RADKE (1969), and GRIGORIEV (1976) regarding the actual radiation exposure (i.e., reduced by shielding) of the astronauts in American and Soviet space flights are summarized in Table 1.9. The majority of the total dose from space flights close to the earth is caused by radiation from the two radiation belts. For example, the higher dose on the Apollo X flight in comparison to Apollo VIII can be attributed to the longer stay in these radiation belts caused by the different route of the flight.

In space far from earth even higher dose rates can be found, caused by protons from sun eruptions because of the missing shielding effect of the magnetic field of the earth.

Table 1.8. Variation of the galactic dose rate and dose equivalent rate with altitude (after O'BRIEN 1975)[a]

Altitude (km)	Absorbed dose rate (μGy·h^{-1})	Dose equivalent rate (μSv·h^{-1})
4	0.14	0.20
6	0.33	0.51
8	0.84	1.35
10	1.75	2.88
12	3.01	4.93
14	4.62	7.56
16	5.92	9.70
18	7.09	11.64
20	7.72	12.75

[a] Values averaged over two geomagnetic latitudes (43° and 55°) and over two periods of solar activity (minimum and maximum)

Table 1.9. Absorbed dose rates of astronauts on space missions (after CURTIS 1974)

Mission or mission series	Launch date	Duration of mission (h)	Type of orbit	Dose $(10^{-5}$ Gy)
Apollo VII	Aug. 1968	260	Earth orbital	120
Apollo VIII	Dec. 1968	147	Circumlunar	185
Apollo IX	Feb. 1969	241	Earth orbital	210
Apollo X	May 1969	192	Circumlunar	470
Apollo XI	July 1969	182	Lunar landing	200
Apollo XII	Nov. 1969	236	Lunar landing	~200
Apollo XIV	Jan. 1971	209	Lunar landing	~500
Apollo XV	July 1971	286	Lunar landing	~200
Vostok 1-6			Earth orbital	2-80
Voskhad 1, 2			Earth orbital	30, 70
Soyuz 3-9			Earth orbital	62-234

Table 1.10. Mean activity concentration of naturally occurring radionuclides in building materials

Type of building material	Mean activity concentration (Bq/kg)		
	^{40}K	^{226}Ra	^{232}Th
1. Sand, gravel	260	15	15
2. Sandstone	190	19	19
3. Other natural stone	480	26	30
4. Limestone	220	19	19
5. Other industrial stone	370	33	30
6. Natural gypsum	70	< 19	< 19
7. Concrete	220	22	26
8. Various additives	220	22	15
9. Basalt	1400	41	52
10. Cement	150	52	52
11. Granite, shale	1480	56	81
12. Bricks	630	67	63
13. Pumice stone	890	81	85
14. Slag-stone	330	81	104
15. Phosphorite	70	520	< 19
16. Lithoid tuff (Italy)	1480	130	120
17. Blast furnace slag	520	120	130
18. Fly-ash	700	210	130
19. Red-mud bricks	330	280	230
20. Concrete containing alum shale (Sweden)	850	1500	70

1.3.2 Radiation Exposure by Use of Special Building Materials with Increased Amounts of Naturally Occurring Radionuclides

Various building materials can increase the natural radiation exposure indoors because of their concentration of naturally occurring radionuclides. These materials can be in their original form like pumice, alum shale, tuff, or granite or as waste products of industrial processes like by-product gypsum, flying ash, and blast furnace slag. In these industrial processes a concentration of poorly soluble radium and thorium compounds is often found, caused by the chemical reactions.

1.3.2.1 Radionuclide Concentrations in Various Building Materials

The radionuclides in building materials which are responsible for increased radiation exposure are potassium 40, radium 226, and thorium 232. Table 1.10 gives the mean activity concentrations of some materials according to KELLER (1980), KELLER et al. (1974), and SWEDJEMARK (1980).

KRISIUK et al. (1971) tried to calculate an acceptable limit for radionuclide concentrations in building materials, using a conservative model. This model is based on infinitely thick walls without windows and doors. The increase in radiation exposure of people staying in these rooms should not exceed 1.5 mGy per year. With these assumptions the following maximum permissible concentrations (MK) for the individual radionuclides can be calculated:

Radionuclide	Maximum permissible concentration
^{40}K	4810 Bq/kg (130 nCi/kg)
^{226}Ra	370 Bq/kg (10 nCi/kg)
^{232}Th	260 Bq/kg (7 nCi/kg)

If all three radionuclides are found in the building material the following relation is valid (C = activity concentration in Bq/kg):

$$\frac{C_{K\,40}}{MK_{K\,40}} + \frac{C_{Ra\,226}}{MK_{Ra\,226}} + \frac{C_{Th\,232}}{MK_{Th\,232}} \leq 1.$$

These very conservative assumptions were later corrected by the authors, in that a finite thickness of walls, windows, and doors was considered through the application of a weighting factor of 0.7 in each case. In this way the maximum permissible concentrations were increased by a factor of 2. That means:

$$\frac{C_{K\,40}}{9620 \text{ Bq/kg}} + \frac{C_{Ra\,226}}{740 \text{ Bq/kg}} + \frac{C_{Th\,232}}{520 \text{ Bq/kg}} \leq 1.$$

Except for concrete containing alum shale, all materials would meet this equation (see Table 1.10). But admittedly only the mean values of all concen-

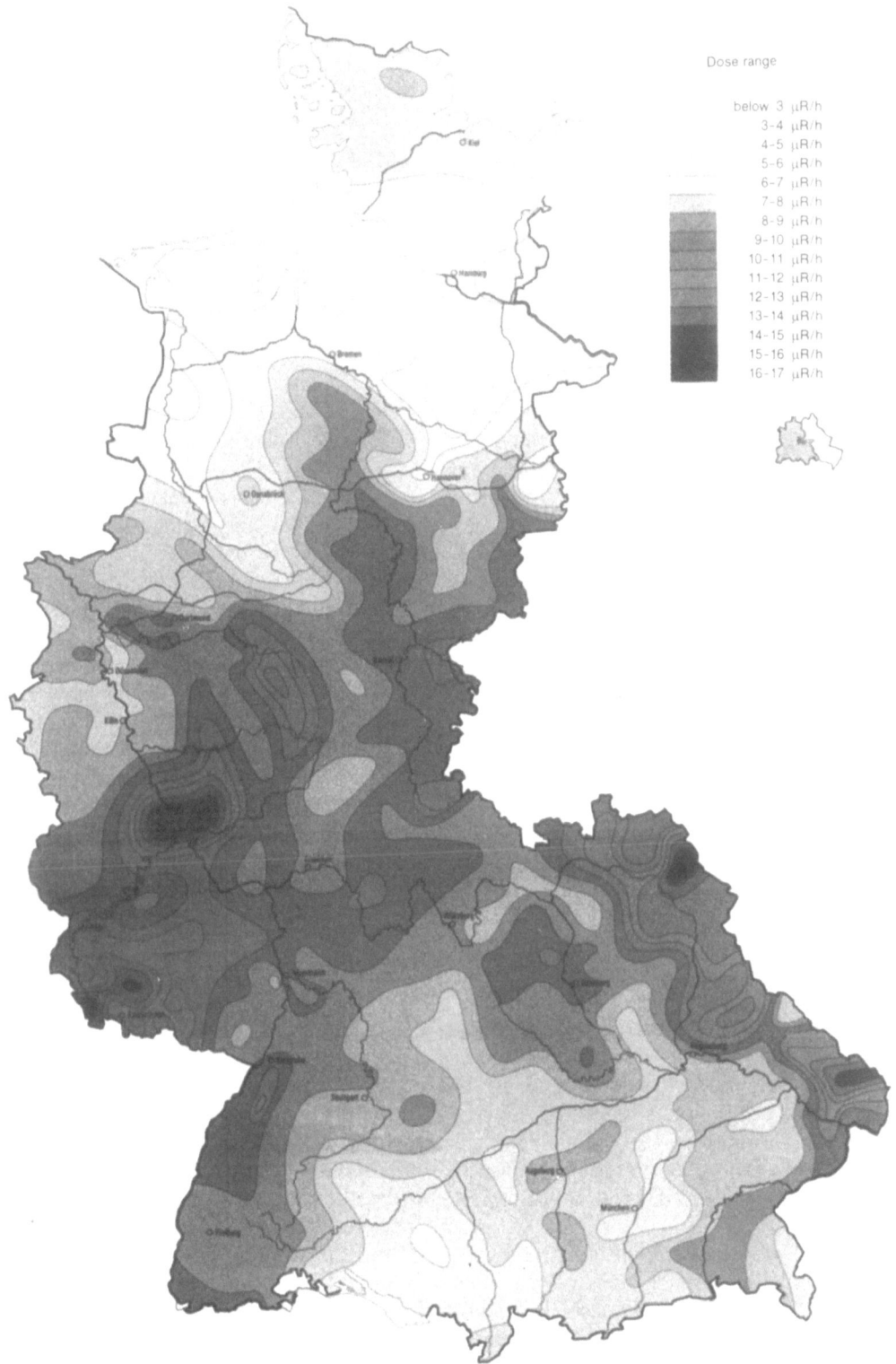

Fig. 1.2. Dose rates from indoor terrestrial γ-irradiation (μR/h) in the Federal Republic of Germany

tration measurements are compared. In individual cases maximum concentrations can be found which do not meet this equation.

1.3.2.2 Radiation Exposure Through γ-Radiation Indoors

The mean dose rate indoors can be calculated from the activity concentration measured in building materials. But as real houses are made of a variety of materials and the geometric parameters vary from room to room, these calculations (KELLER and OBERHAUSEN 1978) are either very elaborate or give unsatisfying results. Therefore in some countries the indoor radiation exposure of the population in comparison to the exposure outside (see Sect. 1.2.1.2) has been determined with special γ-ray dose rate meters. In the Federal Republic of Germany such measurements were carried out by various scientific institutions. The study was based on an analysis of 30000 individual measurements in rooms. Figure 1.2 shows the geographic distribution of the measured dose rates of terrestrial radiation indoors. Comparing the indoor and the outdoor radiation exposure, a considerable difference is found in some regions, which might be related to a regional preference for particular building materials.

The increased γ-ray dose rates found in parts of the Saarland, Hessen, and Rheinland-Pfalz are due to the common use of particular materials in those regions, like pumice, blast furnace stag, and by-product gypsum. In prefabricated wooden houses a lower level of terrestrial radiation was found in comparison to outdoors, which is attributed to the shielding of radiation from the soil and the low radioactivity of the building material itself. For the Federal Republic of Germany a mean dose rate of 8 $\mu R/h$ indoors and 6 $\mu R/h$ outdoors is found. In comparison, the mean dose rate in Swedish houses, which are built using concrete containing alum shale with increased radium concentrations, is 18 $\mu R/h$ according to MJÖNES and SWEDJEMARK (1976).

The measured mean increase in radiation exposure of the population through building materials with a higher content of naturally occurring radionuclides is statistically significant in the Federal Republic of Germany; it is, however, within the range of unmodified exposure to natural radiation. Except for some extreme cases this is undoubtedly true for most countries of the world.

Table 1.11. Distribution of the difference in the mean annual dose to the gonads (D_H-D_F) between indoor and outdoor exposure of the German population, assuming that exposure to be continuous

D_H-D_F ($\mu Sv/y$)	Percent of the population
0	23
0- 50	23
50-100	21
100-200	15
200-250	9
250-300	5
300-450	3
>450	1

1.3.2.3 Mean Dose to the Gonads

Assuming a mean occupancy of 80% indoors and 20% outdoors, a mean annual dose to the gonads of 0.48 mSv is received by the population of the Federal Republic of Germany from terrestrial background radiation and the radiation caused by the radioactivity of building material. A permanent stay outdoors would lead to a mean dose to the gonads of 0.38 mSv per year. Table 1.11 shows the distribution of the annual dose difference to the population of the Federal Republic of Germany between indoors and outdoors. This difference is mainly determined by the radioactivity of the building materials used, the amount of material used, the size of the dose rate outdoors, and the shielding of the houses from the terrestrial and cosmic radiation outdoors.

1.3.3 Radiation Exposure by Inhalation of Radioactive Noble Gases Radon and Thoron and Their Short-Lived Decay Products

In the previous section about the increase in the γ-ray dose rate indoors and the use of building materials with increased naturally occurring radionuclides, the concentrations of ^{226}Ra and ^{232}Th were of special significance. The radioactive gases radon (^{222}Rn) and thoron (^{220}Rn) are decay products of ^{226}Ra and ^{224}Ra, which is a decay product of ^{232}Th. These α-emitting nuclides diffuse out of the building material and get into the room air, which increases the ^{222}Rn or ^{220}Rn concentration in indoor air in comparison with outdoors. Inhalation of these gases and their radioactive decay products leads mainly to radiation exposure of the lung.

Table 1.12. Radioactive decay properties of ^{226}Ra and its daughters (LEDERER and SHIRLEY 1978; ERDTMANN and SOYKA 1979)

Radionuclide	Historical name	Half-life	Major radiation energies and intensities					
			α		β		γ	
			MeV	%	MeV	%	MeV	%
$^{226}_{88}$Ra	Radium	1.6 10³ a	4.60 4.78	6 94			0.186	3.3
$^{222}_{86}$Rn	Emanation Radon (Rn)	3.823 d	5.49	100				
$^{218}_{84}$Po	Radium A	3.05 min	6.00	~100				
99.98% 0.02%								
$^{214}_{82}$Pb	Radium B	26.8 min			0.67 0.73 1.02	48 42 6	0.295 0.352	19 37
$^{218}_{85}$At	Astatine	~2 s	6.65 6.69 6.76	6 90 3.6	?	~0.1		
$^{214}_{83}$Bi	Radium C	19.7 min	5.45 5.51	0.012 0.008	1.0 1.51 3.26	23 40 19	0.609 1.12 1.764	46 15 16
99.98% 0.02%								
$^{214}_{84}$Po	Radium C'	164 µs	7.69	100				
$^{210}_{81}$Tl	Radium C''	1.3 min			1.3 1.9 2.3	25 56 19	0.296 0.795 1.31	80 100 21
$^{210}_{82}$Pb	Radium D	22.3 a			0.015 0.061	81 19	0.047	4
$^{210}_{83}$Bi	Radium E	5.01 d			1.161	~100		
~100% 0.0001%								
$^{210}_{84}$Po	Radium F	138.4 d	5.305	100				
$^{206}_{81}$Tl	Radium E''	4.2 min			1.53	100		
$^{206}_{82}$Pb	Radium G	Stable						

1.3.3.1 General Considerations

Tables 1.12 and 1.13 show the uranium, radium, and thorium decay series according to LEDERER and SHIRLEY (1978) and ERDTMANN and SOYKA (1979). The noble gases ^{222}Rn and ^{220}Rn, which escape from the soil into the air, decay further via short-lived products ^{218}Po, ^{214}Pb, and ^{214}Bi or ^{216}Po, ^{212}Pb, and ^{212}Bi into stable lead isotopes ^{206}Pb or ^{208}Pb. These radioactive heavy metal isotopes attach themselves to the aerosols because of their high diffusion coefficients.

These particles enter the respiratory tract by inhalation; because of their short half-life they are retained there and decay mainly where they are deposited before they can be eliminated by the self-cleansing process of the lung. As a measure of the concentration of ^{222}Rn or ^{220}Rn decay products in air, the concentration of the potential α-energy $E_{pot,\alpha}$ is used. The potential α-energy of an atom in

Table 1.13. Radioactive decay properties of ^{228}Th and its daughters (LEDERER and SHIRLEY 1978; ERDTMANN and SOYKA 1979)

Radionuclide	Historical name	Half-life	Major radiation energies and intensities					
			α		β		γ	
			MeV	%	MeV	%	MeV	%
$^{228}_{90}$Th	Radiothorium	1.913 a	5.34	27			0.084	1.2
			5.43	73			0.216	0.3
$^{224}_{88}$Ra	Thorium X	3.66 d	5.45	6			0.241	3.9
			5.68	94				
$^{220}_{86}$Rn	Emanation Thoron (Tn)	55 s	6.29	100			0.55	0.1
$^{216}_{84}$Po	Thorium A	0.15 s	6.78	100				
$^{212}_{82}$Pb	Thorium B	10.64 h			0.331	83	0.239	43
					0.569	12	0.300	3.2
$^{212}_{83}$Bi	Thorium C	60.6 min	6.05	25	1.55	5	0.040	1.1
			6.09	10	2.26	55	0.727	11.8
							1.620	2.8
$^{212}_{84}$Po	Thorium C'	304 ns	8.78	100				
$^{208}_{81}$Tl	Thorium C''	3.05 min			1.28	23	0.511	23
					1.52	22	0.583	86
					1.80	51	0.860	12
							2.614	100
$^{208}_{82}$Pb	Thorium D	Stable						

(Decay scheme at left: $^{212}_{83}$Bi branches 64% and 36% to $^{212}_{84}$Po and $^{208}_{81}$Tl, both leading to $^{208}_{82}$Pb.)

the decay series of radon or thoron means the sum of all α-energies emitted during the decay of this atom until ^{210}Pb or ^{208}Pb. Table 1.14 shows the potential α-energies of radon, thoron, and their short-lived decay products.

The potential α-energy concentration $C_{pot,\alpha}$ of any mixture of short-lived decay products of radon or thoron is the sum of the potential α-energies of all decay products which are in the volume of air under consideration. If given in SI units:

$$1 \; J \cdot m^{-3} = 6.24 \cdot 10^{12} \; MeV \cdot m^{-3}.$$

"Working Level" (WL) is a unit used in radiation protection for the potential α-energy concentration:

$$1 \; WL = 1.3 \cdot 10^{5} \; MeV \cdot l^{-1}.$$

1 WL corresponds, for example, to the potential α-energy concentration of short-lived ^{222}Rn decay products (or ^{220}Rn decay products) which are in a radioactive equilibrium with a ^{222}Rn concentration of 3.7 Bq$\cdot$l^{-1} (or with a ^{220}Rn concentration of 0.28 Bq$\cdot$l^{-1}). Usually ^{222}Rn or ^{220}Rn decay products are not in equilibrium with their mother nuclides radon or thoron. An "equilibrium factor" F is de-

fined as the ratio of the total actually existing potential α-energy of a given decay product concentration to the total potential α-energy concentration of the decay products, if they are in equilibrium with their mother substance. Therefore F is the ratio of the equivalent equilibrium concentration C_{eq} of radon or thoron to the actual ^{222}Rn or ^{220}Rn concentration C_0 in air, i.e., $F = C_{eq}/C_0$.

1.3.3.2 Radon and Thoron Exhalation from Building Materials

During the α decay of ^{226}Ra or ^{224}Ra the atoms of the gases radon or thoron receive a recoil energy, which pushes them over a distance of $3 \cdot 10^{-6}$ cm in the matrix of the stone. Through this mechanism and by diffusion, a portion of the gas atoms reach the pores of the material. This process is called emanation. The diffusion of the ^{222}Rn or ^{220}Rn atoms to the surface of the material into the air is called exhalation. The exhalation rate in Bq$\cdot$m$^{-2}\cdot$h^{-1} describes how much radon or thoron activity leaves a certain surface per time unit. (This whole process, from the origin of the radon or thoron atoms, through emanation, diffusion, exhalation, further

Table 1.14. Calculated potential α-energy of radon and thoron and their short-lived decay products

Radionuclide	Potential α-energy per				Potential α-energy concentration (C_{pot}) per unit of activity concentration
	atom $E_{pot,at}$		unit of activity $E_{pot,at}/\lambda_j$		
	MeV	10^{-12} J	MeV·Bq^{-1}	10^{-10} J·Bq^{-1}	10^{-6} WL (Bq·m^{-3})$^{-1}$
^{222}Rn	19.2	3.07	9 150 000	14700	
^{218}Po	13.7	2.19	3620	5,79	27.8
^{214}Pb	7.69	1.23	17 800	28.6	137
^{214}Bi	7.69	1.23	13 100	21.0	101
^{214}Po	7.69	1.23	0.002	0.000003	0.000016
Total (rounded)[a]			34 500	55.4	266
^{220}Rn	20.9	3.34	1660	2.65	
^{216}Po	14.6	2.34	3.32	0.00532	0.0256
^{212}Pb	7.8	1.25	431 000	691	3320
^{212}Bi	7.8	1.25	40 900	65.6	315
^{212}Po	2.78	1.41	0.00000305	0.0000000062	0.00000003
Total (rounded)[a]			472 000	757	3640

[a] The total is the sum of the potential α-energies of the daughters only

Table 1.15. Mean activity concentration of radium and thorium in different building materials and their radium and thorium exhalation rates normalized to a thickness of 10 cm (KELLER 1980; FOLKERTS 1983)

Building material	Concentration (Bq/kg)		Exhalation rate (Bq/m²·h)	
	^{226}Ra	^{232}Th	^{222}Rn	^{220}Rn
1. Sandstone	10	10	1.0	170
2. Porphyry	40	22	3.3	150
3. Limestone	10	15	0.9	90
4. Brick	50	15	0.2	30
5. Pumice stone (natural)	60	50	1.5	180
6. Pumice stone (industrial)	70	55	0.7	150
7. Blast furnace slag	75	20	0.6	110
8. Concrete	50	10	1.1	70
9. Gas concrete	20	15	1.0	60
10. Natural gypsum	5	15	0.2	30
11. Industrial gypsum				
- Apatite	20	15	0.4	150
- Phosphorite	260	15	24.1	80

decay into daughter atoms, and adsorption on aerosols and wall surfaces, to inhalation and retention in the respiratory tract, is very complex; details are to be found in the literature, e. g., FOLKERTS 1983.)

In Table 1.15 the measured radon and thoron exhalation rates of some building materials are listed, assuming 10-cm thick walls, together with the mean concentrations of radium and thorium in these materials. The differences between the ^{222}Rn and ^{220}Rn

exhalation rates are generally due to the different decay constants ($\lambda_{thoron} : \lambda_{radon} = 1 : 6 \cdot 10^3$).

If the exhalation rate e is known, the following approximation can be used, according to WICKE (1979), to calculate the expected radon or thoron air concentration indoors C_i:

$$C_i = \frac{e \cdot F \cdot V^{-1} + L \cdot C_a}{\lambda + L}$$

with

C_i = ^{222}Rn or ^{220}Rn indoor air concentration in Bq·m^{-3}.

C_a = ^{222}Rn or ^{220}Rn outdoor air concentration in Bq·m^{-3} (mean for radon ~3.7 Bq·m^{-3}; mean for thoron ~3.7 Bq·m^{-3}).

$F \cdot V^{-1}$ = ratio of surface to volume of the room in m^{-1} (mean ~2 m^{-1}).

L = exchange rate of room air in h^{-1} (mean ~0.3 h^{-1}).

λ = decay constant in h^{-1} (for radon = $7.56 \cdot 10^{-3}$ h^{-1}; for thoron = 45.78 h^{-1}).

e = exhalation rate in Bq·m^{-2}·h^{-1}.

Assuming an exhalation rate for the building material of e = 2.5 Bq·m^{-2}·h^{-1} for a wall thickness of 24 cm and using the previously stated mean rates for the other parameters, for a "standard room" a quite realistic radon concentration $C_i = 20$ Bq·m^{-3} is calculated.

Table 1.16. Mean concentrations in living rooms and outdoors of radon and its decay products in different countries (UNSCEAR 1988; KELLER et al. 1982)

Country	Concentration ($Bq \cdot m^{-3}$)		
	In living rooms		Outdoors
	^{222}Rn	^{222}Rn decay products	^{222}Rn
Austria	24	12	7.0
Canada	14	8	–
Denmark	10	–	–
Finland	18	–	2.3
Hungary	40–240	–	–
Norway	52	–	–
Poland	12– 34	6	–
Sweden	132	66	–
United Kingdom	26	13	3.3
USA	30	15	3.0
Federal Republic of Germany	23–33	6	5.9

1.3.3.3 Concentrations of the Noble Gases Radon and Thoron and Their Decay Products In- and Outdoors

In general, the concentration of the radioactive gases radon and thoron is higher indoors than outdoors. This is mainly due to the ^{222}Rn and ^{220}Rn exhalation from the building materials. Furthermore, in some regions of the earth diffusion out of the surrounding soil and from tap water or natural gas contributes to the increase in indoor air concentrations. Table 1.16 gives information on the mean or median concentrations of ^{222}Rn and its short-lived decay products indoors and the outdoor ^{222}Rn concentrations in different countries, based on the UNSCEAR report (1982) and the findings of KELLER et al. (1982a, b). These results are based on extensive measurements; the minimum and maximum values can differ from the mean by a factor of up to 10.

Only a few results are available on the concentrations of thoron and its decay products, because these studies are very difficult owing to the short half-life ($T_{1/2} = 55.6$ s) of this gas. Based on our own estimates and measurements, the concentration of thoron indoors in the Federal Republic of Germany is about 10 $Bq \cdot m^{-3}$, while that of the decay products is about 0.07 $Bq \cdot m^{-3}$. Outdoors the ^{220}Rn concentration was calculated at about 0.02 $Bq \cdot m^{-3}$.

During the investigations of the concentrations of radon, thoron, and their decay products in- and outdoors, dependency of the results on various parameters was found, e.g., the time of day, the season, and regional or geologic and meteorologic conditions. Figure 1.3 gives the cumulative frequen-

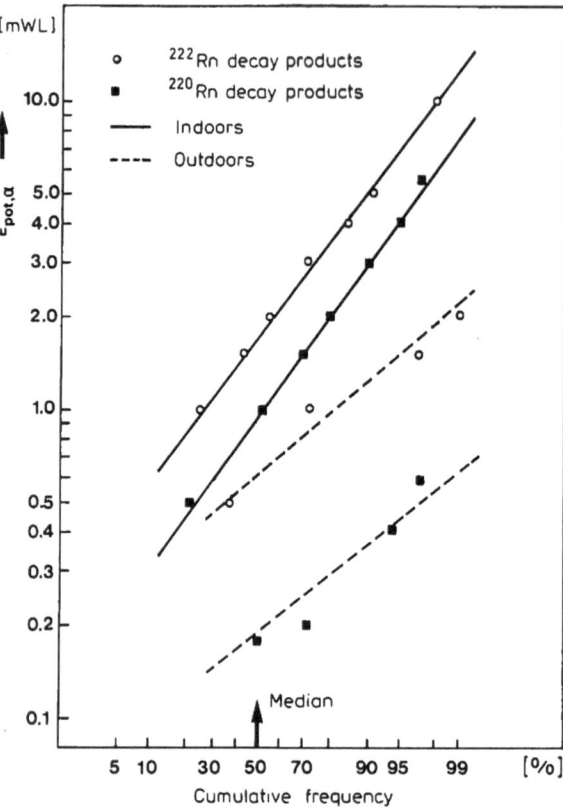

Fig. 1.3. Log-normal cumulative frequency plot of radon and thoron daughter concentrations in German homes and outdoors (KELLER et al. 1982)

cy distribution of radon and thoron decay products in- and outdoors, based on our own systematic measurements in Southwest Germany (KELLER et al. 1982a).

Room Air Exchange Rate (Ventilation Rate). Part of the activity exhaled from building materials leaves the room through natural and artificial ventilation. Artificial ventilation, such as opening windows and doors, forced ventilation by ventilators and air conditioners, etc., has a greater influence than natural ventilation, such as air exchange through pores of masonry or drafts through gaps in windows and doors. Ventilation is an important factor for the indoor concentration of radon and its decay products. For thoron, in contrast to its decay products, the influence of ventilation is very small owing to its short half-life.

To describe the exchange rate of room air, the ventilation rate L is given in units of the reciprocal time, i.e., L describes the proportion of the room volume exchanged by outdoor air per unit of time. In the literature, ventilation rates between 0.1 and

3 h^{-1} have been reported. In "normal" living rooms the ventilation rate is probably between 0.1 h^{-1} and 0.5 h^{-1}, with a mean of about 0.3 h^{-1}. The dependency of radon concentration indoors on the exhalation rate e and the ventilation rate L is shown in Fig. 1.4 according to WICKE (1979).

The Equilibrium Factor in Room Air. The equilibrium factor F is defined as the ratio of the equivalent equilibrium concentration of radon or thoron to the actual radon or thoron concentration (see Sect. 1.3.3.1). Its exact value is important if the measured gas concentrations are to be used to calculate the lung dose, which mainly depends on the decay products. The equilibrium factor F indoors depends on the ventilation, the adsorption on aerosols, and the sedimentation on walls and furniture.

Except for cellars and basements (F = 0.5–0.6), the equilibrium factor indoors is generally below 0.5. According to our own measurements the mean equilibrium factor F of ^{222}Rn and its decay products in living rooms is about 0.3 ± 0.1. Figure 1.5 shows the frequency distribution of equilibrium factors of about 120 measurements for the Federal Republic of Germany (KELLER et al. 1982 a, b).

1.3.3.4 Mean Resulting Lung Dose and Expected Lung Cancer Risk for the Population

Inhalation of short-lived radon and thoron decay products leads to inhomogeneous irradiation of the respiratory tract by α-particles. The resulting dose depends on parameters like the mean respiration rate, the type of respiration (mouth or nose), the size distribution of carrier aerosols, the ratio of adsorbed to free particles, the geometric factors of different regions in the lung, the transport and cleansing processes acting on deposited decay products in the lung, and, of course, the activity concentration in air. Different dosimetric models of the lung allow the calculation of the dose in dependency on these parameters (FOLKERTS 1981; JACOBI 1980). The basal cells of the lung epithelium in the upper bronchi receive the highest doses from the radon decay products, while the terminal bronchioli and the alveolar surface of the lung are exposed to the maximum radiation doses from the thoron decay products.

Table 1.17 shows, according to JACOBI (1982), mean equivalent equilibrium concentrations of radon and thoron decay products, based on studies in several northern countries.

Assuming a mean of 80% of time spent indoors and 20% outdoors, and a mean respiration rate of

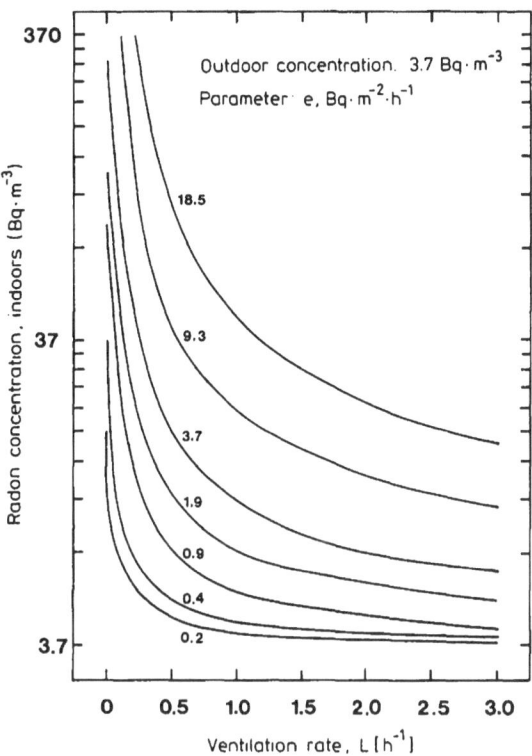

Fig. 1.4. Dependence of the radon concentration on the ventilation rate at different exhalation rates (WICKE 1979)

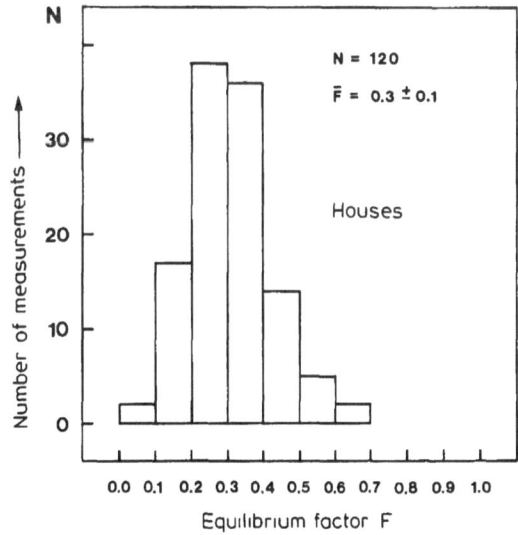

Fig. 1.5. Distribution of the equilibrium factor in German houses (KELLER et al. 1982)

10 l·min^{-1} indoors and 17 l·min^{-1} outdoors, the mean annual respiration volume outdoors is 1800 m^3 and that indoors, 5400 m^3. Using a special lung model, relevant parameters (e. g., quality factor for α-particles Q$_\alpha$ = 20; activity median aerodynam-

Table 1.17. Average equilibrium equivalent concentrations of radon and thoron decay products indoors and outdoors (JACOBI 1982)

	Equilibrium equivalent concentration (Bq·m^{-3})	
	Indoors	Outdoors
^{222}Rn decay products	10-20	~1.8
^{220}Rn decay products	4-11	~0.9

Table 1.18. Estimated median exposure from inhaled ^{222}Rn and ^{220}Rn decay products (JACOBI 1982)

Inhaled radionuclides	Annual dose equivalent (mSv)		Annual effective dose equivalent (mSv)
	Basal cells of the bronchial epithelium	Lung	
^{222}Rn decay products	8.7-17	1.3-2.5	0.6-1.2
^{220}Rn decay products	1.5-3.0	0.4-0.8	0.1-0.2
Total	10-20	1.7-3.3	0.7-1.4

Table 1.19. Estimated lung cancer risk of the population from inhalation of radon decay products (JACOBI 1982)

	Absolute risk R_{Rn} (%)	Relative risk[a] $R_{Rn}/R_{observed}$ (%)
Nonsmokers	0.05-0.2	3 -20
Smokers	0.1 -0.4	0.5- 3
Total	0.06-0.26	1 - 6

[a] The observed lung cancer risk of smokers is about 0.1%-0.2% and that of nonsmokers, about 0.01%-0.02%. 20%-30% of the population are smokers

ic diameter: $AMAD_{indoors} = 0.15-0.2$ μm and free activity $f_{p,indoors} = 0.03$, $AMAD_{outdoors} = 0.1$ μm and $f_{p,outdoors} = 0.005$; weighting factor for the trachial and bronchial region W_{TB} and for the pulmonary region W_P with $W_{TB} = W_P = 0.06$) and the concentrations of ^{222}Rn and ^{220}Rn decay products (see Table 1.17), JACOBI (1982) calculated the mean values for the natural radiation exposure of the population through inhaled radon decay products (Table 1.18). [According to our own measurements (KELLER et al. 1982 a, b) the mean effective equivalent dose by inhalation of Rn decay products for the Federal Republic of Germany is 0.4 mSv/y.]

With a mean life span of 70 years the doses in Table 1.18 lead to a cumulative equivalent dose of about 1 Sv (0.7-1.4 Sv) in the basal cells of the bronchi and about 0.18 Sv (0.12-0.23 Sv) in the pulmonary regions.

In calculating the risk to the population of lung cancer after a mean life span of 70 years, generally a linear relation between dose and effect without a threshold is assumed. UNSCEAR (1977) and EVANS et al. (1981) calculated for nonsmokers a lung cancer risk by inhalation of radon and thoron decay products of 0.05%-0.2% for the general population. The synergistic or cocarcinogenic influence of smoking on radiation-induced lung cancer is taken into account by a risk factor of 2 between the smoking and the nonsmoking sections of the general population (about 20%-30% of the population are smokers).

Table 1.19 shows the estimated lung cancer risk for the population by inhalation of radon and thoron decay products only, according to JACOBI (1982). The total observed lung cancer risk for the population is about 4%-6%. According to this calculation, about 1%-6% of all occurring lung cancers may be due to inhalation of radon and thoron decay products.

1.3.4 Radiation Exposure by Using Phosphate Ores

In 1977 the worldwide production of phosphate ores was about 130 million tons (UNITED NATIONS 1979). The phosphates from sediments, which in total account for about 80% of the total production, e. g., from Florida and Morocco, contain a relatively high concentration of uranium 238, which is in radioactive equilibrium with its decay products, like radium 226. Only in the case of the phosphates of vulcanic origin (like from the Kola peninsula, USSR) no increase in ^{238}U concentration is to be found. The specific activity of ^{232}Th and ^{40}K in the phosphates is generally comparable to rates found in the soil. The ^{226}Ra concentrations in phosphates from sediments average 1500 Bq/kg, with maximal concentrations up to 5000 Bq/kg. Phosphate ores are mainly used to produce fertilizers. They are also used for recovery of uranium and for the production of phosphoric acid.

In addition the chemical gypsum, which is a by-product of processing phosphate ores, is increasingly used as a building material, leading to enhanced radiation exposure indoors (see Sect. 1.3.2 and 1.3.3).

1.3.4.1 The Use of Phosphate Fertilizers

About three-quarters of the phosphate ore production is used in the fertilizer industry. Depending on the amount of P_2O_5 in fertilizers, there are great fluctuations in the ^{226}Ra concentration. The ^{40}K content is mainly dependent on the percentage of nitrogen in the fertilizer. According to PFISTER et al. (1976) and KELLER et al. (1974), the mean concentrations per kg phosphate fertilizer (minimum and maximum values in parentheses) are about 1400 Bq/kg (405200) for potassium 40, about 30 Bq/kg (15-60) for thorium 232, and about 350 Bq/kg (10-850) for radium 226.

Considering the distribution of phosphate fertilizers, only a small contribution to the natural background radiation results even if possible accumulation in the soil is taken into account. The mean additional dose rate of γ-radiation is about 0.1 µR/h, i.e., about 1% of the normal natural background radiation.

Only in various working areas, such as production, storage, or transport of the phosphates or fertilizers, higher occupational radiation exposures can be measured. In extreme cases, the dose rate may be increased to ten times the mean natural dose rate.

1.3.4.2 Production of Phosphoric Acid and Uranium Recovery

The production of phosphoric acid and the recovery of uranium from phosphate ores contributes only a small amount of the total natural radiation exposure to the general population. Only in direct vicinity of these production areas increased concentrations of ^{238}U, ^{230}Th, and ^{226}Ra were found (UNSCEAR 1982).

1.3.4.3 The Use of By-product Gypsum as a Building Material

Calcium dihydrate is a by-product of the production of phosphoric acid, which is chemically processed and marketed as a building material called by-product gypsum (phosphorite). The relatively high ^{226}Ra concentration (see Sect. 1.3.2.1) of this material causes an increase in the γ-radiation dose rate in buildings, where this material has been used. Due to a voluntary reduction of the use of by-product gypsum by the building industry and by mixing it with natural gypsum or apatite, the ^{226}Ra concentrations have been reduced in these building materials in recent years.

1.3.5 Radiation Exposure Through Energy Production

Like nearly all natural materials, the raw materials used to produce energy contain traces of naturally occurring radionuclides (see Sect. 1.2.1.2). Raw materials like coal, oil, gas, and geothermic hot water are transported from the depths of the earth to its surface. In the production of energy the naturally occurring radionuclides are released from these raw materials to the environment, in some cases even enriched. This may lead to an additional natural radiation exposure of the population.

1.3.5.1 Coal-Fired Power Plants

The majority of the coal mined is used for production of electricity; only 2% is used directly for heating of rooms. The main pathway of radiation exposure from coal-fired power plants is the emission of gases and ash particles through the chimneys.

The activity concentration in coal varies, according to BECK et al. (1980), worldwide from 0.7 to 70 Bq/kg for ^{40}K, from 3 to 520 Bq/kg for ^{226}Ra, and from 3 to 320 Bq/kg for ^{232}Th. For coal, mean values of 50 Bq/kg for ^{40}K and 20 Bq/kg for ^{226}Ra and ^{232}Th can be assumed, and radioactive equilibrium between ^{238}U or ^{232}Th and their decay products can be expected.

When burning coal in the plant, naturally radioactive substances are increased in the fly ash as compared to the coal; this increase depends on the individual radionuclide, the burning temperature in the plant, and the type of coal used. Measurements by several authors in the Federal Republic of Germany have shown that in the case of coal plants with a burning temperature of about 1700 °C, the specific activities in fly ash for ^{238}U, ^{232}Th, and ^{226}Ra are about ten times higher than in the coal before burning, while the figures for ^{210}Pb and ^{210}Po are ca. 100 and 200 times higher, respectively. In lignite plants with a burning temperature of about 1100 °C, the accumulation factor for the ash emission for ^{238}U, ^{232}Th, and ^{226}Ra is about 3-5 times higher compared to lignite before burning, while that for ^{210}Pb and ^{210}Po is about ten times higher.

Table 1.20 gives the calculated mean emissions of natural radioactive substances from modern coal-fired power plants per 1 GW·a of electric energy produced (STRAHLENSCHUTZKOMMISSION 1981).

If the radiation exposure is estimated at the most critical point around a coal-fired power plant, an annual effective dose equivalent of about

0.007 mSv is found for the emissions from a modern coal-fired power plant producing 1 GW·a of electric energy. For a comparable modern lignite plant the effective dose equivalent is lower by a factor of about 5. The additional radiation exposure from modern coal-fired power plants in the Federal Republic of Germany is therefore about 1‰–1% of the average unmodified natural radiation exposure, and small in comparison to the range of natural radiation exposure. Thus in terms of the overall radiation exposure of the population emissions of naturally radioactive substances into the air from coal-fired power plants are insignificant when modern cleaning techniques are used.

Table 1.20. Mean emissions of natural radionuclides from coal-fired power plants, normalized to 1 GW electrical energy per year (10^6 Bq)

Nuclide	Bituminous coal plant	Brown coal plant
^{238}U	400	120
^{234}U	400	120
^{230}Th	400	80
^{226}Ra	400	80
^{210}Pb	4000	200
^{210}Po	8000	400
^{232}Th	200	40
^{228}Th	200	40

1.3.5.2 Oil, Natural Gas, and Geothermic Energy

In the production of electricity, oil, natural gas, and, to a small extent, geothermic energy are still used. Oil-fired power plants have significantly lower emissions of naturally radioactive substances than coal-fired power plants and therefore contribute only a negligible amount to the overall radiation exposure of the population. When using natural gas and geothermic energy, only the release of the radioactive noble gas ^{222}Rn (see Sect. 1.3.3) into the atmosphere has to be considered. According to VAN DER HEIJDE et al. (1976), in Germany natural gas as an energy source for cookers and ovens in households has a specific ^{222}Rn activity between 40 and 350 Bq·m^{-3}. However, in other countries maximum values of 40000 Bq·m^{-3} were measured. Ignoring the extreme values and at normal levels of consumption in a household, natural gas again makes only a very small contribution to the radiation exposure of the population.

Hot water from the depth of the earth as a source of energy contributes less than 3% to the total energy production. At the few places where this form of energy is available, significant amounts of the noble gas ^{222}Rn are released from the deep waters to the surface of the earth. According to MASTINU (1980) an Italian 400 MW power plant using geothermic energy releases a ^{222}Rn activity of $110 \cdot 10^{12}$ Bq per

Table 1.21. Consumer products in the Federal Republic of Germany (WEHNER 1978)[a]

Type of consumer product	Produced in the Fed. Rep. of Germany			Number of pieces imported into the Fed. Rep. of Germany
	Number of pieces or weight	Total activity and radionuclide used	Exported	
Radioluminous timepieces				
Devices containing scales or dials with luminous paint	14 10^6	40 TBq ^{3}H 10 TBq ^{147}Pm	50%	8 10^5 (^{3}H) 1 10^5 (^{147}Pm)
Electronic and electrical devices				
High-pressure mercury lamps	7 10^6	15 GBq ^{232}Th	20%	
Ignition devices for fluorescent lamps	26 10^6	3 TBq ^{85}Kr	50%	
Electronic components containing	40 10^6	200 TBq ^{85}Kr		
radioactive substances	11 10^6	10 TBq ^{3}H or ^{147}Pm	40%	3 10^4
	3 10^6	0.2 GBq ^{232}Th		
Electronic tubes	7 10^5	^{3}H, ^{60}Co, ^{63}Ni ^{147}Pm, ^{226}Ra		
Antistatic devices	?	^{210}Po		
Smoke detectors	1 10^5	^{226}Ra, ^{241}Am		
Ceramic, glassware, alloys, etc. containing uranium or thorium				
Articles with uranium paints	3 10^5	0.6 GBq ^{238}U	50%	1 10^6
Glassware containing uranium	4 10^3 kg	2 GBq ^{238}U	50%	3 10^5
Glassware containing thorium	16 10^3 kg	7 GBq ^{232}Th	10%	

[a] The data refer to the years 1973 or 1975, depending on the product

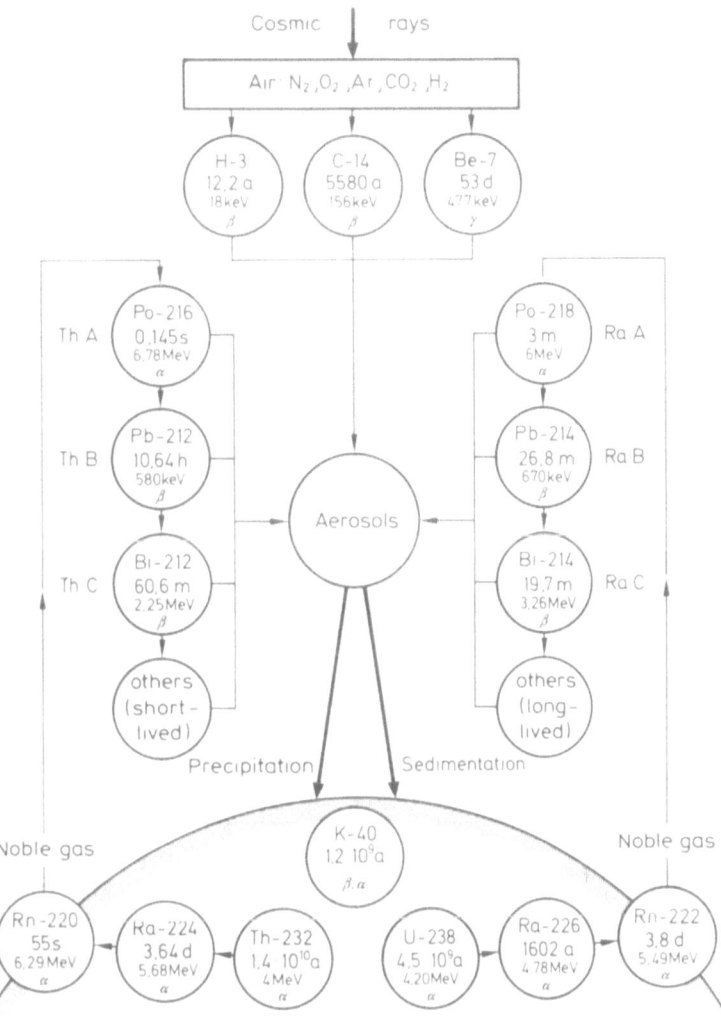

Fig. 1.6. Sources, distribution, and decay of natural radionuclides (after HUBER and STIEVE 1974)

year in an area of 50 km². Despite this considerable radon emission, the contribution of geothermic energy to radiation exposure of the population is not significant, due to the small number of these power plants and the rapid dilution of the radon concentration in the atmosphere.

1.3.6 Radiation Exposure Through Consumer and Industrial Products Containing Naturally Occurring Radionuclides

Many products in technical, scientific, and private use contain small amounts of naturally radioactive substances. Some of these radionuclides are essential for the performance or function of the products. As under some conditions the equipment and products concerned are not subject to compulsory licensing, it is impossible to draw up a complete inventory of them. Table 1.21 lists some of the products in

the Federal Republic of Germany (WEHNER 1978) which contain naturally occurring radioactive substances, giving information on their numbers or weight and the total activity of the radionuclides they contain.

The mean additional radiation exposure of the population from these products is certainly below 0.01 mSv per year, of which the highest proportion is due to luminescent watches. Although the actual radiation exposure is low, some control seems necessary, considering the variety of products and their quickly increasing number.

1.4 Summary

The effective dose equivalent of unmodified natural radiation exposure, averaged worldwide, is about 1.1–1.2 mSv per year in regions of "normal" envi-

Table 1.22. Estimated annual effective dose equivalents from natural sources of radiation in areas of "normal" background (UNSCEAR 1982)

Source	Annual effective dose equivalent (Sv)		
	External irradiation	Internal irradiation	Total
Cosmic rays			
- Ionizing component	$2.8 \cdot 10^{-4}$	-	$2.8 \cdot 10^{-4}$
- Neutron component	$2.1 \cdot 10^{-5}$	-	$2.1 \cdot 10^{-5}$
Cosmogenic nuclides	-	$8.0 \cdot 10^{-6}$	$8.0 \cdot 10^{-6}$
Primordial nuclides			
^{40}K	$1.2 \cdot 10^{-4}$	$1.7 \cdot 10^{-4}$	$2.9 \cdot 10^{-4}$
^{87}Rb	-	$4.0 \cdot 10^{-6}$	$4.0 \cdot 10^{-6}$
^{238}U series			
^{238}U–^{234}U		$9.6 \cdot 10^{-6}$	
^{230}Th		$7.4 \cdot 10^{-6}$	
^{226}Ra	$8.0 \cdot 10^{-5}$	$8.3 \cdot 10^{-6}$	$3.3 \cdot 10^{-4}$
^{222}Rn–^{214}Po[a]		$9.6 \cdot 10^{-5}$	
^{210}Pb–^{210}Po		$1.3 \cdot 10^{-4}$	
^{232}Th series			
^{232}Th		$7.2 \cdot 10^{-6}$	
^{228}Ra–^{224}Ra	$1.2 \cdot 10^{-4}$	$1.2 \cdot 10^{-5}$	$2.4 \cdot 10^{-4}$
^{220}Rn–^{208}Tl[a]		$1.0 \cdot 10^{-4}$	
Total	$6.2 \cdot 10^{-4}$	$5.5 \cdot 10^{-4}$	$1.2 \cdot 10^{-3}$

[a] The dose from inhalation of ^{222}Rn, ^{220}Rn, and their short-lived decay products was only considered for the time spent continuously outdoors

Table 1.23. Estimated annual effective dose equivalents from enhanced natural sources of radiation (KELLER 1988)

Source	Annual effective dose equivalent (Sv)
Cosmic rays (air traffic)	$<1 \cdot 10^{-5}$
Use of building material with high concentration of natural radionuclides	$\sim 1 \cdot 10^{-4}$
Inhalation of ^{222}Rn, ^{220}Rn, and decay products indoors	$0.4-1 \cdot 10^{-3}$
Use of phosphates	$<1 \cdot 10^{-5}$
Energy production - Coal-fired plants	$<1 \cdot 10^{-5}$
- Oil, natural gas, geothermic energy	$<1 \cdot 10^{-5}$
Consumer goods	$<1 \cdot 10^{-5}$
Total	$0.5-1.1 \cdot 10^{-3}$

ronmental radioactivity. Figure 1.6 provides information on the origin, distribution, and decay of natural radionuclides which are relevant for the natural radiation exposure of man (HUBER, see STIEVE 1974).

The individual contributions of radiation from unmodified natural sources are listed in Table 1.22 according to UNSCEAR (1982). The annual effective dose equivalent from inhalation of ^{222}Rn, ^{220}Rn, and their short-lived decay products is only considered for the time spent outdoors as, in our opinion, the contribution of the additional inhalation dose through living indoors has to be regarded as part of the effective dose equivalent of radiation from modified natural sources (see Table 1.23). The total effective dose equivalent of radiation from unmodified natural sources comprises to about an equal extent the two components of natural radiation, with $6.2 \cdot 10^{-4}$ Sv·a^{-1} external and $5.5 \cdot 10^{-4}$ Sv·a^{-1} internal irradiation.

With regard to the radiation from modified natural sources, only the use of building materials with an increased concentration of natural radionuclides

and the inhalation of ^{222}Rn, ^{220}Rn, and their short-lived decay products indoors contribute significantly to the effective dose equivalent of the population. If several unfortunate factors come together like, for example, in some regions of Sweden (an increased radionuclide concentration in soil, in drinking water, and in building materials plus a low ventilation rate indoors, caused by low outdoor temperatures), the contribution of natural radiation exposure increased by human interference can be even higher than that of unmodified natural exposure.

The inhalation of radon and thoron decay products indoors makes the chief contribution towards the effective dose equivalent of radiation exposure from natural sources modified by man. In view of this, the often proposed drastic reduction of the ventilation rate in modern buildings in order to save energy should be reconsidered. In future alternative possibilities should be studied, taking into account on the one hand desirable steps to save energy but on the other, the need to keep radiation exposure as low as is reasonably achievable.

References

Aurand K, Gans I, Rühle H (1974) Vorkommen natürlicher Radionuklide im Wasser. In: Aurand K et al. (eds) Die natürliche Strahlenexposition des Menschen, Thieme, Stuttgart, pp 30-50

Beck HL, Gogolak CV, Miller KM (1980) Perturbations of the natural radiation environment due to the utilization of coal as an energy source. In: Natural Radiation Environment III, CONF-80422, vol 2, pp 1521-1558

Bundesminister des Innern (1978) Die natürliche Strahlenexposition von außen in der Bundesrepublik Deutschland durch natürliche radioaktive Stoffe im Freien und in Wohnungen. Bericht über ein vom Bundesminister des Innern gefördertes Forschungsvorhaben, Bonn

Curtis SB (1974) Radiation physics and evaluation of current hazards. In: Tobias CA, Todd P (eds) Space radiation biology and related topics. Academic, New York

English RA, Bailey JV, Brown RD (1975) Application of Apollo cosmic radiation dosimetry to lunar colonization studies. In: Natural radiation environment, chap 5. United States Energy Research and Development Administration Report CONF-720805-P1

Erdtmann G, Soyka W (1979) The gamma rays of the radionuclides. Tables for applied gamma ray spectrometry. Verlag Chemie, Weinheim

Evans RD, Harley JH, Jacobi W, McLean AS, Mills WA, Stewart CG (1981) Estimate of risk from environmental exposure to radon-222 and its decay products. Nature 290: 98-100

Folkerts KH (1981) Messungen der Aktivitätskonzentrationen der natürlichen Radionuklide ^{222}Rn, ^{218}Po, ^{214}Pb, ^{214}Bi and ^{212}Pb/^{212}Bi in Luft von Wohnungen und im Freien und Abschätzung der daraus resultierenden Strahlenexposition. Diplomarbeit, Saarbrücken-Homburg

Folkerts KH (1983) Theoretische und experimentelle Untersuchungen über Diffusion und Exhalation der natürlich radioaktiven Edelgase Radon und Thoron aus Baustoffen und deren praktische Bedeutung für die Strahlenexposition in Wohnräumen. Dissertation, Universität des Saarlandes, Saarbrücken

Gopal-Ayengar AR, Sundaram KB, Mistry KB (1970) Studies in the high background areas in Kerala state: definition of the population and preliminary dosimetric data. In: Health Physics Division report. Bhabha Atomic Research Centre, United Nations document A/AC.82/G/L 1343

Gopal-Ayengar AR, Sundaram K, Mistry KB (1972) Evaluation of the long-term effects of high background radiation on selected population groups of the Kerala coast. In: Peaceful uses of atomic energy, vol 11. Proceedings of the Fourth International Conference, Geneva. Published by United Nations and the International Atomic Energy Agency, pp 31-51

Grigoriev Ju G (1976) Problems of space radiobiology. In: Kuzin AM (ed) Problems of radioecology and biological effects of low doses of ionizing radiation. Academy of Sciences USSR, Sykryvkar, pp 9-16

International Civil Aviation Organization (1978) Annual report of the council

International Commission on Radiological Protection (1975) Report of the Task Group on Reference Man. ICRP Publication 23. Pergamon, Oxford

Jacobi W (1979) Blei-210, Wismut-210, Polonium-210; Natürliche Radioaktivität, interne Dosimetrie und Dosisfaktoren bei Ingestion und Inhalation, GSF-Bericht, München, p 586

Jacobi W (1980) Zur Strahlenexposition der Bevölkerung durch Inhalation von Radon (^{222}Rn), Thoron (^{220}Rn) und ihrer kurzlebigen Zerfallsprodukte. Personal communication

Jacobi W (1982) Lung dose and lung cancer risk by inhalation of radon daughters. Proceedings of the 30th Annual Meeting of the Radiation Research Society, Salt Lake City, Utah (USA)

Kaufman S, Libby WF (1954) The natural distribution of tritium. Phys Rev 93: 1337-1344

Kaul A, Oberhausen E, Roedler HD, Werner E (1974) Interne Strahlenexposition des Menschen durch ^{40}K. In: Aurand K et al. (eds) Die natürliche Strahlenexposition des Menschen. Grundlage zur Beurteilung des Strahlenrisikos. Thieme, Stuttgart, pp 103-111

Keller G (1980) Der Gehalt natürlicher Radionuklide und die Radon-Exhalationsrate von Baustoffen und Messungen der Gammastrahlendosisleistung in Wohnungen in der Bundesrepublik Deutschland. In: The radiological burden of man from natural radioactivity in the countries of the European Community. CEC, Luxembourg, pp 193-208

Keller G, Oberhausen E (1978) Untersuchung über den Zusammenhang zwischen der spezifischen Aktivität von Baustoffen und der Dosisleistung. In: Die natürliche Strahlenexposition von außen in der Bundesrepublik Deutschland durch natürliche radioaktive Stoffe im Freien und in Wohnungen. Bericht über ein vom Bundesminister des Innern gefördertes Forschungsvorhaben, Bonn

Keller G, Schmier H, Muth H (1974) Externe Strahlenexposition durch terrestrische Strahlung in Gebäuden. In: Aurand K et al. (eds) Die natürliche Strahlenexposition des Menschen. Grundlage zur Beurteilung des Strahlenrisikos. Thieme, Stuttgart, pp 70-79

Keller G, Folkerts KH, Muth H (1982a) Activity concentrations of ^{222}Rn, ^{220}Rn and their decay products in German dwellings, dose calculations and estimate of risk. Radiat Environ Biophys 20: 262-274

Keller G, Folkerts KH, Dudler R, Muth H (1982b) Die Radonexposition in Wohnräumen und im Freien und die Abschätzung der resultierenden Lungendosis für die Bevölkerung. In: Strahlenschutz-Meßtechnik, Fachverband für Strahlenschutz, ISSN 0721-1694, München, pp 403-406

Keller G (1987) Die natürliche Strahlenexposition des Menschen und ihre Beeinflussung durch menschliche Tätigkeiten. Habilitationsschrift, Universität des Saarlandes, Homburg

Kobzev VA, Kolomeets EV, Shabansky VP (1975) Generation of continuous fluxes of protons, electrons and nuclei with $Z \geqq 2$ during the different periods of solar activity. In: 14th International Cosmic Ray Conference, Conference papers, vol 2. Max-Planck-Institut für Extraterrestrische Physik, München, pp 764-767

Kolb W (1974) Radionuclide concentrations in ground level air from 1971 to 1973 in Brunswick and Tromsö. Physikalisch-Technische Bundesanstalt Report PTB-Ra-4, Braunschweig

Krisiuk EM, Tarasov SI, Shamov VP (1971) A study on radioactivity of building materials. Leningrad Research Institute for Hygiene, Leningrad

Lederer CM, Shirley VS (1978) Tables of isotopes, 7th edn. Wiley, New York

Mastinu GG (1980) The radiological impact of geothermal energy. In: The radiological burden of man from natural radioactivity in the countries of the European Community. CEC, Luxembourg, pp 437-445

Mjönes L, Swedjemark GA (1976) Investigation of gamma-radiation in Swedish dwellings. Preliminary results from seven countries, National Institute of Radiation Protection Report 1976-09-01, Stockholm, Sweden

Muth H (1974) Bilanz der externen und internen natürlichen Strahlenexposition. In: Aurand K et al. (eds) Die natürliche Strahlenexposition des Menschen. Grundlage zur Beurteilung des Strahlenrisikos. Thieme, Stuttgart, pp 129-139

Oberhausen E (1963) Die Altersabhängigkeit des Kalium und

Caesium-137 Gehaltes des Menschen. Biophysik 1: 135-142

O'Brien F (1978) Human dose from radiation of terrestrial origin, Paper presented at the US Department of Energy Symposium on the Natural Radiation. Environ Biophys 13: 147-261

O'Brien K (1975) The cosmic ray field at ground level. In: The national radiation environment II. US energy Research and Development Administration report CONF-720805-P1

Pfister H, Phillip G, Pauly H (1976) Population dose from natural radionuclides in phosphate fertilizers. Radiat Environ Biophys 13: 147-261

Radke G (1969) Solar flare dose rates in a near earth polar orbit. In: Janni JF, Holly RE (eds) The current experimental approach to the radiobiological problems of space flight. Aerosp Med 40: 1495-1503

Savun OI, Senchuro IN, Shavrin PI (1973) Distribution of radiation dose in the radiation belts of the earth in the year of maximum solar activity. Kosm Issled 11/1: 119-123

Schaefer HJ (1974) Das Höhenprofil der kosmischen Strahlendosis. In: Aurand K et al. (eds) Die natürliche Strahlenexposition des Menschen. Grundlage zur Beurteilung des Strahlenrisikos, Thieme, Stuttgart, pp 1-9

Spiers FW (1968) Radioisotopes in the human body: physical and biological aspects. Academic, New York

Stieve FE (1974) Strahlenschutzkurs für Ärzte. Hoffmann, Berlin

Strahlenschutzkommission beim Bundesminister des Innern (1981) Stellungnahme zum Vergleich der Strahlenexposition der Bevölkerung durch Emission radioaktiver Stoffe aus Kohlekraftwerken und aus Kernkraftwerken. Personal communication

Swedjemark GA (1979) Indoor measurements of natural radioactivity in Sweden. In: The radiological burden of man from natural radioactivity in the countries of the European Community. CEC, Luxembourg, pp 271-296

Swedjemark GA (1980) Terrestrial and cosmic radiation in Scandinavia. In: The radiological burden of man from natural radioactivity in the countries of the European Community. CEC, Luxembourg, pp 125-147

Telegada K (1971) The seasonal atmospheric distribution and investigations of excess carbon-14 from March 1955 to July 1969. In: Health and Safety Laboratory Fallout Program Quarterly Summary report HASL-243, New York, pp I-2-I-87

United Nations (1979) 1978 Statistical yearbook. New York

UNSCEAR (1977) Sources and effects of ionizing radiation. United Nations Scientific Committee on the Effect of Atomic Radiation, Report 77, New York

UNSCEAR (1982) Sources and effects of ionizing radiation, United Nations Scientific Committee on the Effects of Atomic Radiation, Report 82, New York

UNSCEAR (1988) Sources and effects of ionizing radiation, United Nations Scientific Committee on the Effects of Atomic Radiation, Report 88, New York

Van der Heijde HB, Beens H, Monchy AR de (1976) Koninklijke Shell Laboratorium report AMSR/0047/76

Wehner R (1978) Legal and practical aspects of radioactivity in consumer products in the Federal Republic of Germany. In: Radioactivity in consumer products. US Nuclear Regulatory Commission Report NUREG/CP-0001, pp 97-105

Wicke A (1979) Untersuchungen zur Frage der natürlichen Radioaktivität der Luft in Wohn- und Aufenthaltsräumen. Grundlagen zur Abschätzung der Strahlenexposition durch Inhalation von Radon- und Thoron-Zerfallsprodukten. Dissertation, Universität Gießen

2 Occupational Risks in the Handling of Radioactive Substances

JOHANNES F. MEISSNER

CONTENTS

2.1 Introduction

Extensive superficial injuries were usually observed in physicians and their staff involved in the application of X-rays or radium during the first decades of this century. It was common practice to hold the barium platinum cyanide screen with one hand close to the tube while the other was placed between screen and tube to control the functioning of the equipment and the X-ray intensity. Such long-term continuous irradiation of the skin frequently caused chronic radiation dermatitis and X-ray ulcers. Furthermore, extensive pigmentations occurred on the forearms, face, and breast (MARCUSE 1896; FLASKAMP 1930; BECK et al. 1959).

Six years after the discovery of X-rays, an employee of an X-ray tube factory suffering from cancroid that had developed from an ulcer and which covered the whole of the back of the hand and had metastasized to the cubital and axillary lymph nodes was presented to the medical society of Hamburg by VON FRIEBEN (1902).

In 1959, HOLTHUSEN et al. published in a book of honor the names and records of 359 persons who had lost their lives assumably as a result of their occupational handling of ionizing radiation. The distribution of the dates of their deaths (MEISSNER 1980) shows an increased tendency until the mid-1930s, reflecting the increased use of radiation but also the latency before manifestation and progression of radiation injury. In the following years, however, mortality decreased due to radiation protection facilities and in spite of the further increase in the number of occupationally exposed persons. Most cases were associated with symptoms of the so-called X-ray hand (VON FRIEBEN 1902; FLASKAMP 1930; LEPPIN and MEISSNER 1984; Table 2.1). In addition relatively frequent cases of leukemia and aplastic anemia occurred. There was a tendency towards a higher incidence of leukemia, aplastic anemia, and bronchial carcinoma among those persons who handled radium sources and compounds.

Proposals for establishing dose limits for occupational exposure to radiation were first made in the 1920s. MUTSCHELLER (1925) developed the concept of a tolerance dose, i. e., a dose below an existing threshold for any biologic radiation effects. The "Mutscheller dose," i. e., 1.25 R per 5-day working week, was effective until at least 1950 in several countries (see International Labour Conference 1958). However, when MULLER (1928) demonstrated the proportionality between radiation dose and rate of mutations down to the lowest accessible doses from his *Drosophila* experiments, stricter limitations on permissible doses had to be considered. Accordingly, as early as 1929 the German Congress of Radiology ("Deutscher Röntgen-Kongreß") recommended limiting the permissible dose for workers handling X-rays or radionuclides to a level which would allow the spontaneous rate of mutations to double during their occupational life. This "doubling dose," which might be accumulated until the end of reproductive age, was estimated to be 40–50 R (BECK et al. 1959).

Table 2.1. Causes of death in persons who lost their lives as a consequence of occupational exposure to radiation up until 1959 (HOLTHUSEN et al. 1959; LEPPIN and MEISSNER 1984)

Total number	359
Number of persons for whom biographic data are available	343

Occupational exposure	Disease or cause of death	
X-rays in diagnostics and therapeutics ($n = 297$)	Cancerous tumors, proceeding from the hands (X-ray dermatitis, X-ray ulcus)	221
	Other cancerous tumors	24
	Leukemia and aplastic anemia	45
	Accidents (high voltage set)	7
Radiotherapeutics, using radium-sealed sources ($n = 21$)	Cancerous tumors, proceeding from the hands	9
	Other cancerous tumors	4
	Leukemia, aplastic anemia	8
Radium compounds; work in laboratories generally ($n = 25$)	Cancerous tumors, proceeding from the hands	4
	Bronchial carcinoma	6
	Leukemia, aplastic anemia	15

Obviously, systematic risk estimates had already been made at a time when most of the known radiation injuries belonged to the category of nonstochastic effects. Accordingly, Table 2.2 lists symptoms and prognoses of the radiation syndrome caused by increasing doses of uniform whole body irradiation. Threshold values are listed in Table 2.3 for exposures of particular organs. The doses indicated are standard values and are useful for classifying effects generally caused by accidentally increased occupational exposures only. They are not applicable when considering the risks of stochastic effects under usual working conditions.

From about 1940 onwards, the characteristic acute and delayed sequelae of radiation exposure were known from epidemiologic studies as well as animal experiments. Only the dose-response relationships remained uncertain for some effects. This was especially true for continuous radiation exposures at low dose rates spread over many years of working life. For this reason, efforts were concentrated on agreeing upon "tolerance" doses (DORNEICH et al. 1948).

A very pressing problem was the association of a particular injury in an individual with exposure to radiation. WEISS (1942) reported on physicians and nurses who claimed to have suffered injuries through their occupational exposures in communal hospitals, 25 of them from X-rays and four from radium sources. Considering the individual cases, the author critically discussed the legal problems involved in approving such cases as "occupational diseases." He pointed out the necessity of special regulations for radiation protection, e. g., a reduction in working hours (already considered obligatory at that time) and detailed proposals for rules concerning the control of exposures and medical supervision, which should be continued for a sufficiently long time in those cases in which doses exceeded the permissible limits. Nevertheless, more than a decade elapsed before his proposals were generally accepted within the ICRP recommendations (ICRP 9, 1966, 1969) and subsequent protection laws (IAEA 1987). Generally, the recommended dose limits for exposure of workers were gradually reduced with time. This shows the difficulty in agreeing upon specific and obligatory regulations. On the one hand, the number and types of applications of X-rays and radioactivity were expanding rapidly, thus requiring extension of the category of affected persons as well as prolongation of the individual exposure time. On the other hand, better knowledge of radiation effects and improved techniques of dosimetry provided the basis for the development of improved facilities for radiation protection, as reflected in statutory rules (RAUSCH 1982). Accordingly, the regulations need to be adjusted continuously to take into account the results of current scientific developments, especially if these affect the relations between the risks and benefits of recognized practices.

This report is mostly concerned with the handling of radioactive materials. Therefore, in the main, external exposures to radionuclides in the course of radiologic procedures and internal exposures due to incorporation will be discussed, particularly with respect to the diagnostic and therapeutic applications in the field of nuclear medicine.

2.2 Dose Limitations for Occupational Exposure: Recommendations of the International Commission on Radiation Protection

2.2.1 Principles of Dose Limitation

The limitation of health risks from handling radioactive material requires suitable dose limits for the workers concerned. Essentially, these limits should

Table 2.2. Health effects induced by single, brief, uniform exposure of the whole body to ionizing radiation of various dose equivalents (GLASSTONE 1957; GLASSTONE and DOLAN 1977)

	Subclinical range: 0-1 Sv		Therapeutic range: 1-10 Sv			Lethal range: >10 Sv	
	0-0.5 Sv	0.5-1 Sv	1-2 Sv	2-6 Sv	6-10 Sv	10-50 Sv	>50 Sv
			Clinical surveillance	Therapy effective	Therapy promising	Therapy palliative	
General subjective criteria, increasing within classified ranges	No perceptible effects	5%→10% of exposed persons: vomiting, nausea, fatigue, but not unfit for work	Slight or no illness up to 25% of exposed persons: vomiting (infrequent→common), nausea, combined with other symptoms of radiation disease, but no deaths	Acute symptoms of radiation disease: malaise, vomiting, nausea, diarrhea, loss of appetite. Increasingly fatality	Serious radiation disease. Survival possible, but at higher doses increasingly improbable.	Most serious criteria of illness, caused by prompt changes in the gastrointestinal tract and in the central nervous system	
Leading organ	—		Hematopoietic tissue			Gastrointestinal tract	Central nervous system
Characteristic findings	—	Slight changes of blood picture	Moderate leukopenia	Severe leukopenia, purpura, hemorrhage, infection Epilation (above 3 Sv)		Diarrhea, fever, disturbance of electrolyte balance	Convulsions, tremor, ataxia, lethargy
Initial phase Onset	—	—	3-6 h	½-6 h	¼-½ h	5-30 min	Almost immediately
Duration	—	—	≤1 day	1-2 days	≤2 days	≤1 day	
Latent phase Onset	—	—	≤1 day	1-2 days	≤2 days	1—0 days	Almost immediately
Duration	—	—	≤2 weeks	1-4 weeks	5-10 days	7—0 days	
Final phase Onset	—	—	10-14 days	1-4 weeks	5-10 days	10—0 days	Almost immediately
Duration	—	—	4 weeks	1-8 weeks	1- 4 weeks	10—2 days	
Critical period post-exposure	—	—	—	1-6 weeks		2-14 days	1-48 h
Therapy	—	Reassurance	Reassurance, hematologic surveillance	Blood transfusion, antibiotics	Consider bone marrow transplantation	Maintenance of electrolyte balance	Sedatives
Prognosis	Excellent		Excellent	Guarded	Guarded	Hopeless	
Convalescence period	—	—	Several weeks	1-12 months	Long	—	
Incidence of death	None	—	None	0%-90%	90%-100%	100%	
Death occurs within	—	—	—	2-12 weeks	1-6 weeks	2-14 days	<1-2 days
Cause of death	—	—	—	Hemorrhage, infection		Circulatory collapse	Respiratory failure, brain edema

Table 2.3. Estimates of the thresholds and symptoms for nonstochastic effects following a single brief exposure of particular organs in adults (GLASSTONE 1957; MEHL 1974; ICRP 26, 1977; ICRP 41, 1984)

Tissue	Total dose equivalent (Sv)	Effects	Presently recommended annual dose equivalent limit (Sv)
Skin	>2	Epilation	0.5
	>3	Dermatitis, type I: dermatitis combustionis erythematosa; similar to light heat combustion or sunburn; time of latency 2–3 weeks	
	>10	Dermatitis, type II: dermatitis combustionis bullosa; transepidermal effects, dry or wet, vesicant; time of latency 1–2 weeks	
	ca. 50	Dermatitis, type III: dermatitis combustionis necroticans; similar to scald or chemical cauterization; painful at once	
Lens	0.5–2.0	Detectable opacities	0.15
	5.0	Cataract	
Ovaries	2.5–6.0	Sterility or decreased fertility	0.2
Testes	0.15	Temporary sterility	0.2
	3.5	Permanent sterility	
Bone marrow	0.5	Depression of hematopoeisis	0.4
	1.5	Fatal aplasia	

be based on the quality and quantity of the effects that can be induced by irradiation. Considerations in this respect are decisively influenced by the increasing awareness of the harm possibly involved and the exposures assumed to be tolerable. In order to secure systematically the required international cooperation, the International Commission on Radiological Protection (ICRP) was established as long ago as 1928, on the occasion of the 2nd International Congress of Radiology. Four appointed committees of experts have met periodically ever since, commenting on actual problems. Committee No. 1 deals with radiation effects, No. 2 with internal exposures to incorporated radionuclides, and No. 3 with external exposures, while No. 4 is responsible for implementing ICRP recommendations. The tasks and organization of the Commission are set down in ICRP 9 (1966, 1969) and in ICRP Statement and Recommendations (1980). Since 1959 the ICRP publications have appeared in a successively numbered series, and since 1977 (from No. 24 on) as periodical "Annals of ICRP" (Pergamon Press). Reference data for all ICRP publications until 1979 (also those reports not included in the numbered series) are listed in the appendix to ICRP 26 (1977).

In consequence of these procedures, ICRP recommendations are generally accepted by all developed nations as a basis for legal regulations for nuclear and radiation techniques. The reliability of the methods of radiation protection is improved through the continuous cooperation of the ICRP with other groups of experts, such as the United Nations Scientific Committee on the Effects of Atomic Radiations (UNSCEAR) and the Committee of the Biological Effects of Ionizing Radiations (BEIR) of the USA National Research Council. Obviously, it is neither possible nor necessary within the scope of this paper to discuss all their conclusions. Therefore, in general only the references from the primary reports are given. Particular publications will be cited only if they are directly relevant to the subject concerned.

With regard to occupational risks of exposures, the ICRP aims (a) to define principles and practices of dose limitation, (b) to propose actual figures for the limits, considering exposures under different operating conditions, and to characterize suitable procedures for dose measurements and calculations, and (c) to quantify the risks of harm within the ranges of recommended dose limits with reference to the different effects involved in exposure to ionizing radiations. The recommended dose limits represent maximum permissible doses per year. According to ICRP 1 (1959), paragraph 45: "The Commission recommends, that all doses are to be kept as low as *practicable,* and that any unnecessary exposure is to be avoided." This statement replaced a formulation from an earlier ICRP publication (1955), recommending that the exposure for any use be kept as low as *possible.* Later recommendations, however, admitted that limitations on permissible doses should not jeopardize the application of a specific radiation technique. Accordingly, ICRP 9 (1966, 1969) specified that all doses have to be kept *as low as readily achievable,* commonly abbreviated

as ALARA. This premise was reemphasized in ICRP 22 (1973) and a system of dose limitations was deduced. This system initiated the working basis for ICRP 26 (1977), discussed in detail below. Major concepts and quantities related to the recommended system are compiled in ICRP 42 (1984) with regard to dosimetry, individuals and populations, and the justification and optimization of dose limitation and radiation protection.

2.2.2 The ICRP System of Dose Limitation

The system of dose limitation recommended by ICRP 26 (1977) contains the following main features:
1. No practice (including the handling of radioactive materials) shall be adopted unless its introduction produces a positive net benefit.
2. All exposures shall be kept as low as reasonably achievable, economic and social factors being taken into account.
3. The dose equivalent to individuals shall not exceed the limits recommended for the appropriate circumstances by the Commission.

In fact, the characteristics of this system do not yet seem sufficient to develop obligatory and general controllable regulations valid for any kind or amount of radiation used in practice. However, it was intended that restrictions on current applications and procedures be avoided and also that future requirements for scientific and technical progress be considered. Consequently, each practical application should be discussed individually by comparing the exposure involved with a cost-benefit analysis. Therefore, the acceptability of a proposed practice requires that risks of harm from exposures remain reasonably low and, furthermore, that the expenditure for optimizing the radiation protection appears justified in relation to the expected benefit.

According to ICRP 26 (1977) this approach may not be free from errors due to subjective evaluations. Therefore, recommendations for the techniques to be selected for individual projects should consider not only the requirements for radiation protection and exposure control but also other procedures and, if necessary, even alternative techniques, taking into account harmfulness as well as economic and social consequences. It is for these reasons that the ICRP recommendations are not based on the gross benefit of the procedures but on their positive net benefit only. This is obtained from the gross benefit by subtracting the basic costs of

production as well as those guaranteeing the required degree of safety (e. g., radiation protection) and those for injuries, which must be taken into account in accordance with the technique in question. Thus, a cost-benefit analysis has to ascertain to what extent a reduction of exposure might result in a net benefit. This, however, can only help to decide systematically whether reduced dose equivalents may be considered "reasonably achievable."

Detailed procedures for estimating and comparing cost equivalents have been proposed by the ICRP only with regard to collective dose equivalents (man-rem and man-Sv respectively; ICRP 22, 1973; ICRP 37, 1983; RAUSCH 1977). ICRP 22 has suggested costs ranging between $10 and $250 per man-Sv. The higher costs were derived from calculations of harm risks by considering the slopes of dose-response relationships at low doses in particular. Furthermore, costs of optimized expenditures per unit of collective dose have been published by WEBB and FLEISHMAN (1984), based on newer cost-benefit calculations. Extrapolation of these figures to individuals is not admissible, however, because benefits and costs cannot be assumed to be distributed homogeneously among all members of the group concerned. Thus, the limits on dose equivalents according to ICRP 26 (1977) for individuals are considered to be independent of differential cost-benefit analyses. For this reason they are obligatory even in those cases where the cost-benefit analysis had resulted in lower collective dose equivalents.

General concepts for the use of cost-benefit analyses have been described in ICRP 37 (1983), balancing cost of guaranteeing optimum levels of radiation protection and cost of the remaining detriment. Accordingly, the net benefit B of the introduction of a practice is shown by

$$B = V - (P + X + Y)$$

where V is the gross benefit of the introduction of the practice; P is the basic production cost of the practice, excluding the cost of radiation protection; X is the cost of achieving a selected level of radiation protection; and Y is the cost of the detriment resulting from the practice at the selected level of radiation protection.

ICRP 37 presents a detailed mathematical account for assessments of the different cost items to be considered for optimization. Furthermore, practical examples are discussed, including, for example, protection against external and internal exposures from various operations, including procedures in diagnostic radiology, radiotherapy, and nuclear medicine.

The incorporation of the ICRP recommendations into radiation protection legislation is carried out by international and national committees. These cooperate by regularly exchanging reports and results, especially at the time of the congresses organized by the International Radiation Protection Association every 4 years (KAUL et al. 1984b; Australian Radiation Protection Society 1988).

Most regulations of the developed nations, e. g., the basic standards of EURATOM (atomic authority of the Council of the European Communities 1980) have adopted the ICRP recommendations. In the Strahlenschutzverordnung of the Federal Republic of Germany (1976, 1989), however, the basic principles of radiation protection are set out using the original wording. According to paragraph 45 resp. 28, it has to be proven that in handling of ionizing radiations the exposures involved are kept "as low as possible." This is supposed to guarantee that radiobiologic considerations are taken into consideration in the planning of facilities with the most suitable and optimized radioprotection (STREFFER 1979; MEISSNER 1979). However, not even the ALARA concept remained free from objections. It is based on an appraisal of costs for protection facilities versus risks of harm and death, and is thus still the subject of controversy (ICRP 37, 1983; KATHREN et al. 1980; AUXIER and DICKSON 1983; SCHLAGER 1984; KATHREN 1984; WEBB et al. 1986).

2.2.3 Dose Equivalent Limits for Occupational Exposure

ICRP 1 (1959) expressed the main concepts of the Commission, laying the foundation for all later specific recommendations. Moreover, this first publication was not restricted to definitions of terms for dose limitations for various groups of persons, but also proposed actual dose limits for these groups. By using the term "maximum permissible dose," both in ICRP 1 (1959) and ICRP 9 (1966, 1969), it was implied that risks are considered acceptable up to a certain level. Limits on risks and doses differ for individuals and for groups selected from the general population.

The acceptance of risks cannot be based solely upon consideration of benefits from and need for the analyzed procedures; rather risks to life and health must be taken into account using different, equally appropriate techniques. Under these cir-

cumstances, satisfactory regulations are difficult to achieve. ICRP 22 (1973) proposed that the terms "acceptable dose" and "permissible dose" should not be used any further in order to avoid misinterpretations. Instead, the later ICRP publications exclusively use the term "dose limit." Apart from these formal corrections, the limits which had been proposed by ICRP 1 (1959) have been retained almost unchanged in all further reports of the Commission and have been adopted worldwide for legal regulations and codes of practice. But, as presented in ICRP 26 (1977), the analyzed procedures remain the subject of critical verification and control with regard to both the principles of dose limitation itself and the data on dose limits.

The figures for limitations on exposure are generally given as dose equivalents (H), expressed as the product of absorbed dose (D) and two dimensionless modifying factors, Q and N:

$$H = Q \cdot N \cdot D$$

The quality factor Q relates to the spatial microscopic distribution of the absorbed dose, which influences the effectiveness of irradiation. Thus it depends on the physical quality of the applied radiation. Q is fixed according to the collision stopping power, i. e., the linear energy transfer (LET) for unrestricted energy transfer processes, at the point of matter in question, standardized normally for water and wet tissue (HÜBNER and JAEGER 1974; ICRP 9, 1966, 1969). For a spectrum of radiation an effective value $(\overline{Q})$ can be calculated according to the International Commission on Radiation Units and Measurements (ICRU 25, 1976). For purposes of radiation protection in practice, ICRP 26 (1977) recommends the following figures for $\overline{Q}$ to be used for both external and internal exposures to various types of primary radiation:

X-rays, γ-rays, electrons, β-rays: $\overline{Q} = 1$

Neutrons, protons, and single-charged particles of rest mass greater than one atomic mass unit of unknown energy: $\overline{Q} = 10$

Particles and multiple-charged particles (and particles of unknown charge) of unknown energy:
$\overline{Q} = 20$

However, these values for $\overline{Q}$, conventionally adopted for practical radiation protection use, are still kept under review. For example, the ICRP Statement (1985) recommended changing $\overline{Q}$ from 10 to 20 for fast neutrons. This was rejected in a memorandum from the British Committee on Radiation Units and Measurements (BCRU) (1986), which suggested that the ICRP proposals were premature, at least.

The factor N represents the product of any further modifying factors which might need to be taken into account. In this context, ICRP 26 mentions, as examples, absorbed dose rates and fractionations. In principle, suggestions are considered that the actual values of the dose equivalent might be reduced for lower dose rates or dose fractionations at corresponding levels of absorbed doses. However, at present the ICRP (26, 1977; 42, 1984) has assigned N a value of 1.

Q and N are dimensionless. Thus, the SI unit of the dose equivalent is equal to that of the absorbed dose (Gray, formerly rad):

$$1 \text{ J/kg} = 1 \text{ Gy} = 100 \text{ rd}$$

However, in order to show that a multiplication with $\bar{Q}$ has been carried out, the special name "Sievert" (Sv, formerly rem) has been introduced for the unit of dose equivalent:

$$1 \text{ J/kg} = 1 \text{ Sv} = 100 \text{ rem}$$

The ICRP uses this unit to establish dose limits for practical radiation protection purposes. Limiting values for dose equivalents must ensure that nonstochastic effects are entirely excluded and that stochastic ones are safely kept below a level characterized by the system mentioned above.

Nonstochastic somatic effects are not to be expected at doses below 0.5 Sv (50 rem). Therefore, this value is recommended by ICRP 26 as the annual limit of dose equivalent for all tissues. In ICRP 41 (1984), however, some restrictions are proposed for the gonads, bone marrow, and lens of the eye. The presently recommended annual limits for these tissues are listed in Table 2.3. With regard to the lens, ICRP 26 concluded that a total dose equivalent of 15 Sv protracted over a working lifetime would be below the threshold for the development of severe opacification interfering with vision. For this reason, an annual limit of 0.3 Sv was proposed for the lens. However, from further observations it was concluded that even accumulated exposures of 15 Sv might impair vision. Thus, the recommended limit was reduced to 0.15 Sv per year by ICRP, Statement and Recommendations (1980). Whether this reduction was indeed necessary seems questionable, in view of the investigations by BENDEL et al. (1978), who compared groups of occupationally exposed persons with unexposed controls and did not find any significant differences.

In practice, the limitations for nonstochastic effects are only relevant for the irradiation of particular organs and tissues. However, it does not matter whether, using a particular procedure, an organ is exposed quite selectively or together with surrounding tissues. Nevertheless, utmost significance is attached to the demand of the ICRP system that this limitation is intended to constrain any exposure that fulfils the limitation for stochastic effects discussed below.

With reference to stochastic effects, the dose equivalent limits set by ICRP 26 (1977) cover the total risk to all tissues exposed to the irradiation in question. This applies especially to the effects from incorporation of radionuclides. Although the dose equivalent limits are defined for uniform irradiation of the whole body, it has to be ensured that for inhomogeneous exposure of particular organs and tissues the estimated effective dose equivalent does not exceed the limit for homogeneous irradiation of the whole body. Accordingly, the risks of stochastic effects within the ICRP limitation are based on uniform whole body exposure, but if necessary they need to be calculated from a weighting of the partial risks to the different tissues involved.

The dose equivalent limits refer to individual persons. They are applied to controlled working conditions (according to ICRP recommendations and final national regulations) as well as to environmental controls. For occupationally exposed persons the limiting values recommended are not free from some arbitrariness. Only for the determination of the rate of risks and for establishing protection rules for the public may the range of variation of natural exposure be used as a suitable standard for judging the tolerability of man-made contributions in addition to natural background exposure. With regard to dose equivalent limits for occupational exposure, the ICRP recommendations are based on comparisons of the hazards from radiation exposure with those associated with other occupational risks. For this purpose professions were selected that are known to have a high degree of safety as characterized by the mean annual mortality due to occupational hazards. The standard value for such industrial plants is proposed to be less than 10^{-4} (i. e., out of one million workers less than 100 per year lose their lives through industrial accidents). For many other industrial branches the annual death rate through accidents is far higher. Detailed data for plants in the different industrial countries are published in ICRP 27 (1977) and ICRP 45

(1985), with reference not only to cases of death but also to less severe injuries which occur much more frequently (see Sect. 2.3.1).

Injuries and acute diseases caused by exposures to ionizing radiation are rare among workers. Here, the risk is essentially characterized by the induction of a fatal malignancy. The recommended dose equivalent limits are based on this kind of estimate, and it is to be expected that if these limits are obeyed the mortality caused by radiation exposure will not exceed that in other occupational branches with comparably high degrees of safety. This basic requirement has to be fulfilled by the annual limit (L) H_L of the dose equivalent for uniform whole body (wb) exposure recommended by the ICRP (Nos 1, 9, 14, 26, 30 I):

$$H_{wb,L} = 50 \text{ mSv (5 rem)}$$

The values for risk limitations must be the same for uniform and nonuniform exposures. Thus, for nonuniform exposures the radiation doses to the different tissues must be determined by integrating their relative contribution to the annual limit H_L which characterizes the total risk of whole body exposure. This procedure seems particularly suitable when proposing limits for the incorporation of radionuclides by considering their specific distribution in the organism.

The total sum of the amounts H_T contributed by the particular tissues T must not exceed the annual limit of the dose equivalent $H_{wb,L}$

$$\sum_T w_T \cdot H_T \leq H_{wb,L}$$

where w_T is the weighting factor representing the contribution of the stochastic risk resulting from tissue T to the total risk for uniform exposure of the whole body (Table 2.4). The sum of the weighted mean values of dose equivalents of the particular tissues and organs is termed "effective dose equivalent" for the whole body (ICRP 42, 1984). Its annual limit $H_{wb,L}$ is obligatory for each individual employee. According to ICRP 26 any influence of age and sex on the values for total risk and thus on the weighting factors in Table 2.4 may be neglected for radiation protection purposes.

The annual dose equivalent limits for workers are composed additively of internal and external exposure. In order to determine those for external irradiations, approximate standardizations are useful. ICRP 26 (1977) recommends an assessment of the maximum value of the dose equivalent that would

Table 2.4. Total stochastic risk for nonuniform whole body irradiation (ICRP 26, 1977): Weighting factors w_T for estimating the contribution of the annual dose equivalent (H_T) in the particular tissues (T) to the annual limit ($H_{wb,L}$) valid for uniform exposure of the whole body

Tissue	w_T
Gonads	0.25
Breast	0.15
Red bone marrow	0.12
Lung	0.12
Thyroid	0.03
Bone surfaces	0.03
5 further tissues in order of decreasing dose equivalents with $w_T = 0.06$ each	0.30

The remaining tissues may be neglected. When the gastrointestinal tract is involved, w_T is 0.06 for stomach, small intestine, upper large intestine, and lower large intestine.

occur at a depth d > 10 mm in the irradiated body, simulated by a 30-cm sphere for exposure to penetrating radiations. From this, a deep dose equivalent index is defined, which, for external exposure, should be limited exclusively to the annual value of 50 mSv. For the layer of tissue surrounding this sphere (0.07 < d < 10 mm) a shell dose equivalent index is defined, for which an annual limit of 500 mSv is proposed for skin (ICRP 26); limits for other tissues possibly concerned are listed in Table 2.3 (ICRP 41, 1984).

The inclusion of internal exposure through incorporation of radioactive substances in the system of dose limitations requires an estimation of the effective dose equivalent of the whole body produced by an annual intake I of each of the radionuclides j mentioned (see Sect. 2.3.4). For external and internal exposures together, the annual limits of dose equivalents need to fulfil the following conditions:

$$\frac{H_{I,d}}{H_{E,L}} + \sum_j \frac{I_j}{I_{j,L}} \leq 1$$

and

$$\frac{H_{I,s}}{H_{sk,L}} \leq 1$$

where $H_{I,d}$ is the annual depth dose equivalent index; $H_{I,s}$ is the annual shell dose equivalent index; $H_{E,L}$ is the annual effective dose equivalent limit; $H_{sk,L}$ is the annual dose equivalent limit of the shell (skin); I_j is the annual intake of radionuclide j; $I_{j,L}$ is the annual limit of intake (ALI) for radionuclide j.

In previous publications (ICRP 9, 1966; ICRP 25, 1976) the annual limits were given as "maximum permissible dose equivalents per year for adults, exposed in the course of their work" and tabulated together with dose equivalent limits for members of the public. For the latter it was recommended that values be kept below those for workers by a factor of 10. Since the types of problems in radiation protection vary considerably, ICRP 26 (1977) introduced a system for the classification of working conditions:

Working condition A: The annual exposure might exceed three-tenths of the dose equivalent limits.

Working condition B: It is most unlikely that the annual exposure will exceed three-tenths of the dose equivalent limits.

With reference to these recommendations, Table 2.5 indicates the annual limits for both categories. The national regulations for radiation protection generally contain these figures. In rough accordance with ICRP 14 (1969), the annual limits are listed in Table 2.5 for various tissues mostly exposed during operations. These values appear helpful in facilitating the techniques of radiation application.

The advantages of this classification are obvious. It allows assessment of the extent and frequency of medical surveillance as well as the regular dose controls of workers. However, this procedure does not generally imply reduced annual limits for workers in category B. Rather it is only a matter of introducing standards to classify operation areas with regard to sufficient adaptation of facilities for radiation protection.

For workers under 18 years of age the limits are restricted to one-tenth of the basic values for category A, i. e., to 5 mSv per year for the whole body. Thus, the limits for juvenile workers correspond to those of the public as recommended in ICRP 25 (1976). However, it should be pointed out that the quite formal system for determining the limitation of dose equivalents for the public did provoke objections. Consequently ICRP 26 completely abandoned recommandations of limits for the population. Some of the national regulations nevertheless continue to set limits of this kind, although basing them on a reasonable degree of realization (ICRP 43, 1985).

According to ICRP 26 arrangements should be made for acting in abnormal situations. Intervention levels and appropriate actions for limited exposures should be subject to operating instructions. Corresponding comments and restrictions have been introduced into the regulations:

1. The annual dose equivalent limit (50 mSv) is valid without any reservations. Therefore it is not admissible to balance any overexposure caused by extraordinary events with lower exposures in previous or following years.

2. In the event of accidents, dose equivalents higher than the annual limits must be accepted if they are unavoidable. But precautions must be taken that for any single operation an individual dose equivalent is limited to 100 mSv at most. Over a working lifetime the cumulative dose equivalent has to be limited to 250 mSv for such tasks. In addition, it is required that the mean value of 50 mSv per year is reached again as soon as possible during subsequent employment.

3. It must be ensured that pregnant women continue in employment only under working condition B.

4. The figures in Table 2.5 refer to exposures common in the professional handling of ionizing radiation. ICRP 26 explicitly does not recommend separate annual dose equivalent limits for individual tissues and organs. If such values are required, they may be obtained by dividing $H_{wb,L}$ (50 mSv/a) by the corresponding values w_T (Table 2.4), assuming, however, that the resulting values have upper limits identical to those for nonstochastic effects (500 mSv).

5. The occupational exposures of workers are additively increased by those resulting from individual medical applications. However, in general, medical exposures should not influence any of the procedures of dose limitation applied to exposure from other sources. In exceptional cases, nevertheless, consequences with respect to the working situation and conditions should be considered for those workers who have undergone radiodiagnosis or ra-

Table 2.5. Annual limits of dose equivalents for adults (> 18 years of age) in the course of their work

Organ or tissue	Annual limits for workers (mSv)	
	In category A	In category B
Whole body, red bone marrow, gonads, uterus	50	15
Skin (except skin of hands, forearms, feet, and ankles)	300	100
Hands, forearms, feet, lower legs, ankles (inclusive of skin)	750	250
Bone, thyroid	300	100
Other single organs	150	50

Table 2.6. Radionuclides for nuclear medical use: dose rate constants (Gγ), tenth value lead thickness (TVT), and skin dose rates ($\dot{D}$); estimated maxima to fingers in contact with nonshielded syringes containing radionuclides (IRCP 21, 1973; 25, 1976)

Radionuclide	Gγ $\frac{\mu Gy \cdot m^2}{h \cdot MBq}$	TVT cm	$\dot{D}$ $\frac{\mu Gy}{s \cdot MBq}$
^{99m}Tc	0.018	1.0	0.045–0.225
^{113m}In	0.042	3.4	ca. 0.675
^{131}I	0.048	2.4	0.63–3.15
^{198}Au	0.057	3.6	0.36–0.9

diotherapy involving heavy partial irradiation of the body.

The handling of unsealed radionuclides, for instance in nuclear medicine, involves considerable risks not only through incorporation but also by contaminations and external exposures, especially of the hands. Quantitative statements for the use of γ-emitters are given in ICRP 25 (1976), although modern devices for handling radionuclides have been introduced in the meantime. The data for the dose rate constants (G γ), the dose rate for skin (D) during the handling of unshielded syringes, and "tenth value thickness" (TVT) (ICRP 21, 1973) of lead shielding for heavily attenuated broad beams for some of the radionuclides frequently used for nuclear medical purposes are listed in Table 2.6.

Exposure rates from shielded and unshielded syringes have been compared for ^{99m}Tc by MAINI et al. (1987). The shields were made of lead 33 mm thick. Thermoluminescent dosimeters were fixed at the surface of the syringes as well as at different sites of a tissue equivalent phantom hand. According to ICRP 25 the exposure rates of unshielded syringes handled over the full volume amount to 0.16 μGy/MBq·s for thumb and forefinger. When using shields, these values are reduced to ca. 0.01%. However, the procedures for handling shielded syringes are so unpractical that the operators tend to avoid them in routine clinical practice. Thus the authors consider it important that a substantial reduction in exposure (by a factor of 15) can be obtained by suitable syringe handling, avoiding positioning of the fingers over the active volume.

ICRP 25 also deals with procedures for controlling radioactive contamination and decontamination of equipment and personnel. Basic manuals for estimating incorporations with regard to dose equivalent limits will be discussed in the following section.

2.2.4 Limits for Incorporation (Ingestion and Inhalation)

In ICRP 1 (1959) a general scheme was formulated for the determination of dose equivalents following the incorporation of radioactive substances during clinical or industrial use. Obviously, their distribution and retention in the organism has to be considered. This depends upon their properties and chemical composition with regard to their routes of intake, which are mainly via the gastrointestinal tract (ingestion) or the respiratory system (inhalation). An extensive synopsis of relevant numerical data was issued in ICRP 2 (1959) – and later in ICRP 6 (1964) – resulting in recommendations for maximum permissible doses for internal irradiation as well as maximum admissible concentrations of radionuclides in environmental water and air. Limits were defined for professional exposure based on a 40-h working week, or a total of 168 h per week. They were estimated separately for ingestion and inhalation of soluble and insoluble compounds. Incorporations due to absorption by the skin in the case of submersions or open wounds, however, were only considered for certain radionuclides.

The values of the maximum permissible concentrations in water and air referred to the standard man (70 kg) with a working life of 50 years and were based on an annual dose equivalent of 50 mSv (5 rem) in the critical organ (ICRP 1, 1959; RAJEWSKY 1956). In ICRP 6 (1964) it was recommended that radioactivity controls of exhaled air and excretions be carried out as soon as incorporation is suspected. In this case, whole body counters were suggested to be most useful in determining the amounts of any radionuclide depositions in the organism.

In order to specify the biologic conditions characterizing the passage of the labeled compounds through the organism more accurately, ICRP 10 (1968) and 10 A (1971) introduced the concept of transportability instead of solubility. Evidently, the physical and chemical definition of solubility did not seem sufficiently suitable for describing the metabolism of the radioactive compounds. ICRP 23 (1975) gives various examples of compounds with quite different properties but comparable solubility as regards their absorption and translocation after inhalation or ingestion. The definition of transportability, however, covers the physical and chemical properties of the compound involved as well as its reactions with body cells and fluids and, furthermore, effects of biologic processes.

The ICRP scheme for the determination of the

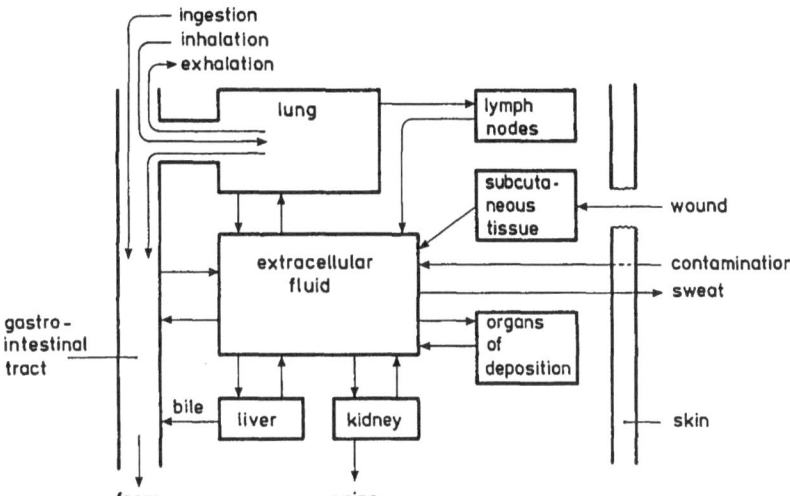

Fig. 2.1. Principal metabolic pathways of radionuclides in the body: model for analysis of incorporation processes. (ICRP 10, 1968; 30 I, 1979)

dose equivalent resulting from incorporation of radionuclides due to occupational exposure requires the quantitative estimation of their distribution within the organism. For this purpose the ICRP recommended that a distinction be made between the intake of radionuclides (i. e., the amount entering the organisms), their uptake (i. e., the amount absorbed into the extracellular fluid), and their deposition (i. e., the specific distribution and retention based on typical radioactivity concentrations in the affected organs and tissues). From experimental results, mathematical models considering the different parameters of the passage have to be derived for each of the relevant radionuclides and their various compounds. The schematic system of the most important metabolic routes dealt with by ICRP 10 is shown in Fig. 2.1. Thus, inhalation, ingestion, absorption through the skin, and entry by wound are possible starting points for incorporation, followed by translocation, deposition, and elimination of the radionuclides involved. The results of studies reported by ICRP 10 and 10A cover the mathematical analysis of metabolic pathways, depending on uptake procedures and distribution mechanisms. The time integrals for the incorporation of various radionuclides over a period of 50 years are determined by retention equations concerning especially the critical organ. Furthermore, practical advice is given for monitoring methods, measurements of incorporation, and calculations of dose equivalents for numerous radionuclides. While ICRP 10 makes general comments on the problems of single incorporations in workers, ICRP 10A refers to recurrent and prolonged uptakes. As mentioned above, the statements in ICRP 10 and 10A are generally

oriented towards the concept of risk of harm induced in the critical organ. According to the definition formulated by ICRP 1 (1959) and 14 (1969), this is the organ for which the ratio of the absorbed dose of the incorporated radionuclide to the dose limit is highest. The basic idea behind this rather formal definition was the attempt to apply the risk of harm for the whole body to that organ in which exposure affects most adversely the health of the individual (RAJEWSKY 1956). This functional association depends not only on the distribution pattern of radioactivity but also on its retention and excretion, and not least on the radiosensitivity of the tissue in question.

However, subsequently this concept was considered to be unsatisfactory for stochastic effects, as it neglects the probability of such effects in other exposed tissues, even at lower concentrations of radioactivity and radiosensitivity. Thus, ICRP 26 (1977) recommended the weighting of different relevant tissues (Table 2.4, see also Sect. 2.2.3) in order to consider the risk of stochastic effects in the whole organism.

The incorporation of radionuclides remains effective for varying lengths of time due to their different metabolic behavior. Consequently, the effective dose equivalents as well as the corresponding risks have to be considered continuously. The corresponding quantity unit introduced by ICRP 26 (1977) is the infinite time integral of the individual dose equivalent rate $\dot{H}$ (t) in an organ or tissue concerned, called "dose equivalent commitment" (H_c)

$$H_c = \int_0^\infty \dot{H} \, (t) \, dt$$

For the purposes of radiation protection with reference to a working life, the term "committed dose equivalent" (H_{50}) is more convenient. It is defined as the integrated dose equivalent rate ($\dot{H}_t$) following a single intake of radioactive material into the body at the time t_0, accumulated over 50 years in a particular organ or tissue:

$$H_{50} = \int_{t_0}^{t_0 + 50\,a} \dot{H}(t)\,dt$$

Thus, the committed dose equivalent represents the quantity convenient for estimating the effective dose equivalent due to intake of radionuclides. It depends on uptake and deposition of the radioactive compound. Corresponding values of the committed dose equivalent for the various tissues are required because these contribute to the effective dose equivalent for the whole body, which must remain within the proposed limits.

With reference to this, the data in ICRP 9 (1966, 1969) for admissible intakes have been replaced by new standards published in ICRP 30 (1979/81). Part 1 of this publication deals with the basic concepts regarding limitations of intake of radionuclides by workers. In parts 2 (1980) and 3 (1981), the annual limits for intake (ALI) are stated for all relevant radionuclides, estimated on the basis of the specific metabolic reactions of their various chemical compounds. Particular supplements (ICRP 30, supplements to part 1, 1979, to part 2, 1981, to part 3, 1981) contain all the details concerning analytical models for body passage and dosimetry necessary for estimating the limits involved. In general, these data are largely restricted to the inhalation and ingestion of radionuclides (see Fig. 2.1). Appropriately, they are also based on the dose equivalent limits for workers given in ICRP 26 (1977) but are now related to the committed dose equivalents for the different tissues (T) by including the weighting factors w_T (Table 2.4):

$$\sum_T w_T \cdot H_{50,T} \leq 50 \text{ mSv for stochastic effects,}$$

$$H_{50,T} \leq 500 \text{ mSv for nonstochastic effects}$$

The standards for controls of the internal dose for workers are derived from anatomic and physiologic data for adults given in ICRP 23, which lists the weights of organs and tissues of the 70-kg reference man (Table 2.7). ICRP 23 and 30 separately consider the effects on organs surrounding the directly exposed tissue in the case of penetrating γ-emitters. Accordingly, Table 2.7 indicates whether the organs

Table 2.7. Weights of organs and tissues of 70-kg reference man (ICRP 23, 1975; ICRP 30, 1979/81)

	Organs	Weight (g)
Both source and target organs	Ovaries	11
	Testes	35
	Muscle	28000
	Red marrow	1500
	Lungs	1000
	Thyroid	20
	Spleen	180
	Kidneys	310
	Liver	1800
	Pancreas	100
	Skin	2600
	Adrenals	14
Source organs exclusively	Stomach content	250
	Small intestine content	400
	Upper large intestine content	220
	Lower large intestine content	135
	Cortical bone	4000
	Trabecular bone	1000
	Bladder content	200
Target organs exclusively	Bone surface	120
	Stomach wall	150
	Small intestine wall	640
	Upper large intestine wall	210
	Lower large intestine wall	160
	Thymus	20
	Uterus	80
	Bladder wall	45

have to be regarded as source or target tissues, or both. Source tissues have been shown to emit considerable amounts of photons into the target tissue. However, for certain source substrates (e.g., intestinal contents), the influence of γ-emitters is negligible for the whole body dose. In general, however, both contributions have to be considered in the determination of the dose equivalent limits.

Controls of dose equivalent limits for workers are derived from the ALI. In ICRP 30 this quantity is defined as the largest value of the annual intake I, which satisfies both of the following inequalities:

$$I \cdot \sum_T w_T \cdot H_{50,T,I_u} \leq 50 \text{ mSv for stochastic effects}$$

$$I \cdot H_{50,T,I_u} \leq 500 \text{ mSv for nonstochastic effects}$$

where I (Bq) is the annual intake of the specified radionuclide either by ingestion or by inhalation, and H_{50,T,I_u} ($Sv \cdot Bq^{-1}$) is the committed dose equivalent in the tissue T from the intake of unit activity ($I_u = 1$ Bq) of the radionuclide by the specified route.

The first of the inequalities ensures that the maximum intake still satisfies the effective whole body annual dose equivalent limit for stochastic effects (50 mSv/a). The second takes into account the ICRP recommendation that the dose equivalent in any of the tissues involved is permitted to exceed the annual limit for nonstochastic effects (500 mSv/a) (see Sect. 2.2.3).

With regard to the route of inhalation, ICRP 30 suggests that the value of ALI is appropriate for limiting the concentration of radionuclide in air C (t) (in $Bq \cdot m^{-3}$) at any time during a single year

$$\int C \ (t) \cdot B \ (t) \leq ALI \ (Bq)$$

where B (t) (in m^3 per unit time) is the volume of air breathed by the worker per unit time, and the limits of integration are spread over a working year. The value of B (t) depends on how heavy the particular work is, as it is not possible to quantify B (t) exclusively from the values for light work given in ICRP 23 in relation to standard man. For heavy work it may be more than twice that value. Therefore, the particular working conditions have to be considered in each case. Standards for "light activity" only are published in ICRP 30 and need to be adapted to particular cases. These values of "derived air concentrations" (DAC) are defined as those concentrations of radionuclide in air (Bq· m^{-3}) breathed by reference man for a working year under conditions of "light activity" which would result in the ALI through inhalation:

Working year: 50 weeks × 40 h × 60 min = $120 \cdot 10^3$ min
Volume of air breathed per minute: 0.02 m^3
$DAC = \dfrac{ALI}{2.4 \cdot 10^3} \ Bq \cdot m^{-3}$

With reference to the metabolism of the various radionuclides, the standard values for ALI and DAC depend on the relevant chemical compounds as well as their physical properties, e.g., solubility, physical state, and intake conditions, especially for colloids. For several substrates the transfer of the radionuclide into the blood is decisive after ingestion or inhalation. In general, however, a quantitative differential analysis of the various pathways, schematically indicated in Fig. 2.1, is extremely complicated. Compartmental models used for the respiratory system and gastrointestinal tract must completely account for the passage of the radionuclides through the organism. Starting with the transfer compartment, organ and tissue compartments

need to be considered and the degree of deposition and excretion for each specific compound has to be determined. Thus, for practical purposes, it is more or less unavoidable that approximate generalizations are resorted to in order to characterize the kinetics of radioactivity in the body.

Concepts and details of applied compartments are given in ICRP 30 I (1979). Dosimetry models are based on the characteristics of the radioactive compounds and their metabolic functions. Thus, models for the respiratory system, the gastrointestinal tract, and, furthermore, osseous tissue have been established, the latter being of particular importance because the cells determining the carcinogenic risk to the skeleton have been identified as the hematopoietic stem cells of the bone marrow, and, of the osteogenic cells, those located on endosteal surfaces as well as certain epithelial cells close to bone surfaces (ICRP 11, 1968). A further model deals with exposure to radioactive clouds of gas, especially with regard to limiting the exposure of workers to inert radioactive noble gases and tritium.

These mathematical models for the kinetics of incorporation of radionuclides have been developed for estimating rates of transfer between the compartments and for determining radioactivity concentrations and equivalent doses in the relevant tissues. ALI and DAC values calculated by this method are given in Table 2.8 for those radionuclides administered most frequently in nuclear medicine. Data for the mother nuclides are also tabulated for generator systems used routinely in nuclear medicine. With regard to the complicated models for the respiratory system, the ICRP recommends the classification of radioactive compounds with respect to their retention in the pulmonary region. Thus, ICRP 30 I distinguishes three classes, D, W, and Y, according to the retention period of the compounds in question after their deposition in the alveoli. The classification in Table 2.8 depends on the chemical and physical form of the inhaled nuclides, especially radiocolloids. Furthermore, in Table 2.8, DAC values for intake after exposure to elemental tritium, ^{85}Kr, and ^{133}Xe are listed.

The ICRP 30 recommendations were commented on by several authors. BAIR (1979), himself a member of the ICRP Committee II, reported on motivation and results, comparing ALI and DAC values with those in ICRP 2 for maximum permissible concentrations. Table 2.9 shows this comparison for data of radionuclides listed in Table 2.8. Obviously, for some of the selected radionuclides, the ICRP 30 values are considerably higher than those in ICRP 2. There are quite a number of reasons for

Table 2.8. Annual limits for intake (ALI: Bq) and derived air concentrations (DAC: $Bq \cdot m^{-3}$) for selected radionuclides used routinely in nuclear medicine, with reference to professionally exposed employees, working a 40-h week (ICRP 30, 1979/81). Inhalation classification D, W, and Y, corresponding to retention half-lives of < 10, 10–100, and > 100 days, standardized for aerosols with median aerodynamic diameter of 1 µm, distributed from compartment of lung parenchyma; ICRP model for clearance from the respiratory system. GRT, group of relative radiotoxicity (ICRU 25, 1976, Council of the European Communities 1980)

Radionuclide (physical half-life)	Oral intake — Metabolic specification (chem. compound, phys. state)	ALI	Intake by inhalation — Chemical compound	Class D ALI	Class D DAC	Class W ALI	Class W DAC	Class Y ALI	Class Y DAC
$^{3}_{1}H$ (12.3 a), GRT 4	Tritiated water	$3 \cdot 10^{9}$	Tritiated water	$3 \cdot 10^{9}$	$8 \cdot 10^{5}$	(Intake by inhalation and skin absorption)			
	Tritium (elemental)	-	Tritium (elemental)	-	$2 \cdot 10^{10}$	Exposure to lungs			
	Organic compounds		No specific values of ALI from ICRP, but they might differ considerably from those of tritiated water; for tritiated thymidine ALI might be as much as 50 times smaller						
$^{11}_{6}C$ (20.38 min), GRT 4	Organic compounds	$2 \cdot 10^{10}$	Organic compounds / CO / CO_2	$2 \cdot 10^{10}$ / $2 \cdot 10^{10}$ / $2 \cdot 10^{10}$	$6 \cdot 10^{6}$ / $1 \cdot 10^{7}$ / $1 \cdot 10^{7}$	Instantaneously uniformly distributed throughout all organs and tissues; retention time indefinite			
$^{14}_{6}C$ (5730 a), GRT 3	Organic compounds	$9 \cdot 10^{7}$	Organic compounds / CO / CO_2	$9 \cdot 10^{7}$ / $6 \cdot 10^{10}$ / $8 \cdot 10^{9}$	$4 \cdot 10^{4}$ / $3 \cdot 10^{7}$ / $3 \cdot 10^{6}$	Instantaneously uniformly distributed throughout all organs and tissues; biological half-life ca. 40 days			
$^{18}_{9}F$ (1.8 h), GRT 3	Soluble fluorides (not differentiated)	$2 \cdot 10^{9}$	Fluorides, classes varying with different metals[a]	$3 \cdot 10^{9}$	$1 \cdot 10^{6}$	$3 \cdot 10^{9}$	$1 \cdot 10^{6}$	$3 \cdot 10^{9}$	$1 \cdot 10^{6}$
$^{22}_{11}Na$ (2.62 a), GRT 2	Not differentiated	$2 \cdot 10^{7}$	All Na compounds	$2 \cdot 10^{7}$	$1 \cdot 10^{4}$				
$^{24}_{11}Na$ (15.05 h), GRT 3	Not differentiated	$1 \cdot 10^{8}$		$2 \cdot 10^{8}$	$8 \cdot 10^{4}$				
$^{32}_{15}P$ (14.3 d), GRT 3	Not differentiated	$2 \cdot 10^{7}$	All P compounds	$3 \cdot 10^{7}$	$1 \cdot 10^{4}$				
			Exception: phosphates of some metals[a]			$1 \cdot 10^{7}$	$6 \cdot 10^{3}$		
$^{35}_{16}S$ (87 d), GRT 3	Inorganic compounds	$4 \cdot 10^{8}$	Sulfates and sulfides[a] either	$6 \cdot 10^{8}$	$3 \cdot 10^{5}$				
			or			$8 \cdot 10^{7}$	$3 \cdot 10^{4}$		
	S (elemental)	$2 \cdot 10^{8}$	S (elemental)			$8 \cdot 10^{7}$	$3 \cdot 10^{4}$		
$^{42}_{19}K$ (12.4 h), GRT 3	Not differentiated	$2 \cdot 10^{8}$	All K compounds	$2 \cdot 10^{8}$	$7 \cdot 10^{4}$				
$^{45}_{20}Ca$ (165 d), GRT 2	Not differentiated	$6 \cdot 10^{7}$	All Ca compounds			$3 \cdot 10^{7}$	$1 \cdot 10^{4}$		
$^{47}_{20}Ca$ (4.5 d), GRT 3	Not differentiated	$3 \cdot 10^{7}$				$3 \cdot 10^{7}$	$1 \cdot 10^{4}$		
$^{51}_{24}Cr$ (27.8 d), GRT 3	Inorganic compounds (3- and 6-valid)	$1 \cdot 10^{9}$	All Cr compounds	$2 \cdot 10^{9}$	$7 \cdot 10^{5}$				
			Exceptions: halides and nitrates			$9 \cdot 10^{8}$	$4 \cdot 10^{5}$		
			oxides and hydroxides					$7 \cdot 10^{8}$	$3 \cdot 10^{5}$
$^{55}_{26}Fe$ (2.6 a), GRT 3	Not differentiated	$3 \cdot 10^{8}$	All Fe compounds	$7 \cdot 10^{7}$	$3 \cdot 10^{4}$				
			Exceptions: oxides, hydroxides, halides			$2 \cdot 10^{8}$	$6 \cdot 10^{4}$		
$^{59}_{26}Fe$ (45 d), GRT 3	Not differentiated	$3 \cdot 10^{7}$	All Fe compounds	$1 \cdot 10^{7}$	$5 \cdot 10^{3}$				
			Exceptions: oxides, hydroxides, halides			$2 \cdot 10^{7}$	$8 \cdot 10^{3}$		
$^{57}_{27}Co$ (267 d), GRT 3	Oxides, hydroxides organic complexed	$2 \cdot 10^{8}$	All Co compounds	$1 \cdot 10^{8}$	$4 \cdot 10^{4}$				
	and other inorganic compounds	$3 \cdot 10^{8}$	Exceptions: oxides, hydroxides, halides, nitrates					$2 \cdot 10^{7}$	$1 \cdot 10^{4}$
$^{58}_{27}Co$ (71 d), GRT 3	Oxides, hydroxides organic complexed	$5 \cdot 10^{7}$	All Co compounds			$4 \cdot 10^{7}$	$2 \cdot 10^{4}$		
	and other inorganic compounds	$6 \cdot 10^{7}$	Exceptions: oxides, hydroxides, halides, nitrates					$3 \cdot 10^{7}$	$1 \cdot 10^{4}$
$^{60}_{27}Co$ (5.27 a), GRT 2	Oxides, hydroxides organic complexed	$7 \cdot 10^{6}$	All Co compounds			$6 \cdot 10^{6}$	$3 \cdot 10^{3}$		
	and other inorganic compounds	$2 \cdot 10^{7}$	Exceptions: oxides, hydroxides, halides, nitrates					$1 \cdot 10^{6}$	$5 \cdot 10^{2}$

Table 2.8 (continued)

Radionuclide (physical half-life)	Oral intake		Intake by inhalation						
	Metabolic specification (chem. compound, phys. state)	ALI	Chemical compound	Class D		Class W		Class Y	
				ALI	DAC	ALI	DAC	ALI	DAC
$^{68}_{31}$Ga (68 min), GRT 3	Citrate (carrier-free)	$6 \cdot 10^8$	All Ga compounds Exceptions: oxides, hydroxides, carbides, halides, nitrates	$2 \cdot 10^9$	$6 \cdot 10^5$	$2 \cdot 10^9$	$8 \cdot 10^5$		
$^{75}_{34}$Se (120 d), GRT 3	Elemental, selenides Other compounds	$1 \cdot 10^8$ $2 \cdot 10^8$	All Se compounds Exceptions: oxides, hydroxides, carbides	$3 \cdot 10^7$	$1 \cdot 10^4$	$2 \cdot 10^7$	$9 \cdot 10^3$		
$^{85}_{36}$Kr (10.7 a), GRT 4	No metabolic model: submersion		DAC for external exposure:	$5 \cdot 10^6$					
$^{89}_{38}$Sr (64 d), GRT 3	Soluble salts SrTiO$_3$	$9 \cdot 10^7$ $1 \cdot 10^8$	Soluble salts SrTiO$_3$	$1 \cdot 10^8$	$4 \cdot 10^4$			$6 \cdot 10^7$	$2 \cdot 10^4$
$^{87m}_{38}$Sr (2.9 h), GRT 4	Soluble salts SrTiO$_3$	$2 \cdot 10^9$ $1 \cdot 10^9$	Soluble salts SrTiO$_3$	$5 \cdot 10^9$	$2 \cdot 10^6$			$6 \cdot 10^9$	$2 \cdot 10^6$
$^{99}_{42}$Mo (67 h), GRT 3	MoS$_2$ All other compounds	$6 \cdot 10^7$ $4 \cdot 10^7$	All Mo compounds Exceptions: oxides, hydroxides, MoS$_2$	$1 \cdot 10^8$	$4 \cdot 10^4$			$5 \cdot 10^7$	$2 \cdot 10^4$
$^{99m}_{43}$Tc (6 h), GRT 4	Not differentiated (for Tc$_2$ S$_7$ listed value probably to high)	$3 \cdot 10^9$	All Tc compounds incl. TcO$_4$ Exceptions: oxides, hydroxides, halides, nitrates	$6 \cdot 10^9$	$2 \cdot 10^6$	$9 \cdot 10^9$	$4 \cdot 10^6$		
$^{99}_{43}$Tc ($2 \cdot 10^5$ a), GRT 3	Not differentiated	$1 \cdot 10^8$	All Tc compounds incl. TcO$_4$ Exceptions: oxides, hydroxides, halides, nitrates	$2 \cdot 10^8$	$8 \cdot 10^4$	$2 \cdot 10^7$	$1 \cdot 10^4$		
$^{113m}_{49}$In (1.7 h), GRT 4	Not differentiated	$2 \cdot 10^9$	All In compounds Exceptions: oxides, hydroxides, halides, nitrates	$5 \cdot 10^9$	$2 \cdot 10^6$	$7 \cdot 10^9$	$3 \cdot 10^6$		
$^{113}_{50}$Sn (118 d), GRT 3	Not differentiated	$6 \cdot 10^8$	All Sn compounds Exceptions: sulfides, oxides, hydroxides, halides, nitrates, phosphates	$5 \cdot 10^7$	$2 \cdot 10^4$	$2 \cdot 10^7$	$9 \cdot 10^3$		
$^{123}_{53}$I (13 h), GRT 3	Not differentiated	$1 \cdot 10^8$	All I compounds	$2 \cdot 10^8$	$9 \cdot 10^4$				
$^{125}_{53}$I (60 d), GRT 2	Not differentiated	$1 \cdot 10^6$	All I compounds	$2 \cdot 10^6$	$1 \cdot 10^3$				
$^{131}_{53}$I (8.05 d), GRT 2	Not differentiated	$1 \cdot 10^6$	All I compounds	$2 \cdot 10^6$	$7 \cdot 10^2$				
$^{132}_{53}$I (2.3 h), GRT 3	Not differentiated	$1 \cdot 10^8$	All I compounds	$3 \cdot 10^8$	$1 \cdot 10^5$				
$^{133}_{54}$Xe (5.65 d), GRT 4	No metabolic model: submersion. DAC for external exposure: $4 \cdot 10^6$ (unlimited cloud). $2 \cdot 10^7$ (in rooms of 100–1000 m^3)								
$^{137}_{55}$Cs (9.7 d), GRT 4	Not differentiated	$8 \cdot 10^8$	All Cs compounds	$1 \cdot 10^9$	$5 \cdot 10^5$				
$^{198}_{79}$Au (2.7 d), GRT 3	Not differentiated	$4 \cdot 10^7$	All Au compounds Exceptions: halides, nitrates oxides, hydroxides	$4 \cdot 10^7$	$2 \cdot 10^4$	$6 \cdot 10^7$	$2 \cdot 10^4$	$6 \cdot 10^7$	$2 \cdot 10^4$
$^{199}_{79}$Au (3.15 d), GRT 3	Not differentiated	$1 \cdot 10^8$	All Au compounds Exceptions: halides, nitrates oxides, hydroxides	$1 \cdot 10^8$	$4 \cdot 10^4$	$1 \cdot 10^8$	$5 \cdot 10^4$	$1 \cdot 10^8$	$5 \cdot 10^4$

Table 2.8 (continued)

Radionuclide (physical half-life)	Oral intake		Intake by inhalation						
	Metabolic specifica-tion (chem. compound, phys. state)	ALI	Chemical compound	Class D		Class W		Class Y	
				ALI	DAC	ALI	DAC	ALI	DAC
$^{197}_{80}$Hg (65 h), GRT 3	Inorganic compounds	$2 \cdot 10^8$	Sulfates	$4 \cdot 10^8$	$2 \cdot 10^5$				
	Methyl-Hg	$4 \cdot 10^8$	Oxides, hydroxides, halides, nitrates, sulfides			$3 \cdot 10^8$	$1 \cdot 10^5$		
	Other organic compounds	$3 \cdot 10^8$	Organic compounds	$5 \cdot 10^8$	$2 \cdot 10^5$				
			Hg vapor	$3 \cdot 10^8$	$1 \cdot 10^5$				
$^{203}_{80}$Hg (47 d), GRT 3	Inorganic compounds	$9 \cdot 10^7$	Sulfates	$5 \cdot 10^7$	$2 \cdot 10^4$				
	Methyl-Hg	$2 \cdot 10^7$	Oxides, hydroxides, halides, nitrates, sulfides			$4 \cdot 10^7$	$2 \cdot 10^4$		
	Other organic compounds	$3 \cdot 10^7$	Organic compounds	$3 \cdot 10^7$	$1 \cdot 10^4$				
			Hg vapor	$3 \cdot 10^7$	$1 \cdot 10^4$				
$^{201}_{81}$Tl (73.5 h), GRT 3	Not differentiated	$6 \cdot 10^8$	All Tl compounds	$8 \cdot 10^8$	$3 \cdot 10^5$				

[a] Further information available from ICRP Task Group on Lung Dynamics

Table 2.9. Comparison of ALI and DAC values (ICRP 30, 1979) with corresponding values of ICRP 2 (1960), according to BAIR (1979)

ICRP	ALI (µCi)				DAC (µCi/cm³)	
	Ingestion		Inhalation			
	2	30	2	30	2	30
$^3_1$H (water)	$2 \cdot 10^4$	$8 \cdot 10^4$	$1.5 \cdot 10^4$	$8 \cdot 10^4$	$5 \cdot 10^{-6}$	$20 \cdot 10^{-6}$
$^{32}_{15}$P	150	500	150	800	$7 \cdot 10^{-8}$	$30 \cdot 10^{-8}$
$^{85}_{36}$Kr					$1 \cdot 10^{-5}$	$15 \cdot 10^{-5}$
$^{99}_{42}$Mo	1500	1500	2000	3000	$7 \cdot 10^{-7}$	$10 \cdot 10^{-7}$
$^{99m}_{43}$Tc	500	500	700	500	$4 \cdot 10^{-7}$	$3 \cdot 10^{-7}$
$^{131}_{53}$I	15	30	20	50	$9 \cdot 10^{-9}$	$20 \cdot 10^{-9}$

this. With respect to tritium, for instance, the differences may be attributed to the different values used for the quality factor (Q = 1, instead of 1.7) and the assumption that the biologic life time is 10 instead of 12 days. Moreover, the water content of the reference man is now taken to be 63 kg wet tissue instead of 43 kg. Lower rates of metabolic turnover in bone have been used in the case of ^{32}P. For ^{85}Kr, the higher DAC value is due to the use of the submersion model, which specifies nonstochastic effects as being restricted to skin instead of whole body exposures. KAUL (1980) compared the results of 187 radioisotopes from 21 elements published in ICRP 30 I with those in ICRP 2. There is agreement within a factor of 3 for 51% of the ALI values, while 25% of the new values are higher and 24% lower than previously.

In practice, compliance with the defined limits for the intake of radionuclides requires suitable methods of control and measurement. This is most important, particularly in the case of accidents resulting in incorporations. Appropriate procedures for working operations using radionuclides have been proposed by SHAMAI et al. (1980); these procedures are suitable for controlling the radioactivity in the whole body in dependence on the time following incorporation, down to the order of 1/20 ALI.

In general, the radiation protection facilities approved for minimizing hazards from internal exposures depend on the amount of activity and the types of radionuclide involved. In order to distinguish operation conditions, the International Atomic Energy Agency (IAEA 1963) proposed a classification of the various radionuclides in practical use according to their radiotoxicity.

In the design of these operation conditions, the ALI values obviously serve as a suitable basis (ICRP 25, 1976). Four groups of radiotoxicity (GRT) have been established: GRT 1, "very high"; GRT 2, "high"; GRT 3, "medium"; and GRT 4, "low." These groups are indicated in Table 2.8 for all radionuclides listed. Nevertheless, this classification has been of some importance in drawing up regulations for radiation protection. Therefore, it is given in the guidelines of the Council of the European Communities (1980). Consequently, obligatory rules for the approval and licensing of the handling of radionuclides were based on the GRTs. Thus, official approval for handling may be renounced for radioactivities in the following amounts:

GRT 1   up to $5 \cdot 10^3$ Bq
GRT 2   up to $5 \cdot 10^4$ Bq
GRT 3   up to $5 \cdot 10^5$ Bq
GRT 4   up to $5 \cdot 10^6$ Bq

These restrictive measures, designated as "free limits of handling," have been introduced into various national regulations, such as the German Strahlenschutzverordnung (1976, 1989). Various problems relating to the implementation of the ICRP dose limitation concept have been discussed. HEALY (1982) expressed general objections to the ALI values, because they do not conform to the ALARA principle, especially as far as incorporations of long-lived radionuclides are concerned. ALI values for ^{125}I and ^{131}I are considered by MAILLIE (1986) as being too high with regard to the assumed risk of thyroid cancer induction. Frequently, the handling of unshielded radionuclides is combined with that of other sources of radiation, thus causing additional exposures in various organs. GILL et al. (1980) have estimated the annual effective dose equivalents for several tissues under working conditions covering nuclear medicine and X-ray diagnosis. Their method depends on the results from two dosimeters, one being worn on the thorax, the other on the coat collar. The difference in the readings is proposed to be due to external exposure exceeding the internal one because of nuclear medical procedures. Although this method is proposed primarily for exposure controls in patients, it also seems adaptable for worker surveillance.

2.3 Concepts for Quantification of Risks from Occupational Exposures

2.3.1 Deleterious Effects and Probability of Detriment

Concepts for the quantification of relevant risks from occupational exposures are based on the dose limits for the various working conditions as stated in Sect. 2.2. Nonstochastic effects can be disregarded, since any manifestations of these defects from exposures during a working life within the range of limits stated (see Sect. 2.2.3 and Table 2.2) are unlikely to appear (ICRP 26, 1977). According to ICRP 26, the detriment to a population is defined mathematically as the "expectation" of cases of harm in dependence on dose equivalents, taking into account the probability of each type of deleterious effect and its severity. Both parameters are required for evaluating risk factors in consequence of occupational irradiation. The ICRP concept of detriment to health (G) in a group of P persons is deduced from

$$G = P \sum_i P_i \cdot g_i,$$

where the probability P_i of the effect i should be acceptably small. The severity of the effect is expressed by a weighting factor g_i. Although merely formal, this definition appears suitable for quantification of the various types of deleterious effect, including the frequency of their occurrence. However, the numerical determination of P_i and g_i is based on more or less arbitrary simplifications. In this connection, the ICRP recommends the use of two statistical procedures. One is based on the comparison of the degree of safety for employees in radiation facilities with that in other industries. The other estimates risks for detriment from experimental data or epidemiologic studies by extrapalating to dose equivalents within the recommended limits. For various industries the number of fatalities is considered to be a suitable safety index for comparing risks due to working conditions. This is usually based on the probability of death caused by industrial accidents. For estimating radiation-induced risks, however, this system seems inadequate in principle as it does not include cases of injury and nonfatal disease, especially those associated with continuous disability. Nevertheless, even cases of death presumably caused by irradiation are not readily comparable with acute deaths from accidents occurring in the course of conventional working operations. Any attempt to attribute the manifestation of malignancy to occupational irradiation remains problematic owing to the long latency period between exposure and disease. Furthermore, fatal accidents in industry can, as a matter of course, be attributed to human or technical failure in most cases. From the stochastic nature of radiation-induced effects, however, it follows that, out of the whole group of employees working under similar conditions, only some will develop cancer, based on an entirely arbitrary statistical selection. This is also true for operations within the prescribed dose limits, although for these exposures the probability of cancer induction is extremely low. Finally, it must be pointed out that a numerical valuation of risks from radiation exposure should not be restricted to the frequency of cancer deaths. Cancer incidence and life span shortening also have to be taken into consideration.

Cancers which may have been induced through working operations, due to occupational exposure to either radiation or chemical carcinogens, are by no means distinguishable from similar ones occurring spontaneously. Moreover, the frequency of diseases is not a sufficient working basis for risk quantification, since their contribution to the total amount of hazard varies greatly under different operating conditions. According to ICRP 27 (1977), in the majority of industries and for the great majority of workers industrial diseases increase by only a few percent the estimate of harm based on the mean period of disability or loss of life from accidents. In certain occupations, however, such as mining or quarrying and some sections of the chemical industry, high incidences of occupational diseases need to be considered to a much greater extent.

In order to compare various industries, deleterious effects due to occupational exposures to ionizing radiations should be expressed as loss of working hours. ICRP 27 (1977) introduced an index of harm based on the loss of working time due to occupational risk in relation to a normal working life. This is defined as man-years lost per working year and 1000 workers. These efforts to develop a comprehensive index of harm have been continued in ICRP 45 (1985). As a basic index criterion, ICRP 45 proposes the total time lost to be the result of all forms of occupational harm, expressed as years of healthy life lost per 1000 worker-years at risk. Basic data suitable for comparison between occupational risks from radiation exposure and those from other fields of work are available for various countries concerning different occupations and activities within different industries. ICRP 45 records occupational injuries (i. e. fatal accidents, accidents causing temporary ore permanent disability), occupational diseases (incidences of nonmalignant or malignant diseases in several occupations) and effects of exposure to radiation (induction of cancers causing death or being cured, inherited abnormalities, effects of exposure during pregnancy and nonstochastic effects). ICRP 45 confirms that no simple numerical index of total occupational harm "can be regarded complete or compelling". That does not affect, however, the importance and necessity of the attempts to define and register the different components of harm from various occupations.

The ICRP proposals, however, are controversial. With regard to the late effects arising from occupational exposure, GONEN (1980) refers to their dependence on the age of the employees concerned. BONNELL and HARTE (1978) suggest the proposed dose equivalent limits to be acceptable in view of the benefit gained from the use of nuclear energy. In accordance with these authors, METCALF and WINKLER (1980) compared the index of harm for radiation exposure for different industrial procedures. ROWE (1980) has pointed out that expenditure on the reduction of risks to well acceptable levels is not exclusively determined by technical possibilities but requires detailed consideration of sociopolitical problems as well.

GONEN (1980) puts forward reasons for expecting fewer late sequelae in practice than are estimated by the product of the ICRP risk factors for collective doses. In particular, attention is drawn to the dependence of harm manifestation on the age of the persons concerned. A similar controversial criticism stems from relevant experiences in radiotherapy. Moreover all the different interpretations involve the uncertainty of extrapolating quantitative values from the range in which approximately nonstochastic effects indeed occur in determining the risk of exposures at low dose equivalents and low dose rates (BEIR III Report 1980). Thus, the results from radiobiologic experiments as well as statistical analyses on groups exposed to anomalously high doses are suspected to be entirely unsuitable for the estimation of risk factors. In light of the normally high rates of spontaneous incidences and their variability due to different conditions, it is even questionable whether any calculations of risk are actually admissible for the low dose range. These procedures were also questioned in respect of the induction or manifestation of leukemia and solid tumors, and the development of genetic effects. OESER and KOEPPE (1981) pointed out that cancer mortality, related to age groups, has remained constant for three generations and is apparently expected to do so in the future. Based on the demographic structure of the Federal Republic of Germany and its prospective changes, KOEPPE and OESER (1982) inferred that from 1976 to 2070 cancer mortality in males will increase by about 20%, and that in females the annual absolute values of deaths caused by cancer will remain approximately constant, although the total number included in the age pyramid will decrease considerably.

In this connection, the authors make no attempts to show the effects of the harmful events involved, e. g., carcinogens or ionizing radiation. Furthermore, causal relations based on estimated findings are inadmissible because cancers caused by radiation cannot be distinguished from spontaneous malignancies (OESER and KOEPPE 1982). However, inspite of all these objections, the Committees on

Radiation Protection are inclined to agree upon the general hypothesis that there is also a real, albeit small probability of induction of malignancy within the low doses ranges (detailed bibliography see ARCHER 1980).

BECK (1982) has pointed out that the dose limits themselves are generally based on working hypotheses. Thus, he rejects any calculations of occupational risk since numerical data obtained in this way may be misinterpreted as a quantification of real harm. BUNGER et al. (1981) reported the results of estimations of risk based on tabulated data for cancer mortality showing that these do not differ significantly between radiation workers and workers in other industries. COHEN (1980a, b, 1981), COHEN and COHEN (1980), and COHEN and LEE (1979) published cost-benefit analyses of risks from occupational exposures in nuclear plants. These calculations have been commented on critically by TAIT (1980) and AXELSSON (1981). Despite their detailed objections, an evaluation of risks from occupational radiation exposures compared with those from other industries seems to be the only practical possibility, notwithstanding the different conditions for nonradiation and radiation operations.

2.3.2 Dose-Response Relationships for the Range of the Annual Dose Equivalent Limits for Occupational Exposure

As pointed out in the last section, the uncertainty of risk evaluations is essentially attributable to the fact that the effects are not specific to irradiation. On the contrary, all relevant responses to exposure may also arise due to different agents and physiologic reactions (ICRP 8, 1966). The Committees for Radiation Protection (ICRP 8, 26; UNSCEAR 1972, 1977, 1982; BEIR Report 1972, 1980) have accepted that there is no alternative to the hypothesis that all effects in the lower dose range are of the same kind and severity per unit dose as those amenable to investigation at higher doses either from experiments or epidemiologc studies on selected groups. Although the conclusions drawn are hardly free from uncertainties, this approach is considered justified as long as no other, more reliable method is available. In view of the complex relationship between the dose received by an individual and any particular biologic effect induced, it is inevitable that simplified assumptions are made for the purpose of radiation protection. Stochastic effects are essentially involved over the range of exposure conditions usually encountered in radiation. In general agreement with the other committees, ICRP 26 (1977) basically recommends that a linear relationship without threshold be used for risk estimation. For various effects studied experimentally, however, the response in this range may indeed be sigmoid. Thus, by making linear extrapolations from data obtained at high doses, the risk from low doses could be overestimated. Nevertheless, the ICRP assumes that the frequency of particular effects per unit absorbed dose induced in the range of about 0.5 Gy is unlikely to lead to an overestimation of such effects in the dose range concerned in radiation protection.

Even so, the ICRP states that the use of linear extrapolations may suffice to assess an upper limit of risk in order to compare the benefit of a practice and the hazard of an alternative one not involving radiation exposure. Thus, the more cautious the assumption of linearity, the more important it becomes to recognize that an overestimation of radiation risks could result in the choice of alternatives that can be more hazardous than those involving radiation exposures. Generally, however, risk quantifications should guarantee that the most unfavorable conditions are considered for any exposure. These requirements appear to be fulfilled only by linear dose-response relationships. Thus, the radiation risks tend to be over- rather than underestimated.

In agreement with ICRP 26, the BEIR III Report (1980) defines the following effects as being of prime concern with regard to consequences arising from low dose exposures: mutagenic effects in germ cells of the gonads, carcinogenic effects in somatic cells, and teratogenic effects in pregnancy. So far, the discussions have not differentiated between the various types of reaction caused by irradiation. About 50 studies concerned with cancer induction by radiation are recorded in the BEIR III Report (1980). The dose-response curves are mainly based on these (RADFORD, 1980a, b).

Nevertheless, different dose-response relationships schematically shown in Fig. 2.2 have been considered. A linear dose-response relationship implies constant effectiveness per dose unit over the total dose range, i. e., even down to the lowest doses. Furthermore, the fact, that the response is independent of the dose rate may also correlate with the linear function. Epidemiologic studies, however, have often suggested that the initial part of the curves become steeper with increasing dose. MAYS et al. (1973) also considered the possibility of a quadratic dose relationship for radiation effects. Objections to this hypothesis are based on the uncertainty

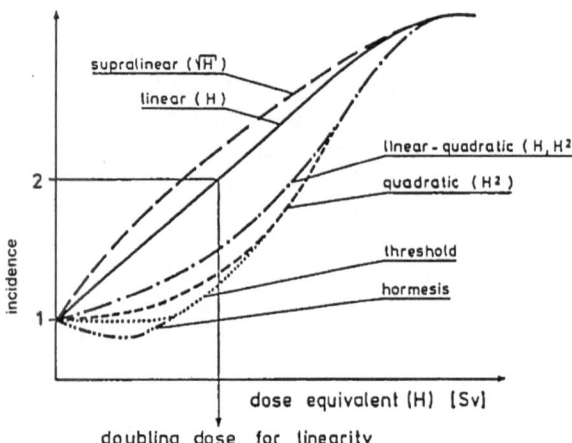

Fig. 2.2. Alternative dose-response curves (schematized) (cf. BEIR III, 1980; BAUM 1973; FEINENDEGEN 1987)

of available data as well as lack of statistical proof that the response curves intersect the risk ordinate horizontally (JACOBI 1974). Some theoretical concepts, however, argue in favor of a quadratic function (KELLERER and ROSSI 1972; ROSSI and KELLERER 1974). Linear-quadratic relationships were first postulated by LEA (1956). MAYS et al. (1973) viewed this functional response as an alternative to the quadratic relationship and argued that the data available were insufficient to decide definitely between these two functions. JACOBI (1974, 1984) also based the quantification of cancer risks from low LET radiation on a linear quadratic relationship.

Whereas ICRP 26 recommends that a linear dose-response relationship should usually be assumed for estimations of carcinogenesis, the BEIR III Report (1980) postulates the functional form of dose response (schematic graph in Fig. 2.3) as follows:

$$E = f(D) = (\alpha_0 + \alpha_1 D + \alpha_2 D^2)e^{-(\beta_1 D + \beta_2 D^2)}$$

where $f(D)$ is the incidence of the relevant effect E (e. g., cancer) at dose D. The parameters α_0, α_1, α_2, β_1, and β_2 have positive values; α_0 is the spontaneous rate of the effect under study (or control for comparison). As differentiated theories of the effects do not exist, the BEIR III response equation is based on empirical derivations of the parameters.

A mathematical analysis of the general function with regard to alternative dose-response curves is reported by UPTON (1977). The linear function is restricted to α_0 and α_1, these being the only parameters relevant for risks at very low doses. With increasing dose the curves become sigmoid and steeper due to the effect of α_2, and at very high

doses exceed a maximum because of the exponential component containing β_1 and β_2. This is generally attributed to cell sterilization in the high dose range.

The BEIR III Report (1980) proposes the use of the linear quadratic function for estimating risks from whole body exposures to low LET radiation. For different tissues, however, the dose-response curves may be approximated either by a linear or a quadratic function (COHEN and COHEN 1980). In addition, genetic effects are presumed to be included and the BEIR III Report suggests that a linear dose response be used for estimations of genetic risk.

The arguments in the BEIR III Report (1980) concerning carcinogenic effects have not been generally approved. Two of the BEIR Committee members have appended separate critical statements to the report (RADFORD 1980a, b; ROSSI 1980a, b). RADFORD is in favor of applying linear dose response functions. ROSSI claims that experimental and statistical findings show the use of linearity to be unjustified. The BEIR III Report as well as ICRP and UNSCEAR do not consider a threshold dose for stochastic effects, be they genetic or somatic. However, the committees concede that experience and experiments are insufficient to allow decisions to be made concerning the existence of a threshold.

Neither experimental results nor theoretical concepts can give rise to safe, empirical conclusions for the low dose range but only to statements on the probability of stochastic effects and their dependence on dose. Therefore, some authors warn against quantitatively correlating cancer manifestations with exposure at low doses without expressly em-

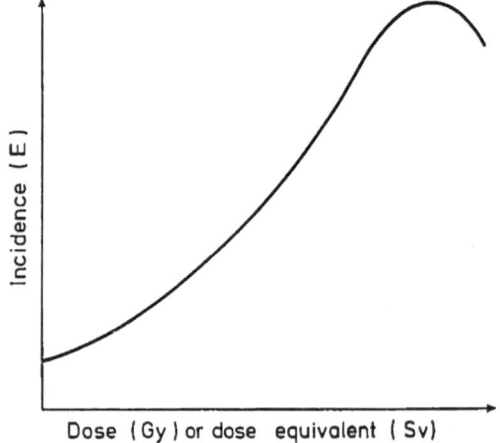

Fig. 2.3. Generalized dose-response relationship (schematized) $E = f(D) = (\alpha_0 + \alpha_1 D + \alpha_2 D^2) \cdot e^{-(\beta_1 D + \beta_2 D^2)}$ (BEIR III, 1980)

phasizing the hypothetical character of the estimation concepts (COHEN and LEE 1979; COHEN 1980b; TAYLOR 1980a,b; BECK 1982; OESER and KOEPPE 1982; HICKEY et al. 1981; FEINENDEGEN 1987). Last but not least, the effects of hormesis at low doses of ionizing radiation should be considered (Fig. 2.2). LUCKEY (1980, 1981, 1982, 1984) attempted to show biopositive effects of low doses of radiation from his own investigations on different biologic objects as well as from numerous references in the literature. In this connection, efforts to demonstrate biopositive effectiveness of balneologic radon applications are worth mentioning (PFALLER 1984). The manifold problems concerning hormesis were the subject of a Conference on Radiation Hormesis (1987) held at Oakland in 1985.

However, even the hypothesis that the linearity of dose exposure guarantees a conservative gradation of risks is not generally accepted. With reference to the results of animal experiments and the studies in Hiroshima and Nagasaki, BAUM (1973) proposed a dose relationship with an exponent smaller than 1 (Fig. 2.2), on the assumption that human populations show a high degree of heterogeneity with regard to cancer disposition. MORGAN (1978, 1979) suggested using this hypothesis in explaining the findings in the Hanford workers (MANCUSO et al. 1977). However, neither ICRP 26 (1977) nor the BEIR III Report (1980) discussed this proposal.

Despite the above-mentioned controversies, UNSCEAR (1977), BEIR III (1980) and ICRP 26 (1977) strongly recommend the use of a linear dose-response relationship as long as more reliable alternatives are lacking. Efforts to develop new concepts for risk assessments, however, seem worthwhile. DUNNING et al. (1984) proposed a new method for assessing potential hazards from low level exposures to radioactive pollutants. They listed risk estimates for about 150 radionuclides and different pathways of exposure resulting in radiation-induced cancer deaths. JONES (1984) presented a general mathematical dose-response mode applicable to human, animal, and cell culture data. This was based on fundamental biologic processes of carcinogenic initiation and promotion as well as competing risk due to other toxic or disease reactions. Consequences of dose relationship for risk estimation based on individual and collective doses were analyzed by PETERSON (1984), especially with reference to protective strategy to minimize the impact on public health either by generally spreading exposures over the whole population or by exposing as few people as possible to an increased dose in spite of greater individual risk.

2.3.3 Estimates of Risks

Risk estimates are mainly concerned with cancer and leukemia and values generally based on a linear dose relationship are given in ICRP 26 (1977) as well as in UNSCEAR (1977) and BEIR III (1980). In particular BEIR III comments on the estimation of basic conditions:

- Cancers induced by radiation are indistinguishable from those occurring naturally. Their existence can be inferred only on the basis of statistical excess above the natural incidence.
- Cancer may be induced by radiation in nearly all tissues of the human body, but tissues and organs vary considerably in their sensitivity to induction.

Moreover, the concepts in BEIR III refer to parameters for natural incidence, type of exposure, latent periods, sex and age factors, interactions between host and encironmental factors, and, not least, to dose - response relationships and their variability under various temporal and spatial conditions of radiation. Further investigations consider interference with various other parameters, such as the relative biologic effectiveness (RBE), dose rate, dose protraction, and dose fractionation as well as the periods to which the involved risk should be related (TOTTER 1980; JABLON 1984; UPTON 1984).

With respect to occupational radiation exposures, ICRP 45 (1985) recommends that the following be considered:

- The frequency of cancer induction and the period of health and life lost owing to curabele and fatal forms
- The number and types of inherited defects
- The incidence of harm to an embryo or fetus resulting from occupational exposure during pregnancy

The irradiation of an embryo or fetus of any pregnant member of a working staff will add to the total harm attributable to occupational radiation exposure. For detailed consideration, ICRP 27 (1977) and 45 (1985) discuss four sources of evidence:

- Frequency of pregnancies
- Risks prior to implantation
- Risks of abnormal development
- Risks of cancer induction in the fetus

Generally it seems very improbable that nonstochastic effects will occur if the limits of occupational exposure are observed. With regard to stochastic teratogenic effects from intrauterine exposures, insufficient human data are available to substantiate the assumption of a dose threshold or even a hormesis (OTAKE and SCHULL 1984; KATO et al. 1987).

The occurrence of cataracts has been investigated experimentally as well as in follow-up studies on the A-bomb survivors (BEIR III 1980). The results are in agreement with sigmoid dose response relationships over the range 0.2-4.5 Gy. An apparent age-dependent dose threshold of about 2 Gy has been suggested for the impairment of vision from cataracts (DODO 1975; MERRIAM and SZECHTER 1975). The influence of radiation on premature ageing and life shortening has been discussed by WALBURG (1975). The follow-up studies on the survivors of Hiroshima and Nagasaki, however, do not show any radiation-induced ageing processes. Thus, ICRP 26 (1977) proposes that any findings of life shortening due to causes other than tumor manifestation are unreliable. A fundamental methodologic assessment of radiation epidemiologic investigations based on the various parameters of radiation exposure and biologic response has been published by BEEBE (1984).

Estimates of risks are based on studies in the following groups:

1. Survivors of the atomic bomb explosions in Hiroshima and Nagasaki, generally with uniform whole body exposures (ISHIMARU et al. 1971; SCHULL 1984; FINCH 1984; TOKUNAGA et al. 1984; PRESTON et al. 1987).

2. Patients given radiotherapy for nonmalignant diseases, e.g., ankylosing spondylitis (COURT BROWN and DOLL 1965; SMITH and DOLL 1978; SMITH 1984), menorrhagia (ALDERSON and JACKSON 1971), thymus hyperplasia (HEMPELMANN et al. 1967, 1975; REFETOFF et al. 1975; SHORE et al. 1984), tinea capitis (ALBERT and OMRAN 1968; MODAN et al. 1974; SHORE et al. 1976; RON and MODAN 1984), and postpartum mastitis (SHORE et al. 1977). The only reference to diagnostic radiology concerns the fluoroscopic exmination of women during the treatment of tuberculosis with pneumothorax (BOICE and MONSON 1977).

3. Persons who are continuously occupationally exposed to ionizing radiation during their working life. This group is described in more detail in Sect. 2.4.

Some basic problems in the use of epidemiologic studies for risk estimations have been considered by BEIR III:

- A selected group of radiation-exposed inidividuals requires a control group, matched as closely as possible.
- Exposures to radiation have mostly occurred in the distant past. Thus, the exact dose of radiation administered is often not adequately known.
- Influence of demographic variables may become

effective during the very long latent periods (30 years and more) that may separate exposure and development of effects in man. For this reason, the true health risk of exposure can only be assessed in extremely long follow-ups of the population concerned.

In order to estimate the carcinogenic effect of exposure at a particular level, BEIR proposed that an absolute and a relative risk model be used alternatively. The absolute risk is defined as the number of excess cancer cases per unit of population, per unit of time, and per unit of radiation dose. The relative cancer risk after the latent period is expressed as a multiple of the natural age-specific cancer risk for the population concerned. Both models will give the same result for lifetime risk only if the epidemiologic follow-up is complete throughout the entire lifetime. As specified by COHEN (1987), the relative risk model assumes that the exposure multiplies the risk to an individual by some factor that remains constant throughout life, while the absolute risk model assumes that exposure adds an annual risk of developing cancer that remains constant thereafter. COHEN (1987) reports both models and considers their applicability as well as the objections to their use.

The estimates of the cancer risk from low dose exposures in ICRP 26 (1977), UNSCEAR (1977), and BEIR III (1980) are based predominantly on the studies of the sample of survivors in Hiroshima

Table 2.10. Cancer mortality risks per 10^4 man-Sv exposure to low LET radiation, sex and age averaged. Comparison of data recommended by ICRP, UNSCEAR, and BEIR (according to COHEN 1981)

Type of cancer	Risk/10^4 man-Sv			
	BEIR 1972	ICRP 26 (1977)	UN-SCEAR (1977)	BEIR III (1980)
Leukemia	25	20	15-25	22
Lung	39	20	25	28
Breast	45	25	~30	11
Bone	6	5	2-5	0.5
GI tract	30		25	19 (incl. stomach)
Thyroid		5	5-15	7
Other	30	50	~25	31
Total (sum) (derived independently)	180	100	120	120

and Nagasaki from 1950 to 1974 (BEEBE et al. 1978a-c). Table 2.10 lists the risk estimates taken from the reports of the three commissions. For purposes of comparison, these were computed by CO-HEN (1981) for cancer of various organs and tissues per 10^4 man-Sv of low LET radiation and in accordance with the linear quadratic dose relationship; the results showed a reasonable level of agreement between the estimates. Especially the data on the incidence of leukemia were considered reliable. Comparative analyses of results from studies of both cities were assumed to supply differing RBE factors, due to their different exposures to neutrons. However, a recent analysis of the γ- and neutron doses has not found any appreciable difference between the two cities (FRY and SINCLAIR 1987).

The ascertainment of received dose from each survivor required first interrogations about distance from hypocenter, shielding from buildings and peculiarities of the terrain. A dose calculated on this basis was proposed first in 1965 as "tentative dose 1965" (TD 65) (MILTON and SHOHOJI 1965) and - being corrected repeatedly - taken as basis for the risk estimations. Accordingly, ROSSI and MAYS (1978) based their analysis of leukemia risk for neutron exposure on those figures. Corresponding studies referred to TD 65-dosimetry of bone marrow (KERR et al. 1977) and other tissues (KERR 1979).

The risk estimates based on the Japanese studies were later questioned by new computations of the doses at Hiroshima and Nagasaki, first published by LOEWE and MENDELSOHN (1981). Considerable differences between the TD65 doses and the new ones, called DS86, became evident. The differences have some influence on the previous risk estimates. According to the new dosimetry, the proportion of neutrons from the Hiroshima bomb is lower than estimated by TD65.

As the revision of the dosimetry to DS 86 is almost completed, the reassessment of its impact for dose-response relationship has now got priority especially concerning chromosome aberrations, brain damage after irradiation in utero and cancer induction. Thus, the new estimates for leukemia and other cancers are expected to be significantly higher than before (FRY and SINCLAIR 1987). The authors address the question, whether the current protection standards require an adaption to the new estimates. In this respect the accuracy of the risk estimates has to be considered, depending not only on correct dosimetry but also on other parameters like appropriateness of the cohort studied, the shape of the dose-response relationship for the various can-

cers, the risk projection models, the RBE factors, and the continuously updated data of cancer mortality (PRESTON et al. 1987). A first extensive analysis by PRESTON and PIERCE (1988) presented a comparison of leucemic and nonleucemic cancer mortality risk estimations under the old and new dosimetries. For both classes of cancer they found thr risk estimates 75-85% higher for DS 86 in air. Not taking the RBE of the small neutron exposure into account the estimates remain essentially unchanged by the dosimetry revision if based on organ doses. As the differences between Hiroshima and Nagasaki are no longer significant, estimates of RBE's are less reliable. Due to the adaption of new risk projection models risk estimates for solid tumors as well as for leukemia increase by a factor of about 2.5 (KELLERER and BRECKOW 1988). It may be argued whether the new estimates require to change the occupational dose limits in radiation protection. In some countries the regulations for occupational exposures have been altered. Usually, the limits for yearly occupational exposures (Table 2.5) remained the same, but they have been further restricted by limiting lifetime exposure. In Great Britain, for instance, the effective dose equivalent as an *average* over the total working life has been limited to 15 mSv/a (National Radiological Protection Board, NRPB 1987). In the Federal Republic of Germany occupational exposures during the *total* working life should not exceed 0.4 Sv (Strahlenschutzkommission 1988). These additional regulations, however, have to be incorporated and unified in the general revision of the basic recommendations of ICRP expected in 1990 (ICRP Statement 1987).

The estimation of risk factors for hereditary effects is based on exposure of the gonads. Genetic response expressed in the following generations is generally induced by mutations resulting from chromosome aberrations in the nuclei of germ cells. For these reactions, a corresponding model should be used as in the case of somatic effects. Similar to the risk estimations for radiation-induced cancer, the hereditary risk estimations require determination of the increase above the natural level of genetic defects as a function of dose. However, even less reliable data are available for radiation-induced hereditary variations than for other risks, and suitable results with reference to subsequent human generations are lacking (BEIR III Report 1980). Even the follow-up studies at Hiroshima and Nagasaki did not demonstrate a significant increase in genetic defects (VOGEL 1984). Since then a reanalysis of the statistical statements has been carried out (ARCHER

Table 2.11. Hereditary risk: Comparison of estimates from UNSCEAR (1982) and BEIR III (1980) on effects of 10^{-2} Gy (1 rad) exposure to low LET radiation, low doses or low dose rates, per 30-year generation for a population of 1 million live-borns

Disease classification	Current incidence	Effect of 10^{-2} Gy per generation			
		UNSCEAR		BEIR III	
		1st generation	Equilibrium	1st generation	Equilibrium
Autosomal dominant and X-linked diseases	10000	20	100	6–65	40–200
Irregular (congenital anomalies, anomalies expressed later, constitutional and degenerative diseases)	90000	5	45		20–900
Recessive diseases	1000/ 1100 (?)	Slight (?)	Slow increase (?)	Very slight	Very slow increase
Chromosomal diseases	4000/6000 (?)	38 (?)	40 (?)	<10	Slow increase
Total	105000 (UNSCEAR) 107100 (BEIR)	63 (?)	185 (?)	15–175	60–1100
Assumed doubling dose		1 Gy		0.5–2 Gy	

1984). The revised doses DS 86 should be determined from this. However, preliminary studies are at present concentrating on chromosome aberration response per unit bone marrow dose in both cities (FRY and SINCLAIR 1987). Thus, estimates of risk factors for hereditary effects expressed in the first two generations could only be based on animal experiments, and sufficient information is required about radiation-induced mutagenicity as well as the reliability of extrapolating from the experimental conditions to the human situation (UNSCEAR, 1972, 1977, 1982; BEIR 1972, 1980), including results on biologic repair (BROYLES and SHAPIRO 1985). VOGEL (1984) considered not only the scientific basis involved but also sociocritical aspects of the estimative procedures. However, obviously no preferable alternative is available.

All estimates of hereditary risks from irradiation in ICRP, UNSCEAR, and BEIR reports assume linear dose-response relationships without thresholds. Thus, any repair mechanisms remain disregarded even in the low dose range for low LET irradiation. ICRP 26 (1977) assumes that the frequency of dominant sex-linked as well as certain chromosomal diseases increases in direct proportion to the dose. The increase in irregularly inherited diseases – congenital or later appearing anomalies as well as constitutional and degenerative defects – is considered to be lower at least in the first two generations. The risk of serious hereditary diseases within the first two generations following irradiation of either parent is taken to be about 10^{-2} Sv^{-1} and the additional damage to later generations is expected to be of the same order. For the purposes of radiation protection, ICRP 26 suggests an average hereditary risk factor of $4 \cdot 10^{-3}$ Sv^{-1} for homogeneous whole body irradiations to individuals of both sexes and all ages. This means that four additional cases of hereditary damage per mSv must be expected in an exposed cohort of 10^6 of either parents. Hereditary risk estimates taken form UNSCEAR (1982) and BEIR III (1980) are given in Table 2.11. The figures show the expected increase in the different classes of genetic disorder among 10^6 live-born people whose ancestors had received additional radiation exposures. There is general agreement between the data from both commissions for autosomal and recessive diseases as well as aberration rates. BEIR III refers explicitly to irregular disorders, this being the largest category of defects listed. For estimates of the number of irregularly inherited disorders induced, present in a state of equilibrium, each disorders requires its own mutational component and each class of disorder its average mutational component. These components are usually difficult to estimate, leading to uncertainties, and are in the proposed range of 20–900 per million live-born at equilibrium (Table 2.11). These mutations would be expected to persist much longer than the simple autosomal dominant ones. Thus, many generations later the incidence of diseases due to gene mutations at equilibrium should have increased by 60-1100, from initially 107100 to around 107160-108200 per million live-born for average exposures of the general population up to 10 mSv per individual and 30-year generation. For occupational exposures, the values for the doubling dose of about 1 Sv, on average (Table 2.11), may be useful in practice.

2.4 Measurement of and Statistical Statements on Occupational Radiation Exposures

2.4.1 Radiation-Induced Diseases of Workers Continuously Exposed During Their Working Lives

The studies on the risks of handling ionizing radiation were generally oriented towards the concepts of dose limitations and correspondingly to the extent of harm induction in exposed workers. Occupational exposures within the scale of recommended dose limits may be of the order of about 5-50 mSv annually. Assuming the complete accumulation of total dose responses during working life, a noticeably higher probability of carcinogenic effects in the group of workers involved must be considered likely. Thus, statistical studies comparing exposed and nonexposed workers generally seem worthwhile. However, the number of employees investigated has been relatively small. Furthermore, statistical results are also essentially lacking in significance since regular dose surveillance shows that dose equivalents usually reach the annual limit only in very few cases, while for the vast majority of the population the annual values of exposure remain below 5 mSv (see Sect. 2.4.2). Thus, stochastic risk estimations for group of workers involved with ionizing radiation will hardly be significant. Employees, however, who are members of a "critical group" (ICRP 42, 1984), representative of those expected to receive occupational doses above the average, should be compulsorily controlled individually. In this connection, MATANOSKI et al. (1984) discussed in general the working conditions in radiologic services. For practical purposes, highly sensitive methods are available for determining the rates of chromosome aberration, usually in peripheral lymphocytes. The range of applications of this method has been extended to occupational irradiation, particularly resulting from industrial accidents (LITTLEFIELD and JOINER 1978; CAO SHU-YUAN et al. 1981).

With reference to the risk of harm to the skin from radioactive compounds used in nuclear medical techniques, LENZ et al. (1979) compared the frequency of dermatitis and skin cancer in particularly exposed staff without, however, specifying the dose values. More detailed pulications are available concerning workers in uranium mines and mills (ICRP 24, 1977; 47, 1986; COHEN 1982; ŠEVC et al. 1984; RADFORD 1984; TIRMARCHE et al. 1984; KHAN et al. 1984). Generally, these papers refer to dose values and risk estimates. During the last 10 years the increased interest in effects induced by incorporated α-emitters (high LET) has resulted in further research (ICRP 32, 1982; 47, 1986; NCRP 77, 1984; STOCKER 1985; MAYS et al. 1985). Recently, data have become available from studies on workers in uranium plants, which have made reference to estimates of doses received (SRIVASTAVA et al. 1986; SINGH and WRENN 1986; FISHER and JACKSON 1986; MORGAN and SAMET 1986; LOMBARD et al. 1986) as well as carcinogenic effects (VOELZ et al. 1984; THOMAS et al. 1985; LEIRA et al. 1986). Occupational exposures in uranium mines and plants are mainly characterized by the inhalation and incorporation of α-emitters. Due to their high LET radiation with a quality factor $\bar{Q}$ of 20, they have considerably more effect than exposures to similar absorbed doses of β- or γ-rays. Thus, risk estimates also need to be taken into account for persons handling radon therapeutically, especially in radon spas. The staff working and living in such spas can be exposed to levels even in excess of limits recommended for uranium miners (ICRP 47, 1986). Experiences gained in this respect have resulted in consideration of the efficiency of repair mechanisms and immunologic reactions (POHL-RÜLING and FISCHER 1979; POHL-RÜLING et al. 1979; TUSCHL and ALTMANN 1979; TUSCHL 1984).

Radium dial painters constitute another group of exposed persons (STEHNEY et al. 1978; POLEDNAK et al. 1978; ROWLAND and LUCAS 1984).

MANCUSO et al. (1977) and KNEALE et al. (1978) attempted statistically to show increased frequencies of different types of cancer, mostly of the lung, pancreas, and bone marrow, as late effects of exposures to workers in the Hanford plutonium plants during their working lives. The reliability of these investigations has been criticized with regard to the admissibility of the statistical methods applied. BEIR III (1980) left these problems unanswered, but the committee refused to accept the results of the Hanford study as cogent or as sufficient for considering modifications of the proposed risk estimates. WILKINSON et al. (1984) reported on the results of the routine medical surveillances of workers in plutonium plants (Manhattan, Los Alamos, Rocky Flats) showing a general plutonium incorporation of between 260 and 8500 Bq. However, dose response in terms of cancer induction could not be observed.

KAUL et al. (1984a) critically evaluated 55 publications with respect to the underlying criteria, number of persons in groups compared, and dose calibration. Out of 33 studies on radiation-induced leukemia, only one met all the requirements for a reliable statistical interpretation; 23 were not sufficient and nine only in restricted extent usable for risk estimations. Likewise, quantitative interpretations were admissible for only four out of 12 surveys of breast cancer: two showed a rough trend towards increased inductions with increasing doses; six, however, did not allow any statistical conclusions to be drawn. Furthermore, only two out of 12 published studies on genetic risks met the requirements for statistical evaluations, the other ten being completely useless. This analysis of papers published in serious scientific journals highlights the difficulties in evaluating risks from statistical studies of groups of persons exposed within the low dose range.

A fundamental methodologic assessment of radiation epidemiologic investigations was published by BEEBE (1984). He drew up the conditions deemed suitable for considering parameters of radiation exposure and biologic response. Epidemiologic studies were based on data from diagnostic and therapeutic radiation, from occupational exposures, from nuclear explosions and fallout as well as from geographic differences in the average level of background radiation. The detection of a relationship between dose and the incidence of the different effects was particularly difficult. BEEBE discussed appropriate control groups to avoid false conclusions with regard to characteristic parameters for the ascertainment of effects and measurement of exposures. Further statements concerned problems related to the choice of exposed subjects, to distribution of effects over time, and to the problem of life time prediction of radiogenic mortality.

2.4.2 Procedures and Results of Dose Controls on Workers and Workplaces

Any procedures for occupational radiologic protection must consider the establishment and maintenance of acceptably safe and satisfactory working conditions (ICRP 35, 1982). As a result, periodic dose measurement are obligatory and must be sufficient for assessment of exposure and individual monitoring and for the recording of results with regard to the ICRP dose limits and the relevant national legislation. Appropriate monitoring programs must include controls of exposure of workers and workplaces to radiation and radioactive materials.

As recommended by ICRP 26 (1977), individual assessments may be restricted to persons employed under working condition A (see Sect. 2.2.3), by considering dose equivalents with respect to external irradiation and committed dose equivalents for internal contaminations as well. This recommendation has been commented by ICRP 35, particularly concerning the general principles of monitoring for radiation protection of workers. This monitoring involves measurements of radiation or radioactivity and specific interpretation of estimates and control of exposures. Appropriate equipment must be provided for the control of individuals, for workplaces, and for special operations, and should include routine programs as well as programs suitable for establishing abnormal or suspectedly abnormal exposures. Accordingly, ICRP 35 deals with the functions of monitoring, models for the interpretation of results, designs for the control of external irradiation, internal exposures, and surface contaminations. Differing conditions of controlled an supervised areas are considered and suitable dosimeters defined.

Obviously, the general standards set out in ICRP 26 and 35 concerning radiation protection facilities offered extensive scope in the establishment of regulations by national bodies because of the different working conditions involved. In principle, however, optimization of protection should be realized on the basis of cost-benefit analyses (see Sect. 2.2.2). Nevertheless, the justification for and the performance and optimization of the protection are generally acknowledged to be elaborations of concepts periodically underlying scientific and organizational developments (ILARI 1986). Accordingly, the effects on the corresponding results remain the subject of discussion. IAEA (1986) published the proceedings of an international symposium on the optimization of radiation protection, with respect to the situation in different countries. KAUL et al. (1986) commented on the problems concerned from the point of view of the Federal Republic of Germany. With reference to the monitoring procedures, they considered three possibilities:

- Statements on the assignment of persons to the various dose classes

- Statements on the number of persons and the mean dose
- Statements on the collective dose, defined as the product of the number of persons exposed and the arithmetic mean of all occurring doses

These statements, of course, imply pronounced simplifications. Nevertheless, the collective dose may be considered a suitable quantity for general optimization procedures.

However, regardless of these problems, the determination of individual and local doses in controlled areas, where continued operation would give rise at least to working condition A, seems reasonable. This is consistently demanded with regard to examinations of the efficiency of operating instructions and installed or mobile protection facilities. For example, in the Federal Republic of Germany appropriate dose controls are carried out compulsorily, supervised by bodies functioning independently of plant management. The summarized results are published annually in general reports on environmental radioactivity and radiation exposure (Bundesrepublik Deutschland 1982–1984). The relevant data for doses are derived from integrated values, obtained under defined exposure conditions as shown by probit analysis. The evaluation procedure has been reported by FÄRBER (1979). The monthly and annual measurements of dose equivalents of exposure of the different groups of persons concerned are suitable for computing the contributions of selected cohorts to the total "genetic significant dose" of the whole population. In this case, the frequency distribution of the grouped doses has to be considered. In case of corresponding average dose levels of comparable dose distributions the response of those with higher doses must be considered more critically. FÄRBER (1979) reported an analysis of the annual statistics for workers handling radioactive materials for medical and research purposes. As the number of controlled persons in two successive years had increased, higher values of

the collective dose as well as of the average individual yearly and monthly doses were obtained. The dose distributions, however, showed a general shift to lower doses.

HOLTHUSEN et al. (1961), BÄUML (1974), and PIETZSCH (1976) have proposed procedures for computing the genetically significant dose from annual mean and collective doses of employees. In the Federal Republic of Germany, the results are published in the above-mentioned annual reports, the last one being available for 1984 (Bundesrepublik Deutschland, 1984, see also STIEVE 1976). Because of their comprehensive data, they will be discussed here in some detail as far as the extent of occupational exposure is concerned. Generally, the reports cover results for external doses as well as for incorporation, acquired by means of body counters or from the measurements of urine samples. Furthermore, all operational problems, accidents, and even peculiar events in connection with the handling of ionizing radiation are listed and interpreted. Plants using X-rays and those with nuclear facilities are listed separately in accordance with the appropriate regulations Röntgenverordnung 1987; Strahlenschutzverordnung 1976, 1989). In particular, the reports represents a collection of data concerning all consequences deriving from operational techniques in practical use. Detailed elucidations of single events are normally excluded. Nevertheless, all kinds of operational problems are discussed at meetings on radiation protection (MESSERSCHMIDT et al. 1980). STIEVE (1980), for example, reported on causes, measures, and countermeasures with reference to all such problems having occurred up to now from supernormal exposures, especially those due to medical procedures.

The following data were taken from the annual reports 1982/1984 to illustrate the situation. The extent of dose controls is listed in Table 2.12 for the individuals as well as for the operating plants involved. Those for medical services, mostly concerning X-ray applications and regulated by the

Table 2.12. Dose-controlled workers and plants handling ionizing radiation. Estimates for the Federal Republic of Germany for mean annual individual doses (mSv) and annual collective doses (man-Sv). (Bundesrepublik Deutschland, 1982, 1983, 1984)

	No. of workers		No. of plants		Mean individual doses (mSv/a)	Collective doses (man-Sv)
	Total	Involved in medical operations	Total	Involved in medical operations		
1982	196694	140634	22274	19035	0.81	158.5
1983	210059	150128	23143	19717	0.75	158.0
1984	217888	157573	23753	20178	0.43	94.7

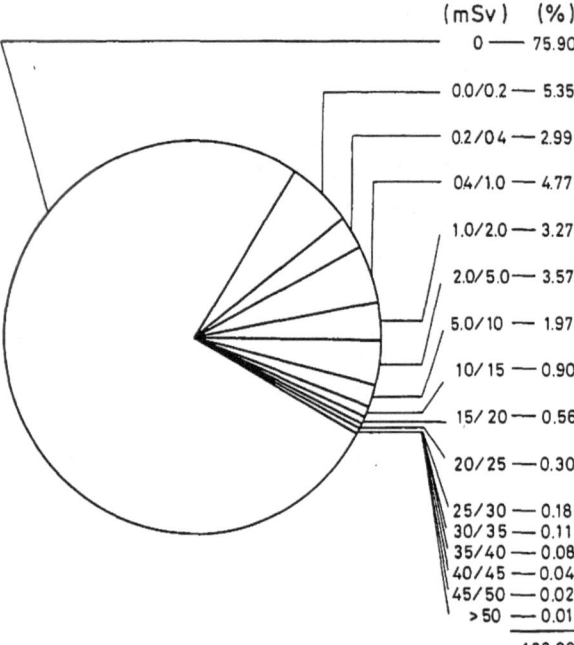

	(mSv)	(%)
	0	75.90
	0.0/0.2	5.35
	0.2/0.4	2.99
	0.4/1.0	4.77
	1.0/2.0	3.27
	2.0/5.0	3.57
	5.0/10	1.97
	10/15	0.90
	15/20	0.56
	20/25	0.30
	25/30	0.18
	30/35	0.11
	35/40	0.08
	40/45	0.04
	45/50	0.02
	>50	0.01
		100.00

Fig. 2.4. Frequency distribution of the annual individual dose: intervals of individual dose (mSv) and corresponding numbers of occupationally exposed employees as a percentage of all regularly controlled workers in the Federal Republic of Germany (Bundesrepublik Deutschland 1983)

Röntgenverordnung 1987, are tabulated separately. Generally, the number of workers regularly controlled is remarkable. The radiation - exposed workers make up approximately 0.8% of the total number of employees (about 25 million) in the Federal Republic of Germany. Both the individual and the collective doses are tabulated for these workers. Figure 2.4 furthermore shows the distribution of persons (as a percentage of all persons controlled) to reasonably selected dose ranges. It should be pointed out that the exposure of more than two-thirds of the involved individuals does not reach a measurable range, and that it exceeds the obligatory equivalent dose limits in less than 0.01% of cases. Even the individuals with exposures between 10 and 50 mSv/a do not represent more than approximately 2% of the total. In general, these statements are reliable and reflect the satisfactory efficiency of the radiation protection facilities. They also indicate the most careful implementation of procedures for control and surveillance of employees (IAEA 1987). Supplementary data to those in Table 2.12 and Fig. 2.4, on occupational exposures particularly in nuclear plants in the Federal Republic of Germany, were reported by MEHL (1979), who listed collective doses for the period from 1962 to 1978.

KOELZER and KIEFER (1980) discussed the mean and individual exposures of employees in the Nuclear Research Center at Karlsruhe from 1968 to 1978 separately for working categories A and B. STÄBLEIN (1980) has commented upon the considerable expenditure necessary in examining about 4000 employees at the Karlsruhe center, which revealed just seven cases in which the dose equivalent limits had been exceeded during the 9 years under consideration.

The results of exposure control in Germany, as mentioned above, should only be regarded as a typical example, but they are an extensive and reliable source of information to the public. Corresponding statements have been published by various other countries and form a basis for a general exchange of ideas on questions concerning the methods and techniques used in occupational radiation control (IAEA and OECD Symposium 1979). IMAHORI (1980), for example, published the annual collective doses and mean individual doses of the workers in nuclear plants in Japan for the period from 1970 to 1978. SHENOY et al. (1980) compared the results of exposures from industrial radiography in India with those in other countries. A statistical analysis of film badge records of workers in Taiwan for the years 1962 to 1983 has been described and interpreted by WENG and CHEN (1985). Corresponding results concerning the users of ionizing radiation in both medical research and industry in Kuwait from 1980 to 1983 have been presented by MUSTAFA et al. (1985) with respect to the efficiency of the adopted protection procedures and the comparable experiences in other countries. All these reports were completed before ICRP 35 (1982) was issued and therefore related to UNSCEAR (1972), which deals with occupational radiation exposures in nuclear plants in appendix D.

Various investigations involve the use of unsealed radionuclides in nuclear medical practice. In particular, iodine isotopes are discussed in terms of their application frequency and the metabolic behavior of iodine. Here, ^{125}I is of particular concern due to the constant increase in radioimmunoassay techniques, resulting in enhanced exposures to laboratory workers. Iodine tends to evaporate in the course of iodination procedures and thus may be easily incorporated via respiration and accumulation in the thyroid. Hence, LAMBERT (1981) suggested control measurements of ^{125}I and ^{131}I be carried out regularly in workers and at shorter intervals whenever required by the working conditions. Results of such controls have been presented by KIVINITTY et al. (1984) in relation to the amount of ^{125}I

activity handled. When activities of approximately 400 MBq (10 mCi) are exceeded weekly, at least monthly controls are advisable, while for smaller activities (less than 40 MBq) annual measurements are proposed.

Industrial procedures in radiopharmaceutical laboratories for the radioiodination of proteins using ^{125}I and ^{131}I in many cases lead to exposures occasionally exceeding the annual dose limits (up to 0.3 Sv/a for iodine isotopes for workers in category A) (ERLENBACH 1980). Thus, precise measuring controls are urgently required for these operations as well as for comparably exposed employees in nuclear medical departments. ANDRE et al. (1980) pointed out that the hands are primarily subject to exposures in radiopharmaceutical laboratories. Resulting injuries were reported by LENZ et al. (1979). ANDRE et al. gave methods for dose control as well as an analysis of results concerning different parts of the hand (fingers, fingertips, etc.). A method for estimating exposures of fingers from badge and ring dosimeters and the use of shielded and unshielded syringes in nuclear medicine has been discussed by MAINI et al. (1987). DANIELLI et al. (1984) carried out control measurements of laboratory air contaminated with ^{125}I. Considering the distribution kinetics of iodine, they estimated the concentrations of ^{125}I in the particular iodine compartments of the body. KRESNIAK et al. (1984) controlled the ^{125}I, ^{131}I, and ^{99m}Tc concentrations in the air of a nuclear medical clinic and calculated the resulting inhalation dose quantities in staff and patients.

Estimations concerning the laboratory use of tritiated compounds have been given in ICRP 30 Part I (1979). Table 2.8 shows the corresponding ALI values for tritiated water and elemental tritium gas. However, no ALI values have been recommended for organic compounds. ICRP 30 only mentioned that these might differ considerably from those of tritiated water, assuming that the value for tritiated thymidine should be as much as 50 times smaller. Corresponding controls have been described by DUIJSINGS et al. (1984). They found values varying between 4 GBq and 4 MBq over the 26-week control period for tritium-labeled thymidine monitored in the laboratories of Dutch universities. As thymidine is metabolized to water to more than 85%, they based their controls on urine samples. They showed that apparently very low fractional intakes can be neglected. Considerably higher exposures must be taken into consideration in the handling of tritium targets in neutron generators. RAS et al. (1980) reported on unexpectedly large incorporations through inhalation. The distri-

bution, transfer, and excretion of these incorporations typically refer to tritium adsorbed on titanium particles. Tritium contaminations arising from the occupational handling of β-radiation have been discussed by KINSLOW et al. (1984). They derived a mathematical model for the urinary excretion of tritium after prolonged multiple exposures resulting in values for the biologic half-lives. Considerable effects of the ambient temperatures on excretion rates of tritium have been reported by CAWLEY et al. (1984).

Particular problems concerning the risks from the use of relatively short-lived radionuclides such as ^{11}C have been discussed by VIALETTES and MOREAU (1980). In practice, activities of 7-20 MBq are usually applied, possibly leading to dose rates of up to 10 Gy/h in workers during syntheses. Thus, operating conditions and restrictions on working time must be considered for reducing exposures.

A survey on problems an procedures for radiation protection in a large and renowned cancer hospital with an associated research institute has been published by TROTT et al. (1980). In particular, routine methods and results of dose equivalent control measurements of the staff handling unsealed sources in therapy, diagnostics, and research applications are explained. Individual total body doses have generally been kept to below 15 mSv/a, corresponding to the annual limit for workers of category B. TROTT (1981) compared the resulting exposure risks to the staff with the benefit to the patients, characterized by the number of clinical treatments and investigations carried out. A detailed synopsis covering the complex problems of the significance of relevant risk-benefit analyses for patients and staff in hospitals and in research has been presented by TROTT (1982) in his "Gastein Lecture" on "the safe and effective use of radiopharmaceuticals." With regard to the general tendency towards economic heating of modern houses - neglecting effective ventilation and thus provoking increased radon levels - TROTT finally stated: "It is a pleasant irony that, while some effort is being devoted ... for guidance in nuclear medicine ... to help us reduce still further the radiation exposure of our staff from this man-made activity, it is possible ... that our exposure to natural sources of radiation may have measurably increased as a consequence of our higher standard of living." No comment!

References

Albert ER, Omran AR (1968) Follow-up study of patients treated by x-ray epilation for tinea capitis. Arch Environ Health 17: 899–950

Alderson MR, Jackson SM (1971) Long term follow-up of patients with menorrhagia treated by irradiation. Br J Radiol 44: 295–298

Andre JJ, Goubert J, Moreau A, Perotin JP (1980) Problèmes de radioprotection recontres dans un laboratoire de controle radiopharmaceutique irradiation des extremites des mains. In: Radiation protection. A systematic approach to safety. Proceedings of the 5th Congress of the International Radiation Protection Society. Jerusalem, March 1980, vol I. Pergamon, Oxford, pp 60–63

Archer VE (1980) Effects of low-level radiation: A critical review. Nuclear Safety 21: 68–82

Archer VE (1984) Reanalysis of genetic effects of atomic bombs: comment and further analysis. Health Phys 46: 1147–1149

Australian Radiation Protection Society (1988) Radiation Protection Practice. Seventh International Congress of the International Radiation Protection Association, Sydney, 10–17 April 1988. Pergamon Press (Australia)

Auxier JA, Dickson HW (1983) Concern over recent use of the ALARA-philosophy. Health Phys 44: 595–600

Axelsson I (1981) Analysis of risks, costs and benefits. Health Phys 40: 255–256

Bair WJ (1979) Review of report of the International Commission on Radiological Protection committee 2: limits for intakes of radionuclides by workers. In: Deutsches Atomforum (ed) Radioökologie. Berichtsband der Fachtagung Radioökologie vom 2.–3. Oktober 1979. Vulkan, Essen, pp 182–195

Baum JW (1973) Population heterogeneity hypothesis on radiation induced cancer. Health Phys 25: 97–104

Bäuml A (1974) Strahlenbelastung beruflich strahlenexponierter Personen in den Bundesländern Hamburg, Schleswig-Holstein, Niedersachsen und Berlin in den Jahren 1967–1974. Institut für Strahlenhygiene des Bundesgesundheitsamtes. St H-Bericht 11/76

Beck HR (1982) Anmerkungen zu den Empfehlungen der Internationalen Strahlenschutzkommission (ICRP 26/27) hinsichtlich der Definition und der Vergleichbarkeit von Schadensrisiken. In: Messerschmidt O, Börner W, Holeczke F, Olbert F, Seyss R (eds) Zur Problematik der Wirkung kleiner Strahlendosen. Strahlenschutz in Forschung und Praxis, vol XXIII. Thieme, Stuttgart, pp 165–175

Beck HR, Dresel H, Melching H-J (1959) Leitfaden des Strahlenschutzes. Thieme, Stuttgart

Beebe GW (1984) A methodological assessment of radiation epidemiology studies. Health Phys 46: 745–762

Beebe GW, Kato H, Land CE (1978a) Life span study report 8. Mortality experience of atomic bomb survivors 1950–1974. In: Radiation Effects Research Foundation Technical Report (TR 1–77). Radiation Effects Research Foundation, Hiroshima, pp 64–67

Beebe GW, Land CE, Kato H (1978b) Studies of the mortality of A-bomb survivors. 6. Mortality and radiation dose. 1950–1974. Radiat Res 75: 138–201

Beebe GW, Kato H, Land CE (1978c) The hypothesis of radiation-accelerated aging and the mortality of Japanese A-bomb victims. In: Late biological effects of ionizing radiation (Proc Symp Vienna 13–17 March 1978). IAEA, vol I, pp 3–27

BEIR Report (1972) National Research Council: Advisory Committee on the Biological Effects of Ionizing Radiations (ed) The effects on populations of exposure to low levels of ionizing radiations. National Academy of Sciences, Washington

BEIR III Report (1980) National Research Council, Committee on the Biological Effects of Ionizing Radiations (ed) The effects on populations of exposure to low levels of ionizing radiations. National Academy of Sciences, Washington

Bendel I, Schüttmann W, Arndt D (1978) Cataract of lens as late effect of ionizing radiation in occupationally exposed persons. In: Late biological effects of ionizing radiation (Proc Symp Vienna, 13–17 March 1978). IAEA, vol I, pp 309–319

Boice JD Jr, Monson RR (1977) Breast cancer in women after repeated fluoroscopic examinations of the chest. J Natl Cancer Inst 59: 823–832

Bond VP, Fliedner TM, Archambeau JO (1965) Mammalian radiation lethality. A disturbance of cellular kinetics. Academic Press, New York

Bonnell JA, Harte G (1978) Risk associated with occupational exposure to ionizing radiation kept in perspective. In: Late biological effects of ionizing radiation (Proc Symp Vienna, 13–17 March 1978). IAEA, vol I, pp 413–425

British Committee on Radiation Units and Measurements (1986) Effective quality factor for neutrons. Radiation Protection Dosimetry 14: 345–346

Broyles AA, Shapiro CS (1985) Biological repair with time-dependent irradiation. Health Phys 49: 701–705

Bundesrepublik Deutschland (1982, 1983, 1984) Umweltradioaktivität und Strahlenbelastung, Jahresbericht: Der Bundesinnenminister, Der Bundesminister für Umwelt, Naturschutz und Reaktorsicherheit, Bonn

Bunger BM, Cook JR, Barrick MK (1981) Life table methodology for evaluating radiation risk: an application based on occupational exposures. Health Phys 40: 439–455

Cao Shu-Juan, Deng Zhicheng, Shou Zhenying, Li Yun-hua, Yu Cui-fang (1981) Lymphocyte chromosome aberrations in personnel occupationally exposed to low levels of radiation. Health Physics in the People's Republic of China: 586–587

Cawley CN, Cannon LA, Moschella JJ (1984) Tritium excretion and ambient temperature. Health Phys 46: 1155–1157

Cohen AF, Cohen BL (1980) Tests of the linearity assumption in the dose-effect relationship for radiation-induced cancer. Health Phys 38: 53–69

Cohen BL (1980a) Society's valuation of life saving in radiation protection and other contexts. Health Phys 38: 33–51

Cohen BL (1980b) The cancer risk from low level radiation. Health Phys 39: 659–678

Cohen BL (1981) Proposals on use of the BEIR-III report in environment assessments. Health Phys 41: 769–774

Cohen BL (1982) Health effects of radon emissions from uranium mill tailings. Health Phys 42: 695–702

Cohen BL (1987) Alternatives to the BEIR relative risk model for explaining atomic bomb survivor cancer mortality. Health Phys 52: 55–83

Cohen BL, Lee I-S (1979) A catalog of risks. Health Phys 36: 707–722

Conference on Radiation Hormesis (1987) Proceedings of conference held August 14–16 1985 in Oakland California. Health Phys 52: 517–680

Council of the European Communities (1980) Council directive of 15 July 1980 amending the directives laying down the basic safety standards for the health protection of the general public and workers against the dangers of ionizing radiation (80/836/Euratom) Official J. European Communites 23 No 246: 1-80

Court Brown WM, Doll R (1965) Mortality from cancer and other causes after radiotherapy for ankylosing spondylitis. Br Med J II: 1327-1332

Danielli C, Gaiba W, Rossi A, Vianello Vos C, Calamosca M (1984) ^{125}I-airborne contamination, levels in vitro radiometry laboratories. In: Kaul A, Neider R, Peński J, Stieve F-E, Brunner H (eds) Radiation-risk-protection. 6th International Congress of the International Radiation Protection Association, Berlin (West), 7-12 May 1984. Compacts publ. by Fachverband für Strahlenschutz e. V. Jülich. Verlag TÜV Rheinland, Köln, pp 829-832

Dodo T (1975) Cataracts. J Radiat Res (Tokyo) 16 (Suppl): 132-137

Dorneich M, Jaeger R, Schaefer H, Muth H, Henschke U, Rajewsky B (1948) Strahlenschutz und Toleranzdosis. In: Rajewsky B, Schön M (eds) Naturforschung und Medizin in Deutschland 1939-1946 (Fiat-Bericht), vol 21/I. Biophysik. DVB, Wiesbaden, pp 177-226

Duijsings JH, Beentjes LB, Coops AJ, van der Jagt PJ (1984) Internal contamination of radiological workers by tritium compounds at two universities in The Netherlands. Health Phys 46: 665-669

Dunning DEJ, Leggett RW, Sullivan RE (1984) An assessment of health risk from radiation exposures. Health Phys 46: 1035-1051

Erlenbach HR (1980) Occupational thyroid exposure due to internal radioiodine contamination in radiation workers handling iodine-131 (125) for non-therapeutic use. In: Radiation protection. A systematic approach to safety. Proceedings of the 5th Congress of the International Radiation Protection Society, Jerusalem, March 1980, vol I. Pergamon, Oxford, pp 72-75

Färber K (1979) Strahlenbelastung beruflich strahlenexponierter Personen: Institut für Strahlenhygiene des Bundesgesundheitsamtes. STH-Bericht 9/1979. Reimer, Berlin

Feinendegen LE (1987) Das Problem der kleinen Strahlendosen - eine Herausforderung für die Beurteilung von Schadenfolgen. In: Gesellschaft für Reaktorsicherheit (ed) Ergebnisse neuer Sicherheitsanalysen. 10 GRS-Fachgespräche, 12-13 November 1986. GRS 64: Köln

Finch SC (1984) Leukemia and lymphoma in atomic bomb survivors. In: Boice JD Jr, Fraumeni JF Jr (eds) Radiation carcinogenesis: epidemiology and biological significance. Progress in cancer research and therapy, vol 26. Raven, New York, pp 37-44

Fisher DR, Jackson PO (1986) Reply to Singh and Wrenn regarding levels of ^{234}U, ^{238}U and ^{230}Th in excreta of U mill crushermen. Health Phys 50: 302-303

Falskamp W (1930) Über Röntgenschäden und Schäden durch radioaktive Substanzen. Sonderbände zur Strahlentherapie XII. Urban und Schwarzenberg, Berlin

Fry RJM, Sinclair WK (1987) New dosimetry of atomic bomb radiations. Lancet II: 845-848

Gill JR, Beaver PF, Dennis JA (1980) The practical application of ICRP recommendations regarding dose-equivalent limits for workers to staff in diagnostic x-ray departments. In: Radiation protection. A systematic approach to safety. Proceedings of the 5th Congress of the International Protection Society, Jerusalem, March 1980, vol I. Pergamon, Oxford, pp 31-34

Glasstone S (ed) (1957) The effects of nuclear weapons. USAEC (United States Atomic Energy Commission), Washington

Glasstone S, Dolan PJ (eds) (1977) The effects of nuclear weapons, 3rd edn. United States Department of Defense and United States Department of Energy

Gonen YG (1980) Risk estimates of stochastic effects due to exposure to radiation. A stochastic harm index. In: Radiation protection. A systematic approach to safety. Proceedings of the 5th Congress of the International Radiation Protection Society, Jerusalem, March 1980, vol I. Pergamon, Oxford, pp 31-34

Healy JW (1982) The ICRP dose limitation system - solution or problem? Health Phys 42: 407-413

Hempelmann LH, Pifer JW, Burke GJ, Terry R, Ames WR (1967) Neoplasms in persons treated with x-rays in infancy for thymic enlargement. A report of the third follow-up survey. J Natl Cancer Inst 38: 317-341

Hempelmann LH, Hall WJ, Phillips M, Cooper RA, Ames WR (1975) Neoplasms in persons treated with x-rays in infancy. Fourth survey in 20 years. J Natl Cancer Inst 55: 519-530

Hickey RJ, Browers EJ, Spence DE, Zemel BS, Clelland AB, Clelland RC (1981) Low level ionizing radiation and human mortality: multi-regional epidemiological studies. Health Phys 40: 625-641

Holthusen H, Meyer H, Molineus W (eds) (1959) Ehrenbuch der Röntgenologen und Radiologen aller Nationen. Sonderbände zur Strahlentherapie, vol 42. Urban & Schwarzenberg, München

Holthusen H, Leetz H-K, Leppin W (1961) Die genetische Belastung der Bevölkerung einer Großstadt (Hamburg) durch medizinische Strahlenanwendung (Schriftenreihe des Bundesministers für Atomkernenergie und Wasserwirtschaft, Strahlenschutz, vol 21). Gersbach, München

Hübner W, Jaeger RG (1974) Strahlenfeldgrößen und Einheiten. Dosisgrößen und Dosiseinheiten. In: Jaeger RG, Hübner W (eds) Dosimetrie und Strahlenschutz, 2nd revised edn. Thieme, Stuttgart, pp 53-83

IAEA (International Atomic Energy Agency) (1963) A basic toxicity classification of radionuclides. IAEA Technical Report Series No. 15, Vienna

IAEA (International Atomic Energy Agency), NEA, OECD (Nuclear Energy Agency of the Organisation for Economic Co-operation and Development) (1979) International symposium on occupational radiation exposure in nuclear fuel cycle facilities. Los Angeles/USA, 18-22 June 1979. IAEA, Vienna

IAEA (International Atomic Energy Agency) NEA, OECD Nuclear Energy Agency of the Organisation for Economic Co-operation and Development) (1986) Proceeding of an international symposium on the optimization of radiation protection, 10-14 March 1986 in Vienna. IAEA, Vienna

IAEA (International Atomic Energy Agency) (1987) Radiation protection in occupational health. Manual for occupational physicians. IAEA Safety series no. 83, Vienna

ICRP Publication 1 (1959) Recommendations of the International Commission on Radiological Protection (adopted 9 September 1958) Pergamon, New York. Reprinted ICRP Publication 2 (1959). IX-XXXII

ICRP Publication 2 (1959) Recommendations of the International Commission on Radiological Protection, Report of Committee II on Permissible Dose for Internal Radiation. Pergamon, New York

ICRP Publication 6 (1964) Recommendations of the International Commission on Radiological Protection, as amended 1959 and revised 1962. Pergamon, New York

ICRP Publication 8 (1966) The evaluation of risks from radiation. Pergamon, Oxford

ICRP Publication 9 (1966, reprint 1969) Recommendations of the International Commission on Radiological Protection, Pergamon, Oxford

ICRP Publication 10 (1968) Evaluation of radiation doses to body tissues from internal contamination due to occupational exposure. Pergamon, Oxford

ICRP Publication 10 A (1971) The assessment of internal contamination resulting from recurrent or prolonged uptakes. Pergamon, Oxford

ICRP Publication 11 (1968) A review of the radiosensitivity of the tissues in bone. Pergamon, Oxford

ICRP Publication 14 (1969) Radiosensitivity and spatial distribution of dose. Reports prepared by two task groups of Committee 1 of the International Commission on Radiological Protection. Pergamon, Oxford

ICRP Publication 21 (1973) Data for protection against ionizing radiation from external sources: supplement to ICRP Publication 15. Pergamon, Oxford

ICRP Publication 22 (1973) Recommendations of the International Commission on Radiological Protection. Implications of Commission recommendations that doses be kept as low as readily achievable. Pergamon, Oxford

ICRP Publication 23 (1975) Report of the task group on reference man. Pergamon, Oxford [Revision and update in preparation, cf. Health Phys (1985) 49: 1015]

ICRP Publication 24 (1977) Radiation protection in uranium and other mines. Ann ICRP 1: no. 1

ICRP Publication 25 (1976) The handling, storage, use and disposal of unsealed radionuclides in hospitals and medical research establishments. Ann ICRP 1: no. 2

ICRP Publication 26 (1977) Recommendations of the International Commission on Radiological Protection (adopted 17 January 1977). Ann ICRP 1: no. 3

ICRP Publication 27 (1977) Problems involved in developing an index of harm. A report prepared for the International Commission of Radiological Protection. Ann ICRP 1: no. 4

ICRP Publication 30 part 1 (1979) Limits for intakes of radionuclides by workers. Ann ICRP 2: no. 3/4

ICRP Publication 30, Suppl to part 1 (1979) Limits for intakes of radionuclides by workers. Ann ICRP 3: no. 1-4

ICRP Publication 30 part 2 (1980) Limits for intakes of radionuclides by workers. Ann ICRP 4: No. 3/4

ICRP Publication 30, Suppl to Part 2 (1981) Limits for intakes of radionuclides by workers. Ann ICRP 5: no. 1/6

ICRP Publication 30 part 3 (1981) Limits for intakes of radionuclides by workers. Ann ICRP 6: no. 2/3

ICRP Publication 30, Supplements A and B to part 3 (1981) Limits for intakes of radionuclides by workers. Ann ICRP 7: no. 1/3 and 8: no. 1/3

ICRP Publication 32 (1982) Limtis for inhalation of radon daughters by workers. Ann ICRP 6: no. 1

ICRP Publication 35 (1982) General principles of monitoring for radiation protection of workers. Ann ICRP 9: no. 4

ICRP Publication 37 (1983) Cost-benefit analysis in the optimization of radiation protection. Ann ICRP 10: no. 2/3

ICRP Publication 41 (1984) Non-stochastic effects of ionizing radiation. Ann ICRP 14: no. 3

ICRP Publication 42 (1984) A compilation of the major concept quantities in use by ICRP. Ann ICRP 14: no. 4

ICRP Publication 43 (1985) Principles of monitoring for the radiation protection of the population. Ann ICRP 15: no. 1

ICRP Publication 45 (1985) Quantitative basis for developing a unified index of harm. Ann ICRP 15: no. 3

ICRP Publication 47 (1986) Radiation protection of workers in mines. Ann ICRP 16: no. 1

ICRP (1980) Statement and recommendations of the 1980 Brighton meeting of the ICRP. Ann ICRP 4: no. 3/4

ICRP (1985) Statement from the 1985 Paris meeting of the ICRP. Ann ICRP 15: no. 3

ICRP (1987) Statement from the 1987 Como meeting of the ICRP. Ann ICRP 17: no. 4

ICRU (International Commission on Radiation Units and Measurements) Report 25 (1976) The conceptual basis for determination of dose equivalent. ICRU Publication, Washington

Ilari O (1986) The concept of optimization of protection in the evolution of ICRP recommendations. In: IAEA and NEA (1986) Optimization of radiation protection (cf. l. c.). Vienna, pp 3-15

Imahori A (1980) Occupational radiation exposure at nuclear power plants in Japan. Health Phys 40: 317-322

International Labour Conference (1958) 43rd meeting, report VI no 1. Protection of employees from radiation effects. Intern. Labour Office Geneva

Ishimaru T, Hoshino T, Ishimaru M, Okada H, Tomiyasu T, Tsuchimoto T, Yamamoto T (1971) Leukaemia in atomic survivors, Hiroshima and Nagasaki, 1 October 1950-30 September 1966. Radiat Res 45: 216-233

Jablon S (1984) Epidemiologic perspectives in radiation carcinogenesis. In: Boice JD Jr, Fraumeni JF Jr (eds) Radiation carcinogenesis: epidemiologica and biological significance. Progress in cancer research and therapy, vol 26. Raven, New York, pp 1-8

Jacobi W (1974) Beziehungen zwischen der Strahlendosis und dem somatischen Strahlenrisiko. Atomwirtschaft-Atomtechnik 19: 278-283

Jacobi W (1983) Strahlung und Risiko. Atomwirtschaft-Atomtechnik 28: 238-248

Jones TD (1984) A unifying concept for carcinogenic risk assessments: comparison with radiation-induced leukaemia in mice and men. Health Phys 47: 533-558

Kathren RL (1984) Comments regarding ALARA editorial. Health Phys 47: 323

Kathren RL, Selby JM, Vallario EJ (1980) A guide to reducing radiation exposure to As Low As Reasonably Achievable (ALARA). U. S. Department of Energy (Report DOE/EV/1830-T5: Washington)

Kato H, Schull WJ, Awa A, Akiyama M, Otake M (1987) Dose-response analyses among atomic bomb survivors exposed to low level radiation. Health Phys 52: 645-652

Kaul A (1980) Inkorporationsgrenzwerte nach ICRP 30. In: Messerschmidt O, Feinendegen LE, Hunzinger W (eds) Industrielle Störfälle und Strahlenexposition, Strahlenschutz in Forschung und Praxis, vol XXI. Thieme, Stuttgart, pp 110-132

Kaul A, Elsasser U, Hinz G, Kossel F, Martignoni K, Nitschke J, Stephan G (1984a) Bewertung ausgewählter epidemiologischer Studien an strahlenexponierten Kollektiven. In: Leppin W, Meißner J, Börner W, Messerschmidt O (eds) Die Hypothesen im Strahlenschutz. Strahlenschutz in Forschung und Praxis, vol XXV. Thieme, Stuttgart, pp 61-90

Kaul A, Neider R, Peńsko J, Stieve F-E, Brunner H (eds) (1984b) Radiation-risk-protection. 6th International Congress of the International Radiation Protection Association, Berlin (West), 7-12 May, 1984, Compacts publ. by Fachverband für Strahlenschutz e. V. Jülich. Verlag TÜV Rheinland, Köln

Kaul A, Aurand K, Bonka H et al. (1986) Possibilities of and limitations to the application of the collective dose: a recommendation of the Radiation Protection Commission of the Federal Republic of Germany. In: IAEA (1986) Optimization of radiation protection (cf. l. c.). Vienna, pp 87-95 ·

Kellerer AM, Rossi HH (1972) The theory of dual radiation action. In: Ebert M, Howard A (eds) Current topics in radiation research, vol 8. North Holland, Amsterdam, pp 85-158

Kellerer AM, Breckow J (1988) Neue Erkenntnisse zur Dosisrelation nach der Revision der Dosimetrie in Hiroshima und Nagasaki und Auswirkungen auf den Strahlenschutz. In: Nüsslin F (ed) Medizinische Physik 1988. Deutsche Gesellschaft für Medizinische Physik, pp 499-513

Kerr GD (1979) Organ dose estimates for the Japanese atomic bomb survivors. Health Phys 37: 487-508

Kerr GD, Jones TD, Hwang JML, Miller FL, Auxier JA (1977) An analysis of leukaemia data from studies of atomic bomb survivors based on estimates of absorbed dose to active bone marrow. In: Proceedings of the 4th International Congress of the International Radiation Protection Association, Paris, vol 3, pp 714-718

Khan AK, Raghavayya M, Soman SD (1984) Radiation exposure of indian uranium miners and estimate of associated risk. In: Kaul A, Neider R, Peńsko J, Stieve F-E, Brunner H (eds) Radiation-risk-protection. 6th International Congress of the International Radiation Protection Association, Berlin (West), 7-12 May 1984. Compacts publ. by Fachverband für Strahlenschutz e. V. Jülich. Verlag TÜV Rheinland, Köln, pp 687-690

Kinslow R, Moschella JJ, Cawley CN (1984) A model of urinary tritium following multiple contaminations. Health Phys 46: 1154-1155

Kivinitty K, Nasman P, Leppaluoto J (1984) Accumulation of ^{125}I in the thyroid glands of laboratory workers. Health Phys 46: 234-236

Kneale GW, Stewart AM, Mancuso TF (1978) Reanalysis of data relating to the Hanford study of the cancer risk of radiation workers. In: Late biological effects of ionizing radiation (Proc Symp Vienna, 13-17 March 1978). IAEA, vol I, pp 387-410

Koelzer W, Kiefer H (1980) Mean and individual radiation exposures of the staff of Karlsruhe nuclear research center 1969-1978. In: Radiation protection. A systematic approach to safety. Proceedings of the 5th Congress of the International Radiation Protection Society, Jerusalem, March 1989, vol I. Pergamon, Oxford, pp 80-83

Koeppe P, Oeser H (1982) Prognose der Krebsmortalität in der Bundesrepublik Deutschland 1976-2070. Lebensversicherungsmedizin 34: 50-60

Krzesniak JW, Schürnbrand P, Porstendörfer J, Schicha H, Krajewski P, Becker KH, Emrich D (1984) Levels of airborne contamination while handling ^{125}I and ^{123}I and ^{99m}Tc unsealed sources in medical diagnostic procedures. In: Kaul A, Neider R, Peńsko J, Stieve F-E, Brunner H (eds) Radiation-risk-protection. 6th International Congress of the International Radiation Protection Association, Berlin (West) 7-12 May 1984. Compacts publ. by Fachverband für Strahlenschutz e. V. Jülich. Verlag TÜV Rheinland, Köln, pp 833-836

Lambert JP (1981) Report of a minor ^{125}I exposure in a research laboratory. Health Phys 40: 746-748

Lea DE (1956) Action of radiation on living cells, 2nd edn. Cambridge University Press, London

Leira HL, Lund E, Refseth T (1986) Mortality and cancer incidence in a small cohort of miners exposed to low levels of α-radiation. Health Phys 50: 189-194

Lenz U, Schüttmann W, Arndt D, Thormann T (1979) Late effects of ionizing radiation on the human skin after occupational exposure. In: Occupational radiation exposure in nuclear fuel cycle facilities. Proceedings of a Symposium of IAEA and OECD in Los Angeles, 18-22 June 1979, pp 321-329

Leppin W, Meissner J (1984) Zur Problematik der Strahlenschutzhypothesen. Vorwort zum Tagungsthema. In: Leppin W, Meißner J, Börner W, Messerschmidt O (eds) Die Hypothesen im Strahlenschutz. Strahlenschutz in Forschung und Praxis, vol XXV. Thieme, Stuttgart, pp XII-XVI

Littlefield LG, Joiner EE (1978) Cytogenic follow-up studies in six radiation accident victims 16 and 17 years post-exposure. In: Late biological effects of ionizing radiation (Proc Symp Vienna, 13-17 March 1978). IAEA, vol I, pp 297-308

Loewe WE, Mendelsohn E (1981) Revised dose estimates at Hiroshima and Nagasaki. Health Phys 41: 663-666

Lombard J, Oudiz A, Zettwoog P (1986) A contribution to optimizing radiological protection in a U mine. Health Phys 50: 473-483

Luckey TD (1980) Hormesis with ionizing radiation. CRC Press, Boca Raton

Luckey TD (1981) Ionizing radiation hormones of non-specific immunity. Microecol Ther 11: 113-123

Luckey TD (1982) Physiological benefits from low levels of ionizing radiation. Health Phys 43: 771-789

Luckey TD (1984) Beneficial physiologic effects of ionizing radiation. In: Leppin W, Meissner J, Börner W, Messerschmidt O (eds) Die Hypothesen im Strahlenschutz. Strahlenschutz in Forschung und Praxis, vol XXV. Thieme, Stuttgart, pp 184-196

Maillie HD (1986) Should the annual limit on intake for ^{125}I and ^{131}I be lowered? Health Phys 50: 425

Maini CL, Cichocki F, Rossi G, Venga L, Marchetti L (1987) The radiation dose to the hand of nuclear medicine operators using shielded and unshielded syringes and a simple method for such estimate from badge and ring dosimeters. Nuc-Compact 18: 302-396

Mancuso TF, Stewart A, Kneale G (1977) Radiation exposures of Hanford workers dying from cancer and other causes. Health Phys 33: 396-385

Marcuse W (1896) Dermatitis and Alopecie nach Durchleuchtungsversuchen mit Röntgenstrahlen. Dtsch Med Wochenschr 22: 481-483, 681-682

Matanoski GM, Sartwell P, Elliott E, Tonascia J, Sternberg A (1984) Cancer risks in radiologists and radiation workers. In: Boice JD Jr, Fraumeni JF Jr (eds) Radiation carcinogenesis: epidemiology and biological significance. Progress in cancer research and therapy, vol 26. Raven, New York, pp 83-96

Mays CW, Lloyd RD, Marshall JH (1973) Malignancy risk to humans from total body γ-ray radiation. In: Snyder WSL (ed) Proceedings of the 3rd International Congress of the International Radiation Protection Association. NT IS, Washington Springfield, pp 417-428

Mays CW, Rowland RE, Stehney AF (1985) Cancer risk from the lifetime intake of Ra and U isotopes. Health Phys 48: 635-647

Mehl J (1974) Strahlenschutz. In: Jaeger RG, Hübner W (eds) Dosimetrie und Strahlenschutz, 2nd revised edn. Thieme, Stuttgart, pp 319-416

Mehl J (1979) Radiation exposure control of nuclear power plants personnel in the Federal Republic of Germany. In: Occupational radiation exposure in nuclear fuel cycle facilities. Proceedings of a Symposion of IAEA and OECD in Los Angeles, 18-22 June 1979, pp 275-283

Meissner J (1979) In: Deutsches Atomforum (ed) Rede und

Gegenrede. Symposion der Niedersächsischen Landesregierung zur grundsätzlichen sicherheitstechnischen Realisierbarkeit eines integrierten nuklearen Entsorgungszentrums, 28-31 March, 2.-3. April 1979. Verlag Deutsches Atomforum, Bonn, pp 231-233

Meissner J (1980) Gesundheitsbelastung durch ionisierende Strahlung und chemische Schadstoffe. Zur Quantifizierung gesundheitlicher Risiken durch Strahlenwirkungen und strahlensynergistische Effekte. Atomwirtschaft-Atomtechnik 25: 93-101

Merriam GR, Szechter A (1975) The relative radiosensitivity of rat lenses as a function of age. Radiat Res 62: 488-497

Messerschmidt O, Feinendegen LE, Hunzinger W (eds) (1980) Industrielle Störfälle und Strahlenexposition. 21. Jahrestagung der Vereinigung Deutscher Strahlenschutzärzte, 29.-31.05. 1980 (gemeinsam mit 14. Jahrestagung des Fachverbandes für Strahlenschutz) in Jülich. Strahlenschutz in Forschung und Praxis, vol XXI, Thieme, Stuttgart

Metcalf PE, Winkler BC (1980) Risk ratio for use in establishing dose limits for occupational exposure to radiation. In: Radiation protection. A systematic approach to safety. Proceedings of the 5th Congress of the International Radiation Protection Society, Jerusalem, March 1980, vol I. Pergamon, Oxford, pp 272-275

Milton RC, Shohoji T (1965) Tentative 1965 radiation dose (T65D). Estimation for atomic bomb survivors, Hiroshima and Nagasaki. Atomic bomb casualty commission. Technical report TR 1-68 Hiroshima

Modan B, Baidatz D, Mart H, Steinitz R, Levin SG (1974) Radiation-induced head and neck tumours. Lancet 1: 277-279

Morgan KZ (1978) Risk of cancer from low level exposure to ionizing radiation. Bull Atomic Sci, Sept 1978: 30-41

Morgan KZ (1979) In: Deutsches Atomforum (ed) Rede und Gegenrede. Symposion der Niedersächsischen Landesregierung zur gesundheitlichen sicherheitstechnischen Realisierbarkeit eines integrierten nuklearen Entsorgungszentrums, 28-31 March, 2-3 April 1979. Verlag Deutsches Atomforum, Bonn, pp 241-243

Morgan MV, Samet JM (1986) Radon daughter exposures of New Mexico U miners 1967-1982. Health Phys 50: 656-662

Muller HJ (1928) The production of mutations by x-rays. Proc Nat Acad Sci 14: 714-726

Mustafa A, Sabol J, Janeczek J (1985) Doses from occupational exposure: a study of radiation doses to workers in Kuwait over a four-year period. Health Phys 49: 1197-1204

Mutscheller A (1925) Physical standards of protection against roentgen ray dangers. AJR 13: 65-70

NCRP (National Council on Radiation Protection and Measurements) (1984) Exposures from the uranium series with emphasis on radon and its daughters. NCRP Report No. 77: Bethesda

National Radiological Protection Board NRPB (1987) Interim guidance on the implications of recent revisions of risk estimates and the ICRP 1987 Como statement. NRPB-Report GS 9 Chilton

Oeser H, Koeppe P (1981) Voraussichtliche Entwicklung der Krebssterblichkeit in der Bundesrepublik Deutschland. Münch Med Wochenschr 123: 706-708

Oeser H, Koeppe P (1982) Kritische Betrachtungen zur Abschätzung von Strahlenrisiken. In: Messerschmidt O, Börner W, Holeczke F, Olbert F, Seyss R (eds) Zur Problematik der Wirkung kleiner Strahlendosen. Strahlen-schutz in Forschung und Praxis, vol XXIII. Thieme, Stuttgart, pp 154-162

Otake M, Schull WJ (1984) In utero exposure to A-bomb radiation and mental retardation, a reassessment. Br J Radiol 57: 409-414

Peterson HT Jr (1984) Regulatory implications of radiation dose effect relationship. Health Phys 47: 345-359

Pfaller W (1984) Subzelluläre Veränderungen nach Radium 222-Einwirkung. In: Leppin W, Meissner J, Börner W, Messerschmidt O (eds) Die Hypothesen im Strahlenschutz. Strahlenschutz in Forschung und Praxis, vol XXV. Thieme, Stuttgart, pp 205-206

Pietzsch W (1976) Approximation zur Berechnung der genetisch-signifikanten Dosis. Institut für Strahlenhygiene des Bundesgesundheitsamtes. Arbeits-Bericht E I 2-3/76

Pohl-Rüling J, Fischer P (1979) The dose-effect relationship of chromosome aberrations to alpha and gamma irradiations in a population subjected to an increased burden of natural radioactivity. Radiat Res 80: 61-81

Pohl-Rüling J, Fischer P, Pohl E (1979) Chromosomenaberrationen nach Inhalation von ^{222}Radon und seinen Zerfallsprodukten. Z Angew Bäder- und Klimaheilkd 26: 437-443

Polednak AP, Stehney AF, Rowland RE (1978) Mortality among women first employed before 1930 in the US radium dial-painting industry. A group ascertained from employment lists. J Epidemiol 107: 179-195

Preston DL, Kato H, Kopecky K, Fujita S (1987) Studies of the mortality of A-bomb survivors. 8. Cancer mortality, 1950-1982. Radiat Res 111: 151-178

Preston DL, Pierce DA (1988) The effect of changes in dosimetry on cancer mortality risk estimates in the atomic bomb survivors. Radiat Res 114: 437-466

Radford EP (1980a) Statement concerning the current version of cancer risk assessment in the report of the advisory committee on the biological effects of ionizing radiations. In: BEIR III report, pp 227-253

Radford EP (1980b) Human health effects of low doses of ionizing radiation: the BEIR III controversy. Radiat Res 84: 369-394

Radford EP (1984) Radiogenic cancer in underground miners. In: Boice JD Jr, Fraumeni JF Jr (eds) Radiation carcinogenesis: epidemiology and biological significance. Progress in cancer research and therapy, vol 26. Raven, New York, pp 225-230

Rajewsky B (1956) Strahlendosis und Strahlenwirkung, 2nd edn. Thieme, Stuttgart

Ras EMM de, Vaane JP, Suetendael W van (1980) Investigation of the nature of a contamination caused by tritium targets used for neutron production. In: Radiation protection. A systematic approach to safety. Proceedings of the 5th Congress of the International Radiation Protection Society, Jerusalem, March 1980, vol I. Pergamon, Oxford, pp 48-51

Rausch L (1977) Das manrem-Konzept.-Formalismus und Realität. In: Rausch L, Messerschmidt O, Möhrle G, Zimmer R (eds) Betrieblicher Strahlenschutz aus ärztlicher Sicht. Grundlagen und Praxis des Strahlenschutzes in der Medizin. Strahlenschutz in Forschung und Praxis, vol XVII. Thieme, Stuttgart, pp 66-82

Rausch L (1982) Mensch und Strahlenwirkung. Strahlenschäden Strahlenbehandlung Strahlenschutz. Piper, München

Refetoff S, Harrison J, Karanfilski BT, Kaplan EL, De Groot LJ, Bekerman C (1975) Continuing occurrence of thyroid carcinoma after irradiation to the neck in infancy and childhood. N Engl J Med 292: 171-175

Ron E, Modan B (1984) Thyroid and other neoplasms following childhood scalp irradiation. In: Boice JD Jr, Fraumeni JF Jr (eds) Radiation carcinogenesis: epidemiology and biological significance. Progress in cancer research and therapy, vol 26. Raven, New York, pp 139-151

Röntgenverordnung (RÖV) (1987) Verordnung über den Schutz vor Schäden durch Röntgenstrahlen. Vom 8. Januar 1987, BGBL I: 114-133

Rossi HH (1980a) Separate statement - critique of BEIR III. In: BEIR III Report, pp 254-260

Rossi HH (1980b) Comments on the somatic effects section of the BEIR III Report. Radiat Res 84: 395-406

Rossi HH, Kellerer AM (1974) The validity of risk estimates of leukaemia incidence based on Japanese data. Radiat Res 58: 131-140

Rossi HH, Mays CW (1978) Leukaemia risk from neutrons. Health Phys 34: 353-360

Rowe WD (1980) Risk assessment perspectives in radiation protection. In: Radiation protection. A systematic approach to safety. Proceedings of the 5th Congress of the International Radiation Protection Society. Jerusalem, March 1980, vol I. Pergamon, Oxford, pp 255-261

Rowland RE, Lucas HF Jr (1984) Radium-dial workers. In: Boice JD Jr, Fraumeni JF Jr (eds) Radiation carcinogenesis: epidemiology and biological significance. Progress in cancer research and therapy, vol 26. Raven, New York, pp 231-240

Schlager KJ (1984) More comments on ALARA. Health Phys 47: 321-323

Schull WJ (1984) Atomic bomb survivors: pattern of cancer risk. In: Boice JD Jr, Fraumeni JF Jr (eds) Radiation carcinogenesis: epidemiology and biological significance. Progress in cancer research and therapy, vol. 26. Raven, New York, pp 21-36

Ševc J, Kunz E, Plaček V, Šmid A (1984) Comments on lung cancer estimates. Health Phys 46: 961-964

Shamai Y, Tirkel M, Schlesinger T (1980) Investigation levels of radioisotopes in the body and in urine. Consequences of the recent recommendations on the annual limits of intake. In: Radiation protection. A systematic approach to safety. Proceedings of the 5th Congress of the International Protection Society, Jerusalem, March 1980, vol I. Pergamon, Oxford, pp 88-91

Shenoy KS, Patel PH, Madhavanath U (1980) Occupational exposure in industrial radiography practice. Health Phys 40: 323-326

Shore RE, Albert RE, Pasternack BS (1976) Follow-up study of patients treated by x-ray epilation for tinea capitis. Resurvey of post-treatment illness and mortality experience. Arch Environ Health 31: 17-28

Shore RE, Hempelmann LH, Kowaluk E, Mansur PG, Pasternack BS, Albert RE, Haughie GE (1977) Breat neoplasms in women treated with x-rays for acute postpartum mastitis. J Natl. Cancer Inst 59: 813-822

Shore RE, Woodward ED, Hempelmann LH (1984) Radiation-induced thyroid cancer. In: Boice JD Jr, Fraumeni JF Jr (eds) Radiation carcinogenesis: epidemiology and biological significance. Progress in cancer research and therapy, vol 26. Raven, New York, pp 131-138

Singh NP, Wrenn McDE (1986) Levels of ^{234}U, ^{238}U and ^{230}Th in excreta of U mill crushermen. Health Phys 50: 300-302

Smith PG (1984) Late effects of x-ray treatment of ankylosing spondylitis. In: Boice JD Jr, Fraumeni JF Jr (eds) Radiation carcinogenesis: epidemiology and biological significance. Progress in cancer research and therapy, vol 26. Raven, New York, pp 107-118

Smith PG, Doll R (1978) Age- and time-dependent changes in the rates of radiation-induced cancers in patients with ankylosing spondylitis following a single course of x-ray treatment. In: Late biological effects of ionizing radiation. Proc Symp Vienna, 13-17 March 1978. IAEA, vol I, pp 205-218

Srivastava GK, Raghavayyayya M, Kotrappa P, Somasundaram S (1986) Radium-226 body burden in U miners by measurement of Rn in exhaled breath. Health Phys 50: 217-221

Stäblein G (1980) Health pyhsics documentation. In: Radiation protection. A systematic approach to safety. Proceedings of the 5th Congress of the International Radiation Protection Society, Jerusalem, March 1980, vol I. Pergamon, Oxford, pp 92-94

Stehney AF, Lucal HF Jr, Rowland RE (1978) Survival times of women radium dial workers first exposed before 1930. In: Proceedings of the symposion of late biological effects of ionizing radiation, vol I. IAEA, Vienna, pp 333-351

Stieve F-E (1976) Kontrolluntersuchungen und Behandlungsverfahren. In: Bünemann D, Meißner J, Rausch L (eds) Medizinische und biologische Probleme der Strahlenexposition. Atomkernenergie 28: 49-53

Stieve F-E (1980) Erfahrungen mit bisherigen Strahlenunfällen - Ursachen, Abläufe, Gegenmaßnahmen. In: Messerschmidt O, Feinendegen LE, Hunzinger W (eds) Industrielle Störfälle und Strahlenexposition. Strahlenschutz in Forschung und Praxis, vol XXI. Thieme, Stuttgart, pp 80-108

Stocker H (ed) (1985) International conference on occupational radiation safety in mining, Toronto, October 14-18. Canadian Nuclear Association, Toronto

Strahlenschutzkommission SSK (1988) Recommendations of the Commission on Radiological Protection 1987. Vol. 10. Fischer Stuttgart

Strahlenschutzverordnung (StrSchV) (1976, 1989) Verordnung über den Schutz vor Schäden durch ionisierende Strahlen. Vom 13. Oktober 1976 BGBl I, pp 2905-2931; revised edn. 30. 6. 1989 BGBl I, pp 1321-1375

Streffer C (1979) In: Deutsches Atomforum (ed) Rede und Gegenrede. Symposion der Niedersächsischen Landesregierung zur grundsätzlichen sicherheitstechnischen Realisierbarkeit eines integrierten nuklearen Entsorgungszentrums, 28-31 März, 2.-3. April 1979. Verlag Deutsches Atomforum, Bonn, pp 171-173

Tait GWC (1980) Cost-benefit analysis: reality or illusion. Health Phys 39: 835-838

Taylor LS (1980a) Dealing with radiation hazards. Perspect Biol Med 23: 325-334

Taylor LS (1980b) Some nonscientific influences on radiation protection standards and practice. The 1980 Sievert Lecture. Health Phys 39: 851-874

Thomas DC, McNeill KG, Dougherty C (1985) Estimates of lifetime lung cancer risks resulting from Rn progeny exposure. Health Phys 49: 825-846

Tirmarche M, Chameaud J, Piechowski J, Pradel J (1984) Enquete epidemiologique francaise sur les mineurs d'uranium: Difficultes et progres. In: Kaul A, Neider R, Peńsko J, Stieve F-E, Brunner H (eds) Radiation-risk-protection. 6th International Congress of the International Radiation Protection Association, Berlin (West), 7-12 May 1984. Compacts publ. by Fachverband für Strahlenschutz e. V. Jülich. Verlag TÜV Rheinland, Köln, pp 574-577

Tokunaga M, Land CE, Yamamoto T, Asano M, Tokuoka S, Ezaki H, Nishimori I, Fujikura T (1984) Breast cancer among atomic bomb survivors. In: Boice JD Jr, Fraumeni

JF Jr (eds) Radiation carcinogenesis: epidemiology and biological significance. Progress in cancer research and therapy, vol 26. Raven, New York, pp 45-56

Totter JR (1980) Some observational bases for estimating the oncogenic effects of ionizing radiation. Nuclear Safety 21: 83-94

Trott NG (1981) Balancing the risk to hospital staff and the benefit to the patient. J Soc Radiol Protection 1: 20-26

Trott NG (1982) The safe and effective use of radiopharmaceuticals, Gastein lecture. In: Höfer R, Bergmann H (eds) Radioaktive Isotope in Klinik und Forschung. Gasteiner Internationales Symposium 1982, vol 15, part 2. Egermann, Vienna, pp 465-479

Trott NG, Anderson W, Davis R, Parker RP, Garden DM, Pearson N, Harbottle E (1980) Some radiation protection problems in a cancer hospital and associated research institute. In: Radiation protection. A systematic approach to safety. Proceedings of the 5th International Radiation Protection Society, Jerusalem, March 1980, vol I. Pergamon, Oxford, pp 56-59

Tuschl H (1984) Die Wirkung niederer Dosen ionisierender Strahlung auf DNA-Reparaturvorgänge. In: Leppin W, Meissner J, Börner W, Messerschmidt O (eds) Die Hypothesen im Strahlenschutz. Strahlenschutz in Forschung und Praxis, vol XXV. Thieme, Stuttgart, pp 197-204

Tuschl H, Altmann H (1979) Untersuchungen über den Einfluß von Radon auf Immunsysteme und DNA-Stoffwechsel. Z Angew Bäder- und Klimaheilkd 26: 391-398

UNSCEAR United Nations Scientific Committee on the Effects of Atomic Radiation (ed) (1972) vol I: Levels, vol II: Effects, report to the general assembly. United Nations Sales Publication, no E. 72. IX. 17 and 18, New York

UNSCEAR United Nations Scientific Committee on the Effects of Atomic Radiation (ed) (1977) Sources and effects of ionizing radiation, report to the general assembly, United Nations Sales Publication, no. E. 77.IX. I, New York

UNSCEAR United Nations Scientific Committee on the Effects of Atomic Radiation (ed) (1982) Sources and biological effects of ionizing radiation, report to the general assembly, United Nations Sales Publications, no. E. 82. IX, New York

Upton AC (1977) Radiobiological effects of low doses. Implications for radiological protection. Radiat Res 71: 51-74

Upton AC (1984) Biological aspects of radiation carcinogenesis. In: Boice JD Jr, Fraumeni JF Jr (eds) Radiation carcinogenesis: epidemiology and biological significance. Progress in cancer research and therapy, vol 26. Raven, New York, pp 9-19

Vialettes H, Moreau A (1980) Les Problemes de radioprotection rencontres dans un laboratoire de marquage de molecules au carbone-11. In: Radiation protection. A systematic approach to safety. Proceedings of the 5th Congress of the International Radiation Protection Society, Jerusalem, March 1980, vol I. Pergamon, Oxford, pp 52-55

Voelz GL, Grier RS, Hempelmann LH (1984) A 37-year medical follow-up of Manhattan project Pu workers. Health Phys 48: 249-259

Vogel F (1984) Gesichertes und Hypothetisches im Bereich der Strahlengenetik. In: Leppin W, Meissner J, Börner W, Messerschmidt O (eds) Die Hypothesen im Strahlenschutz. Strahlenschutz in Forschung und Praxis, vol XXV. Thieme, Stuttgart, pp 144-168

von Frieben A (1902) Cancroid des rechten Handrückens nach langdauernder Einwirkung von Röntgenstrahlen. Fortschr. Roentgenstr. 6: 102

Walburg HE Jr (1975) Radiation-induced life shortening and premature aging.In: Lett JT, Adler H (eds) Advances in radiation biology, vol 5. Academic Press, New York, pp 145-179

Webb GAM, Fleishman AB (1984) Optimization of protection of radiation workers. In: Kaul A, Neider R, Peńsko J, Stieve F-E, Brunner H (eds) Radiation-risk-protection. 6th International Congress International Radiation Protection Association, Berlin (West), 7-12 May 1984. Compacts publ. by Fachverband für Strahlenschutz e. V. Jülich. Verlag TÜV Rheinland, Köln, pp 637-640

Webb GAM, Oudiz A, Lochard J, Lombard J, Croft JR, Fleishman AB (1986) Development of a general framework for the practical implementation of ALARA. In: IAEA (1986c) Optimization of radiation protection. (cf. l. c.): Vienna, pp 123-136

Weiss K (1942) Die Röntgenschäden der letzten 20 Jahre in den gemeindlichen Krankenhäusern Deutschlands. Strahlentherapie 72: 307-329

Weng P-S, Chen T-C (1985) Occupational radiation exposures in Taiwan 1962-1983. Health Phys 49: 411-418

Wilkinson GS, Voelz GL, Acquavella JF, Tietjen GL, Wiggs L, Waxweiler M (1984) Health effects amoung plutonium workers. In: Kaul A, Neider R, Peńsko J, Stieve F-E, Brunner H (eds) Radiation-risk-protection. 6th International Congress of the International Radiation Protection Association, Berlin (West), 7-12 May 1984. Compacts publ. by Fachverband für Strahlenschutz e. V. Jülich. Verlag TÜV Rheinland, Köln, pp 570-573

3 Occupational Radiation Carcinogenesis

KLAUS-RÜDIGER TROTT and CHRISTIAN STREFFER

CONTENTS

3.1 Introduction

On 21 October 1902, Dr. FRIEBEN demonstrated to the Medical Society of Hamburg the first recognized case of radiation-induced cancer. A worker at the local X-ray tube factory had used his hands for testing the output of the tubes for 4 years when he developed severe pigmentation of the arms, face, and chest as well as chronic radiodermatitis which progressed to chronic ulceration and cancer metastasizing to the regional lymph nodes within a few months.

Skin cancer arising from chronic radiodermatitis, mostly of the hands, was very common among radiation workers in the first decades of this century. Of the 343 radiation victims listed on the memorial at the Allgemeines Krankenhaus Sankt Georg in Hamburg, 234 died from occupational skin cancer originating in the hands (HOLTHUSEN et al. 1959). Until the middle of this century, perception of radiation carcinogenesis was essentially limited to skin cancer developing after long exposure to very high radiation doses and secondary to chronic radiodermatitis.

When, in 1928, the first "recommendations for X-ray and radium protection" were published by the International Congress of Radiology, creating the International Commission on Radiological Protection (ICRP), they were primarily concerned with the dangers of overexposure of radiation workers to guard them against "(a) injuries to the superficial tissues and (b) derangement of internal organs and changes in the blood." Working hours were limited and the need for shielding X-ray tubes and rooms was established. Further recommendations improved the standard of radiologic protection of radiation workers during the following decades as a consequence of better understanding of the various radiobiologic effects of radiation until, nearly 50 years after the first recommendation, in 1977 the ICRP published their recommendation No. 26 setting limits of radiation exposure for radiation workers. These limits are "intended to prevent non-stochastic effects and to limit the occurrence of stochastic effects to an accepted level" (§ 103). "The Commission believes that for the foreseeable future a valid method for judging the acceptability of the level of risk in radiation work is by comparing this risk with that for other occupations recognised as having high standards of safety, which are generally considered to be those in which the average annual mortality due to occupational hazards does not exceed 10^{-4}" (§ 96).

In addition to epidemiologic studies on increased cancer incidence after various levels of radiation exposure among the survivors of the atomic bomb explosions in Hiroshima and Nagasaki, the risk estimates of ICRP publication 26 and the recommended dose limits derived from them were also based on some epidemiologic studies on various radiation workers who had received radiation doses substantially exceeding those established as dose limits by ICRP in 1977. These studies will also be reported in this chapter; however, more emphasis is laid on those studies concerning radiation workers who received their exposure in the decades since 1945 and most of whom accumulated occupational radiation doses well within the generally accepted limits. These studies have been performed either because of a widely expressed feeling of unease about the safety of present regulations, or, sometimes, with the purpose of proving that these regulations are indeed safe.

3.2 Epidemiology of Occupational Radiation Hazards

To study the possibility of increased morbidity or mortality among occupationally exposed people, epidemiologists may use a variety of methods which can be described as being either cohort studies, case control studies, or nested studies.

3.2.1 Cohort Studies

In a cohort study or follow-up study, groups of people who have been occupationally exposed to different levels of radiation but who are otherwise comparable are identified. These groups are then followed over a period of time to determine the frequencies of disease occurrence. These frequencies are then compared among the categories of exposure or with those of a comparable nonexposed population, using national or regional health registries.

3.2.2 Case Control Studies

In a case control study, groups of individuals are selected who have the disease under study. From a suitable group of people, several individuals are then selected for each diseased person who are as similar as possible with regard to age and other known risk factors except for the factors which are to be investigated. In a case control study, a specific hypothesis is usually tested, for example that a connection exists between occupational radiation exposure and a specific type of cancer, e. g., leukemia or lung cancer. However, case control studies can look at more than just on type of possibly harmful exposure in relation to one disease whereas cohort studies evaluate different diseases in relation to one exposure.

3.2.3 Nested Studies

The nested design incorporates features of both the cohort and the case control studies. Once a cohort

has been identified and followed up it is possible to conduct a case control study within the cohort. All individuals with the disease of interest are matched to nondiseased controls from the pool of the study on the basis of well-defined stratification criteria, and detailed exposure histories to radiation and other possibly toxic agents are then obtained only for the diseased cases and their controls.

3.2.4 Comparison Groups

In both cohort and case control studies, a basis for comparison is required to evaluate whether the morbidity or mortality experienced by the occupationally exposed group differs from what would have been expected if they had not been at any increased risk from that exposure. Ideally, the compared subjects should be as similar as possible with respect to all variables that could affect their probability of contracting the disease under study except for the exposure of interest. This, however, is rarely possible.

In a cohort study, comparison groups are drawn either from within the study group (nonexposed workers of the same establishment) or from the general population. However, since the incidence of most diseases varies according to age, sex, race, and socioeconomic status, the disease rates in such a comparison group must be adjusted or standardized for these variables so that comparison with the study cohort can be made.

In most of the studies reported here, lower morbidity and mortality from most causes of death are observed in the study groups compared with the comparison group drawn from the general population. This is known as the "healthy worker effect" and is usually ascribed to selection of a particularly healthy subpopulation at employment.

3.2.5 Analytical Methods

Morbidity and mortality of the study group and the comparison group are compared by calculating the standardized mortality ratio (SMR) by relating the observed effect to that expected for a similar (i. e., stratified with regard to known confounding factors, especially age, sex, and calendar year) but nonexposed group. If several exposure categories are available, the relationship of SMR to exposure dose is analyzed to see whether there is a trend towards an increase in SMR as exposure dose increases.

Alternatively, the expected rate can be calculated from the total study group (rather than the nonexposed subgroup or the general population) stratified to mirror the various exposure groups. The resulting SMR values for the different exposure groups are then tested for trend in relation to radiation dose.

3.2.6 Confounding Factors and Other Problems

Radiation is a known carcinogen but by no means the only one or the most potent one. For some specific types of cancer, factors are known to be associated with an increased risk and they should be controlled for in the study design by adequate stratification. However, for some other cancers, except for obvious factors like age, sex, and ethnic group (and related life-style and habits), no association with specific carcinogenic factors is known; this makes stratification (except for age, sex, and calendar year) impossible and one has to rely on random chance.

Of great importance for stratification is the temporal pattern of incidence and mortality for various types of cancer, e. g., the continuing decrease in the incidence of cancers of the stomach and of the cervix over the last 30 years and the increasing frequency of cancers of the lung and the large bowel. Moreover, dramatic changes in cure probability, e. g., in childhood leukemia or Hodgkin's disease, from a few percent to well over 50% within that same period may influence results if they are based on mortality rather than on incidence. Therefore, comparison groups have to be stratified or corrected for calendar years.

A problem common to most reported studies is the very skewed distribution of exposure doses. Even in some very large studies, only a few subjects have accumulated radiation doses which are well above those doses which may represent the upper limit of radiation exposures from natural sources. Since background external and internal radiation exposure may vary between individuals by more than one order of magnitude and is not known for the individuals at risk (in practically all studies), problems may arise in the interpretation of results regarding the effect of occupational radiation doses which are smaller than those within the range of natural radiation exposure.

Most of these problems arise in several of the reported studies and will be discussed whenever necessary.

3.3 Miners

Long before Becquerel detected natural radioactivity in ores, occupational lung cancer had been described in hard rock miners which, however, only much later could be definitely ascribed to exposure of the bronchial mucosa to radioactivity originating from decay products of radon in the mines. In a very extensive article covering 68 pages, HÄRTING and HESSE (1879) described the details of one of the first comprehensive studies in industrial medicine, which compares favorably with the famous studies by POTT (1775) on cancer of the scrotum and by VIRCHOW (1849) on the epidemic among Silesian weavers.

In the cobalt mines of Schneeberg, which was part of Saxonia in those days, many miners succumbed to a fatal lung disease. In 20 autopsied cases HÄRTING consistently found the cause of death to be cancer which he described as lymphosarcoma but which today would probably be classified as small cell lung cancer. Not a single miner had tuberculosis, the most common fatal lung disease in the nineteenth century. Of the total workforce of 650 men, 63 died from lung cancer in the 3-year period 1869–1871. Primary lung cancer was the cause of death in 75% of all deaths among miners in Schneeberg. All miners who did not die in an accident or had an infectious disease died from lung cancer. Death occurred after a minimum working life of 20 years but rarely as late as 50 years after commencement of mining (which was usually at the age of 16 years). Very detailed investigations of the working conditions in the mines, which take up most of the publication, suggested that a toxic substance associated and inhaled together with the drill dust was causing cancer. It was suggested that this was arsenic. To decrease the concentration of the supposed carcinogenic substance in the air, HESSE initiated improvements in the ventilation which appear to have resulted in a drop in the number of deaths from lung cancer to 40 miners in the period 1875–1877.

Soon after Becquerel detected radioactivity in 1895, it was suggested that inhalation of radioactive radon 222, which is produced in the rock by decay of ^{226}Ra, was causing lung cancer. However, radon, being a noble gas which is not retained in the body, is relatively nontoxic. In 1955, BALE and SHAPIRO as well as AURAND et al. demonstrated that lung cancer of the miners was due to α-irradiation from inhaled aerosols containing short-lived daughter products of the decay of radon.

Table 3.1. Data from epidemiological studies on uranium minera (Jacobi and Paretzke 1985)

Quantity	Colorado	CSSR	Ontario	France
Initial number of miners	3366	2433	ca. 13 400	1957
Observation period	1950–77	1948–75	1955–81	1947–83
Average follow-up period	19	26	15	25.9
Surviving fraction at end (%)	72	–	ca. 80	81
Median age at start of mining	30	35–40	ca. 25	ca. 30
Average working period in mine	9	10	ca. 2	11.4
Person · years at risk (PYR)	62 556	ca. 60 000	202 795	50 784
Mean cumulated exposure (WLM)	820	310	60 ± 25	–
Chronic cigarette smokers (%)	ca. 70	ca. 70	50–60	ca. 70
Number of lung cancer cases ⎫ observed	194	ca. 250	82	36
during follow up ⎭ expected	40	ca. 60	57	18.8
Observed/expected cases	4.8	ca. 4.2	1.45	1.9

Epidemiologic studies on men working in uranium mines in the United States (WHITTEMORE and MCMILLAN 1983), Czechoslovakia (KUNZ et al. 1979), Ontario (MÜLLER et al. 1985), and France (TIRMARCHE et al. 1985) confirmed the close association between increased lung cancer risk and high levels of exposure to radioactivity in the mines. There seems no question that extended exposure in some uranium mines as well as in some other mines – especially fluorospar, lead, zinc, and iron – which have high radon levels is associated with an increase in lung cancer incidence. There is evidence that the incidence is correlated to the product of the length of time worked and the average concentration of radon and its daughter products during that time. JACOBI and PARETZKE (1985) summarized these four studies on altogether more than 22 000 uranium miners in whom about 570 lung cancer deaths were observed, compared with less than 180 expected cases. The details of these four studies are tabulated in Table 3.1. The observed excess of about 400 cases of lung cancer is similar to the observed excess of all cancers in the lifespan study of Hiroshima and Nagasaki survivors comprising a four times larger group of people. This reflects the very high radiation exposure of the respiratory tract of the miners.

Epidemiologic studies of miners relate the cancer incidence to the individual cumulative exposure to radon in the air, called the Working Level Month (WLM). One Working Level is defined as a ^{222}Ra activity concentration of 100 pCi per liter air, which can be readily measured in mines. However, conversion of the radon concentration in air to radiation dose to the bronchial mucosa requires additional information on equilibrium of radon with the short-lived radon daughters, on particle size distribution of the aerosol, etc. and has to be based on models of breathing physiology. This has been dis-

cussed in some detail in various reports (e. g., UNSCEAR 1982). ICRP 32 recommended (1982), on the basis of recently developed models, annual limits of exposure and derived air concentrations of radon and radon daughters to conform with the general system of dose limitations to radiation workers.

The studies of both the Colorado miners and the Czechoslovakian miners provided evidence for a linear relationship between excess lung cancer cases and cumulative exposure (WLM) (Fig. 3.1). With short follow-up periods, small cell lung cancer was the most frequent histologic type whereas after long latency times epidermoid bronchogenic cancer was the dominant type of lung cancer (NCRP Report 78, 1984). Therefore, the original suggestion that exposure to radon daughters led predominantly to small cell lung cancer is unjustified and was due to insufficient follow-up.

NRCP Report 78 (1984) gives a detailed summary of seven studies on hard rock miners, stressing the difficulties of data collection, differences in the age structure and follow-up periods of the various exposure groups, etc. Above all, epidemiologic methods are not strictly comparable in the various studies, making any comparison of them doubtful.

Nevertheless, the two most important studies, on the US and Czechoslovakian miners, gave very similar results, which are summarized by NCRP Report 78 as follows:

A review of epidemiological data on underground miners indicates that an excess lung cancer mortality exists above cumulative exposures of about 100 WLM. Radon daughter exposures appear to be more efficient in inducing lung cancer when cumulative exposures are below 1000 WLM.

In the reported analyses of lung cancer among these mining groups, a number of important factors involved in the relationship between radon daughter exposure and resultant lung cancer in man are becoming apparent. The latent period seems to vary inversely with age at first exposure, with

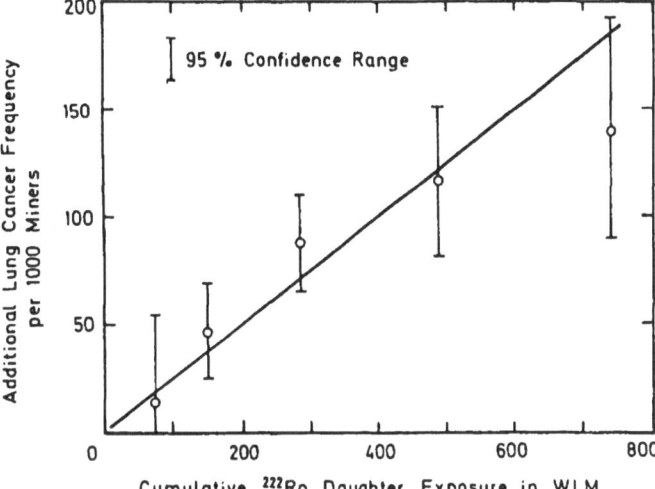

Fig. 3.1. Observed excess lung cancer frequency among uranium miners in Czechoslovakia as a function of their potential α-energy exposure to short-lived ^{222}Rn-daughters (data from KUNZ et al. 1979)

amount of cigarette smoking, and with total exposure and/or exposure rates. That is, the shortest latent periods are found among those men who are elderly at start of mining, who smoke heavily, and who have the most intense exposures. The latent period has a large range of about 7 to 50 years. Mean values are usually considered to be between 20 and 30 years.

The risk calculated from these studies varies by more than one order of magnitude between the groups, with a rounded average value of 10^{-5} per person per WLM.

3.4 Radiologists

The mortality patterns of the members of two radiologic societies have been compared with those of other medical societies or the population at large.

COURT BROWN and DOLL (1958) analyzed the mortality of 1381 male members of the British Institute of Radiology and the Faculty of Radiologists from 1897 to 1957. The radiologists were divided into two groups, those who joined the professional society before 1921 and those who joined after 1921, this being the year when the British X-ray and Radiation Protection Committee was formed and issued its first recommendations, which appeared to have had a swift effect in reducing the exposure of radiologists to radiation. Three hundred and two deaths were recorded in the pre-1921 population and 161 deaths in the post-1921 population. Altogether 79 cancer deaths were identified while 86 were expected if the general population were taken for comparison or 61-71 when doctors belonging to other medical societies served as the control group. Whereas no increased cancer mortality was ob-

served among the post-1921 group (24 cases observed with 26-38 expected in the different standardized comparison groups) there was a definite excess of cancer deaths in the pre-1921 population, with 55 observed while between 35 and 47 were expected. This excess cancer mortality was mostly due to skin cancer (six observed while 0.6-1 were expected), leukemia (two observed while 0.3-0.5 were expected), and cancer of the pancreas (six observed while 1.6-1.9 were expected).

The excess mortality from skin cancer is in keeping with the numerous reports of the development of tumors in skin showing evidence of radiation damage. In addition to the six cases noted above, in another two cases skin cancer was an associated cause of death and in another 15 cases severe damage to the skin was reported which probably included skin cancers, as well. While the increase in leukemia is significant, it is small and in keeping with expectations. No explanation could or can be given for the increased mortality from cancer of the pancreas.

SMITH and DOLL (1981) followed this group of British radiologists until 1977. The number of evaluable deaths increased to 319 in the pre-1921 group and to 411 in the post-1921 group. SMRs were calculated against three different comparison groups: the general population of England and Wales, the general population of social class 1, and general practitioners. Compared with general practitioners, the most striking effect was a 75% increase in cancer mortality ($P < 0.001$) among the pre-1921 radiologists but no difference in those radiologists who entered the profession after 1921. The most noteworthy change in the cancer mortality pattern of the early radiologists was the very late occur-

Table 3.2. Deaths from major causes and leukemia in American radiologists, 1935-1958 (SELTSER and SARTWELL, 1965)

Cause of death	Age group (years)	Observed deaths in RSNA	Expected no. of deaths at AAOO rates	Excess of observed over expected	Mortality ratio
All causes	35–49	79	61.5	17.5	0.3
	50–61	339	271.5	67.5	1.2
	65–79	438	295.0	143.0	1.5
	35–79	856	628.0	228.0	1.4
Leukemia	35–49	2	1.9	0.1	1.0
	50–64	8	1.1	6.9	7.3
	65–79	9	4.7	4.3	1.9
	35–79	19	7.7	11.3	2.5
All other cancer	35–49	9	7.3	1.7	1.2
	50–64	54	32.0	22.0	1.7
	65–79	72	47.6	24.4	1.5
	35–79	135	86.8	48.2	1.6
Cardio-vascular-renal disease	35–49	33	31.6	1.4	1.0
	50–61	209	189.1	19.9	1.1
	65–79	292	209.8	82.2	1.4
	35–79	534	430.6	103.4	1.2
All other causes	35–49	35	20.8	14.2	1.7
	50–64	68	49.2	18.8	1.4
	65–79	65	32.7	32.3	2.0
	35–79	168	102.7	65.3	1.6

rence of two additional cases of leukemia, adding up to a total of four cases against 0.65 expected. The intervals between entering the profession and leukemia were 9, 30, 52, and 57 years.

SELTSER and SARTWELL (1965) compared the mortality of 3697 members of the Radiological Society of North America, of 7052 members of the American College of Physicians, and of 6059 members of the American Academy of Ophthalmology and Otolaryngology. Only male doctors practising in the United States between 1915 and 1958 were analyzed. Between 1935 and 1944 radiologists had the highest death rate in eight of the nine age groups, and this pattern was similar in the subsequent time periods up to 1958. The data, classified into 15-year age groups, are tabulated in Table 3.2. Life table analysis showed a mean age of death of 71.4 years for radiologists, of 73.4 for physicians, and of 76.2 for ophthalmologists. Interestingly, age-adjusted death rates for all causes, for cardiovascular diseases, and for cancer were highest in radiologists above the age of 50. Yet the most pronounced

excess of mortality was found for leukemia when radiologists were compared with ophthalmologists. Nineteen leukemias were observed in radiologists, where only 7.7 would have been observed in a similar group of ophthalmologists. The relative excess for all cancers was smaller, with 135 cases observed while 87 were expected.

Although there are some differences in the observations of the two major studies on the mortality of radiologists which might be ascribed to the selection of different comparison groups, both studies showed a pronounced increase in leukemia mortality, especially in those radiologists who started their practice in the first decades of the century; other cancers also appeared more frequently than in other comparable professional groups.

AOYAMA et al. (1983) studied the occupational cancer risk in a group of 2028 Japanese radiologic technologists born before 1933 and related it to the accumulated radiation dose which could be ascertained in 599 persons. Between 1969 and 1982 131 died, compared with an expected number of 182 based on the general population in Japan. This "healthy worker effect," however, was not found for cancer mortality. Fifty-one deaths from cancer were observed. A breakdown into 17 types of malignancy did not show any significant excess for 15, including leukemia (one observed while 1.8 were expected), pancreas (two observed while 2.2 were expected), myeloma (none observed while 0.34 were expected), and skin cancer (none observed while 0.23 were expected). Only the four urogenital tumors and two brain tumors significantly exceeded the expectations. Of the 51 cancer deaths, radiation exposure could be determined in 22, 13 exceeding 0.5 Sv and 5 exceeding even 1 Sv. Yet no correlation between cancer mortality and radiation dose was apparent.

This conclusion changed considerably after more deaths had been recorded in this group (AOYAMA et al. 1987). A total of 137 cancer deaths were observed while 108 were expected based on the general population cancer risk. The estimated mean exposure dose was exceptionally high and was estimated to be about 0.55 Sv. There was a trend for an increase in the ratio of observed versus expected cases with increasing exposure dose for all types of cancer ($P=0.05$) and for stomach cancer ($P=0.06$). However, this analysis of trend could only be performed on a small subset of the whole group for which information on radiation exposure was available, comprising only 55 cancer deaths. As in the first publication, no indication of an increased risk of leukemia was found.

3.5 Radium Dial Workers

Up to about 1926, the luminescent paint used in the dial painting industry contained considerable amounts of ^{226}Ra and ^{228}Ra in varying proportions. Being unaware of the potential hazards, the women who painted the dials with a fine brush incorporated considerable amounts of radium over an extended period of time. From external measurements of the γ-rays emitted from incorporated radium, the radium burden and the radiation dose to the skeleton can be estimated (ROWLAND et al. 1978).

Epidemiologic studies in the American (ROWLAND et al. 1978; STEHNEY et al. 1979) and the British (BAVERSTOCK et al. 1981) dial painters have been published. The American study population consisted of 1474 female dial workers first employed before 1930. Body burdens of radium were measured in 759 women; 235 had died by the end of 1976, 38 from bone sarcomas and 17 from squamous cell carcinomas of the paranasal sinuses. No radium-related malignancies were observed in workers who had entered the industry after 1925. No osteosarcoma was diagnosed in the 644 dial painters with a systemic intake of less than 3.7×10^6 Bq, whereas there was a massive increase in the risk of bone sarcomas at higher intake levels. Half of the 38 who had incorporated more than 3.7×10^7 Bq radium (both isotopes combined but corrected for the shorter half-life of ^{228}Ra) died from bone sarcoma. Similarly no cancers of the paranasal sinuses were found in 625 patients with a systemic intake of less than 10^6 Bq ^{226}Ra (^{228}Ra has not been considered as contributing to any significant degree to this type of malignancy), but a steep rise again occurred with higher radium intake. For both types of malignancy induced by incorporated radium ROWLAND et al. (1978) suggest a dependence of the incidence on the square of intake or the square of mean skeletal radiation dose. Latency to cancer varied very greatly. The first bone sarcoma appeared 5 years after first exposure to radium but others took 50 years to develop; the mean latency time was 20 years. Latency to cancer of the paranasal sinuses was even longer. The first case appeared after 19 years, and the mean latency was 35 years (STEHNEY et al. 1979).

The study of BAVERSTOCK et al. (1981) on 1110 women who worked between 1939 and 1961 for the UK luminescing industry with paint containing radium did not consider any large amount of radium incorporation: the mean uptake in 470 assessed women was about 7×10^4 Bq, the maximum being about 3.7×10^6 Bq in one woman. External radiation exposure to these women from γ-rays was greater than doses from incorporated radium. No excess for bone sarcomas was observed, but there was a significant increase in the number of breast cancers in those 632 women who were under the age of 30 when they started work in the luminescing industry and who received external radiation doses in excess of 0.2 Gy (with a mean dose of 0.5 Gy). Twelve women died from breast cancer while 5.4 were expected.

3.6 The Nuclear Industry

Due to the principles on which the limits of radiation exposure of occupationally exposed persons are based, it is generally impossible to detect, with epidemiologic methods, health defects as a result of radiation exposure if the limits for occupation in the nuclear industry are carefully controlled. This is very distinctively pointed out by BERAL et al. (1985) in the introduction to her study: "Before the investigation began it was recognized that the study would not have sufficient power to detect an increase in radiation related cancers unless the existing figures of the International Commission on Radiological Protection underestimated risk by a factor of about 20 or more." This restriction applies to all following studies.

3.6.1 Hanford

In 1943 a big industrial complex was built in Hanford, USA. Its first task was the production of plutonium for manufacturing bombs. Later on, it was mainly a reprocessing plant. Everybody who has ever worked in Hanford and was monitored for occupational radiation exposure has been documented with the doses read from the film badges, some of them for more than 40 years. In 1964 the USAEC (now DOE) commissioned a study on the causes of death of these workers, to find out whether there were any peculiarities and to provide evidence especially on the frequency of leukemia and cancer. MANCUSO et al. (1977) published an analysis of 3520 deaths up to 1974 among the 25000 workers and demonstrated an increased cancer risk among these occupationally exposed people. As this was in contrast to the generally accepted perception of occupational radiation risk, this publication caused widespread discussion and set in motion further investigations, which are summarized in this chapter. However, the Hanford study, its original interpretation by MANCUSO, and further interpretations will be reviewed in more detail, first.

It has to be realized that the mortality of the workers in Hanford is lower than that of the general population of the USA. The SMR of all workers who were employed longer than 2 years at Hanford is 75%. Obviously the workers at Hanford are healthier than the average American. These results are not surprising, as the "healthy worker effect" is found whenever employees of other branches of industry are investigated. Reasons for the higher life expectancy are, in particular, the selection of healthy people at employment, a generally better medical care system, systematic preventive health care, and often a higher socioeconomic status of workers in large industries. Nevertheless, hidden beneath these reassuring results one may still find a higher rate of specific illness, especially illness specific for an occupation. Analyzing the spectrum of causes of death in more detail, a higher percentage of deaths from cancer is found in Hanford than in the general population. The SMR of malignant disease of 85% among those employed for more than 2 years at Hanford is considerably higher than the SMR of other diseases. This is not peculiar to Hanford either, as in such studies, in which the general mortality is lower than for the general population, cancer mortality is often relatively increased compared to other causes of death, as it is absolutely, less affected by the healthy worker effect than other causes of death, e. g., infections or chronic diseases.

To look for a connection between radiation exposure and cancer deaths, MANCUSO et al. (1977) used a method of statistical analysis which they called the "comparative mean dose method": they added up the accumulated radiation doses of all workers dying from cancer and calculated their mean accumulated radiation dose and then added up the accumulated radiation doses of the workers dying from other causes and calculated their mean radiation dose. They found that the workers dying from cancer had accumulated a distinctly higher radiation dose (20 mSv) than the others (16 mSv). They interpreted this as an indication that a causal relation might exist. A more detailed breakdown of the different types of cancer yielded some more surprises: While the mean radiation dose of those dying from myeloid leukemia was lower than that of those dying from all malignant tumors, some tumors showed a particularly high mean radiation dose, especially myeloma (78 mSv), carcinoma of the pancreas (25 mSv), and brain tumors (22 mSv).

Compared to the spectrum of cancer death cases in the general population, the rate of myelomas, leukemias, and tumors of the kidneys, liver, and gallbladder is increased. The proportion of the observed to the expected cancer deaths (according to the different tumor types in the general population) from those tumors, with a mean radiation dose above average, was 1.25, whereas with a mean dose below average it was 0.78.

The on average higher radiation dose of the workers dying from cancer was also evident if the workers were classified according to time of exposure. MANCUSO et al. (1977) placed special emphasis on the fact that the mean annual radiation dose of the workers dying from cancer and exposed between the ages at 20 and 40 was lower than that of the workers dying from other diseases. It is only in the older age groups that the mean radiation dose of the workers dying from cancer is higher than for the other diseases. They attributed this to a remarkably higher radiosensitivity of the workers over 40. Finally they showed that with increasing, cumulative radiation dose the proportion of cancer cases among the deaths in each dose group would increase from about 17% at less than 2 mSv to 22% at 2-10 mSv, 22% at 10-50 mSv, and 25% at more than 50 mSv. This increase was significant for all cancer deaths, as well as after grouping into five age groups. If there were no correlation between the radiation exposure of the workers and any other carcinogenic substance or factor, these results would indicate a causal association between radiation and cancer. But precisely this kind of correlation does exist. The accumulated radiation dose increases with duration of employment. Likewise there is a probability that with increasing length of employment in a job of particularly low average mortality, the relative risk of dying from nonmalignant disease decreases. This may lead to a relative increase in the frequency of malignant diseases with increasing length of employment. In addition an increase in the mortality for certain types of cancer has been observed during the last 20 years: for example lung cancer was twice as common in 1970 as in 1950. Inevitably the number of lung cancer cases will be higher among those who died later and were therefore exposed to a higher radiation dose than among those who died earlier. Finally, MANCUSO et al. (1977) calculated doubling doses and the latency of the different tumors. The mean period between highest radiation exposure and death from cancer was calculated as 9-12 years. The calculated doubling dose for leukemia was 8 mSv, for neoplasms of the reticuloendothelial system 25 mSv, for lung cancer 61 mSv, for carcinoma of the pancreas 74 mSv, and for all cancers 122 mSv. The doubling dose decreased from infinity between 30 and 40 to 700 mSv at the age of 45, to 180 mSv between 50 and 60, to 160 mSv at 70, and to 10 mSv at the age of 75. In 1979 KNEALE et al. presented a continuation of this analysis and confirmed the significantly higher mean radiation dose of the workers dying from cancer up to 1977.

Looking at the Hanford data, it is noteworthy that only the occupational radiation exposure from low dose rate external γ-radiation is included in the calculation. In fact the natural radiation exposure with similar doses and dose rates in the two groups should be compared as well. Proceeding from the assumption of an annual natural radiation exposure of 1 mSv, the mean *natural* radiation exposure of all workers dying from cancer is 60.6 mSv, while it is 60.2 mSv for all workers dying from other causes; the total radiation exposure (natural and occupational) is therefore 74.4 mSv for all workers dying from cancer and 70.1 mSv for all those dying from other causes. Among the cancer cases the increase in the mean accumulated radiation dose of 4.3 mSv is 6%. According to MANCUSO, this increase in radiation exposure should lead to an increase in the "spontaneous" cancer mortality risk of 4%–26%, depending on the analytic procedures used. If MANCUSO et al. had included in 1977 the natural radiation exposure, they would have come to the conclusion that all cancer deaths in the population of the USA were caused by natural radiation exposure *exclusively*. The only alternative would be the assertion that the natural, external γ-radiation exposure acts qualitatively and quantitatively differently than the occupational external γ-radiation exposure with comparable doses and dose rates.

GILBERT and MARKS (1979) published an analysis of the Hanford data, considering all white male workers employed before 1964 (20842) and all registered death cases until 1 April 1974 (3994). From the start 7767 workers were excluded from further analysis, as they were employed in Hanford for less than 2 years. Out of the remaining 13075 workers, more than three-quarters (9902) had only accumulated a radiation dose of less than 50 mSv during their entire occupational radiation exposure (i.e., less than the natural radiation exposure); 1107 had accumulated a radiation dose between 50 and 100 mSv, 742 between 100 and 200 mSv, 711 between 200 and 400 mSv, and only 188 workers had a radiation dose of more than 400 mSv. The mean radiation exposure of the workers rose with increasing length of employment. As the workers at Han-

ford showed a lower mortality than the general population, the correlation with the radiation exposure was tested by dividing the radiation workers into four dose groups: 0–20 mSv (7607 people), 20–50 mSv (2294 people), 50–150 mSv (1567 people), and more than 150 mSv (1211 people). The mortality for the various diseases in the different age groups was defined independently of radiation exposure and, according to the number of people in the four dose groups, the expected number of deaths was calculated. The expected rates for all age groups, types of occupation, and times of employment were added up for the various causes of death and compared with the observed rates. A statistical test was performed to determine whether the ratio of observed to expected deaths rose with increasing dose and to work out the significance of this tendency. A significant trend was found only for myeloma ($P = 0.01$) and cancer of the pancreas ($P = 0.03$). This was based on two deaths from myeloma instead of the expected 0.4 in the radiation dose group receiving more than 150 mSv, and on four deaths from cancer of the pancreas instead of the expected three in the groups receiving more than 50 mSv. No significant trend was observed for any of the other tumors.

These studies led to a number of further publications about the methods used in the analysis of the Hanford data, most of which concluded that the mathematical models used by MANCUSO et al. (1977) were not suitable for proving a causal relation between cancer and radiation (e. g., ANDERSON 1978). Also noteworthy is a letter by KNEALE et al. published in 1979, in which they point out that they "did not claim that cancer was a major hazard of the nuclear industry or even that the cancer mortality of Hanford workers was significantly raised but rather that there is prima facie evidence of a relationship between cancer mortality from certain specific cancers and radiation even at the low doses received by Hanford workers".

GOFMAN (1979) did another analysis of the Hanford data, following a different statistical approach. Based on the fact that the accumulated doses do not follow a normal distribution and that therefore the statistical procedures employed by MANCUSO are open to criticism, he compared the ratio of cancer deaths to other causes of death in two groups: workers who had accumulated less than 100 mSv (mean dose less than 10 mSv for 3193 deaths) and workers who had accumulated more than 100 mSv (mean dose 230 mSv for 115 deaths). This ratio was significantly ($P = 0.047$) higher for those exposed to a higher radiation dose. The calculated doubling dose was 435 mSv (with a 95% confidence limit from 290 mSv to infinity). However, no significant correlation with the radiation dose could be established for the individual types of cancer. GOFMAN rejected the claim by MANCUSO that sensitive organs and age groups could be identified. The latest analysis of the Hanford workers by TOLLEY et al. (1983) was based on 2500 death cases until 1979, including 512 cancer deaths. The trend towards a higher ratio of observed to expected deaths was still significant for myeloma ($P=0.01$), although there had been no further deaths from myeloma since the first publications, but no longer for carcinoma of the pancreas, as there were two new cases in the lower dose groups.

In summary, it has to be realized that examination of the causes of death of the Hanford workers only provides evidence for a higher incidence of myeloma on the basis of three observed cases whereas just one was expected, in the group receiving more than 50 mSv.

3.6.2 Oak Ridge National Laboratory

Oak Ridge National Laboratory in Tennessee was founded in 1943 as part of the Manhattan project, but after the war it developed into an important center for all fields of nuclear research. Between 1943 and 1972 about 20000 people were employed there. The study by CHECKOWAY et al. (1985) was limited to 8375 male white workers who were not employed in any other radiation works. In this group 966 deaths were identified, and the cause could be ascertained in 97% of cases. In addition to the data on external radiation exposure, information was gained on the incorporation of radioactive substances by means of measurements in urine and feces. The mortality for different causes of death was compared with that of the white male population of the USA between 1943 and 1977. The workers at Oak Ridge were divided into four dose groups: no radiation dose, under 10 mSv, 10-50 mSv, and more than 50 mSv. The highest dose was 1 Sv; 297 workers had accumulated a dose of more than 100 mSv. The mean dose of all workers was less than 2 mSv. Relatively high values were found for 115 workers at the incorporation control. A close correlation was observed between incorporation and external dose. As for the Hanford workers, it was found that the SMR was lower than that for the male white US population, with values of 73% for all causes of death and 78% for all cancers. Breaking down the mortality ratio into

the various tumor types, the ratios for carcinoma of the prostate, leukemia, and Hodgkin's disease were higher than 1, which, however, was not significant in any of these cases. While in Hanford there was a definite correlation of myeloma with increased radiation exposure, only one death from myeloma was found among the Oak Ridge workers. When breaking down the SMR according to radiation dose or incorporation, no significant trend towards dependency on the radiation dose was found either for all deaths or for all cancer deaths. As radiation exposures during the last few years before death are unlikely to have a great influence on radioinduced cancer, it is reasonable not to consider the radiation exposure during the 10 years preceding death. Even after an appropriate correction of the data no correlation was found between the cancer mortality ratio and the radiation dose for all cancer cases. However, in the leukemia group a definite increase in the standard leukemia mortality was found up to a dose of 50 mSv. This correlation was based on a relatively small number of only 13 leukemia cases, of which seven were found at an accumulated dose of 10-50 mSv. No cases of leukemia, however, were found among the 64 deaths of workers who had received a radiation exposure of more than 50 mSv. A similar result appeared for the correlation with the amount of incorporated radioactivity.

The study came to the conclusion that apart from the correlation with the leukemia mortality, which is not statistically significant, there is no sign of specific radiation damage among the workers at Oak Ridge.

3.6.3 Sellafield

The study by SMITH and DOUGLAS (1986) reported on the causes of death of all people who worked in Sellafield from 1947 until 1975 and who died before the 31 December 1983. Overall the study dealt with 14327 employees. It drew a distinction between workers who were paid weekly and those who were paid monthly (i. e., clerical, managerial, and scientific workers).

Analysis of the medical reports of each worker was made possible by the fact that all Sellafield workers were recorded in the central register of the National Health Service. In the event of death, the death certificate with the diagnosis of the cause and the time of death was made available for the study. In addition, information was received through the Department of Health and Social Security (DASS) and the pension scheme of BNFL. In this way it was possible to follow up a high percentage of workers. On 1 January 1984 72.9% of the workers were still alive, 2.9% had emigrated, and 2277, i. e., 15.9% had died. Of the 14000 staff at Sellafield, 27.5% had not worn dose meters. For the 10000 workers and employees in Sellafield who had worn dose meters the mean accumulated radiation dose was 90 mSv. 3469 accumulated a dose of more than 100 mSv, 542 more than 500 mSv, and 48 more than 1000 mSv. The highest accumulated dose in a worker (the person in question had been employed at Sellafield for 28 years) was 1760 mSv. The mean length of employment at Sellafield was 9 years.

Comparison of the causes of death among the Sellafield workers with those in the population of England and Wales gave an SMR of 98%. Thus, no "healthy worker effect" was observed, in contrast to the two American studies. The SMR of cancer deaths was 95%, and therefore lower than for other causes of death. The mortality for nonmalignant causes of death was lower for the radiation workers than for those who were not classified as radiation workers, while the rate of cancer deaths was the same for both groups. The usual fluctuations were reflected in a detailed breakdown of the SMR for the various tumors. Only one group was significantly above national average, i. e., the ill-defined and secondary tumors among the nonradiation workers. The rates of cancers of the liver, gallbladder, and lung and of Hodgkin's disease among the radiation workers were significantly below the national averages, while leukemia was underrepresented among nonradiation workers. If the SMR is based on the local population of Cumbria instead of the population of England, the overall rate decreases from 98% to 91%. There was no dependency of the SMR on the length of employment at Sellafield.

The various cancer deaths were divided into different dose groups, the relative risk was calculated, and alteration in the relative risk with increasing dose was determined. Negative as well as positive trends were found. However, the only significant trend was for leukemia and myeloma, if the radiation dose accumulated until 15 years before death was taken into account.

In summary, the "healthy worker effect," which usually appears in this sort of study and which was found for the Sellafield group in an earlier publication by CLOUGH (1983), did not appear in this study; the authors trace this back to the longer observation period. Distinct differences were found between radiation-exposed and non-radiation-ex-posed workers in terms of the spectrum of nonmalignant diseases that caused death; in particular the mortality from bronchitis was significantly lower among the radiation-exposed workers than among the others. This could be because people with chronic bronchitis would not be hired for a radiation job, because radiation workers have a higher socioeconomic status, and, above all, because there is a lower percentage of smokers among radiation workers. These differences have to be taken into account when interpreting the differences in mortality for radiation and other workers. A distinct negative correlation between the frequency of death from lung diseases and the radiation dose was established, i. e., the higher the radiation dose, the lower the probability of dying from chronic bronchitis. One explanation may be that workers suffering from chronic bronchitis (or any other chronic disease) are usually taken out of the radiation area and, therefore, accumulate a lower radiation dose during their time of employment. Since in all other studies leukemia has usually appeared within 15 years after radiation exposure, it was an unexpected and radiobiologically inexplicable finding that there was no significant correlation between dose and relative risk of leukemia if the total radiation dose until death was used as a criterion but that there was a significant correlation if the calculations were based on the radiation dose until 15 years before death. This correlation, however, was due to just one single patient, who died of a stem cell leukemia at the age of 75 years, 10 years after his retirement, after he had accumulated a radiation dose of 767 mSv during the entire length of his employment.

There was also a correlation between the relative risk of death from myeloma and increasing radiation dose if only the dose until 15 years before death was taken into account. Although there was no significant increase in the overall mortality from myeloma among the workers at Sellafield compared to the national mortality (seven observed instead of 4.2 expected cases of death from myeloma), the authors concluded that in the light of the material from Hanford, myeloma is probably the only malignant disease which appears more frequently in dependence on the accumulated radiation dose among workers at reprocessing plants. Among the seven deaths from myeloma, two were in workers who had accumulated radiation doses of 565 and 865 mSv during 26 and 31 years (respectively) of employment at Sellafield. For the other five workers the radiation doses were below 120 mSv, i. e., in a dose range which is within the range of variation of natural radiation exposure.

The high probability of this trend was exclusively due to the aforementioned two cases. While the authors assumed that the correlation between radiation dose and myeloma was possibly a causal one, they believed that the correlation between bladder cancer and dose was a chance result. They justified this assumption with the argument that, for statistical reasons, one significant result has to be observed by chance if such a high number of correlations are tested. Since no correlation was found with bladder cancer in other studies, it is more likely that the correlation with bladder cancer was coincidental, rather than the one with myeloma, as this correlation was observed in Hanford, too.

3.6.4 UKAEA

The second English study was by BERAL et al. (1985). The study examined the employees of the Atomic Energy Authorities of Harwell, Culham, London, Dunray, and Winfrith between January 1946 and December 1979. The examined group consisted of 39 546 people, of whom 19 164 were radiation workers. The mean radiation dose of the radiation workers was 32 mSv; in 84 the dose was more than 500 mSv, and in 1675 it was between 100 and 500 mSv. Therefore the radiation dose in this study was significantly lower than that of the Sellafield workers. In contrast to the Sellafield study a significant deficit of deaths was found among the employees of the Atomic Energy Authorities in comparison with the national cancer mortality, with SMRs of 76% for radiation workers and 78% for nonradiation workers.

When breaking down the mortality ratios into the various types of cancer, a significantly lower cancer mortality was found for lung cancer among both the radiation workers and the nonradiation workers. The same result was found for stomach cancer, bladder cancer, and brain tumors. Relating the relative risks to the accumulated radiation doses, a significant trend toward disease with increasing radiation dose was observed only for cancer of the prostate. The 25 men who had died from cancer of the prostate at ages between 49 and 90 years were listed with their accumulated radiation dose and the result of incorporation measurements. In only seven of them was the accumulated radiation dose more than 100 mSv, and in none of them was it higher than 300 mSv. The relative mortality risk was especially high in the age group under 60 years, in which eight cases were observed while only three would have been expected according to the nation-

al cancer statistics. This increase in the mortality for carcinoma of the prostate was based on an increased mortality of younger men, who also had incorporated an undetermined amount of tritium. This group is going to be studied more carefully, in order to identify possible causes of this cluster of carcinomas of the prostate. BERAL et al. (1985) discussed the possibility that the incorporation of tritium might be but an indicator for incorporation of other chemical carcinogens which may cause specifically cancer of the prostate, although none such are known. In summary an increased mortality from carcinoma of the prostate was found, which was limited to a small group of young men, working in one of the five research laboratories, who had also incorporated some tritium.

3.6.5 Shipyard Workers

A good example for the problems associated with epidemiologic studies which are initiated because of a suspected increase in the incidence of certain diseases and which then search for further cases is the study of the leukemia mortality among the workers at the Portsmouth shipyard in New Hampshire, USA, which specializes in building and overhauling nuclear submarines. The study of NAJARIAN and COLTON (1978) was initiated when a patient, treated for leukemia in Boston, told his doctor that he was a radiation worker at the Portsmouth shipyard and that several of his younger colleagues had died of cancer. Among the death certificates for the years 1959–1977 from the three States of Maine, New Hampshire, and Massachusetts, where the shipyard workers came from, 1722 death certificates of former shipyard workers were identified. The relatives of 592 were contacted to find out whether the deceased had been classified as a radiation worker. The causes of death of 146 former radiation workers were compared with those of 379 nonradiation workers, and the SMRs were calculated in comparison to the white male population of the USA.

The most remarkable result was the increase by a factor of five in the number of leukemia deaths among the radiation workers, to six instead of the expected 1.1. Data on the accumulated radiation doses were not available. The authors warned against taking this result as proof of an increased leukemia risk, but stated that it should give rise to a systematic epidemiologic study, which was indeed performed and published 3 years after the pilot study (RINSKY et al. 1981): 24 545 workers were investigated, who had worked in the shipyard between 1952 and 1977. Among them were 7615 who had an accumulated external radiation dose of up to 914 mSv. The mean occupational radiation exposure was 28 mSv. It was possible to identify the causes of death for 96% of all deceased radiation workers. The SMR of all workers was 89% (94% for malignant diseases). Among the radiation workers it was 78% (92% for malignant diseases). Seven cases of leukemia were observed among the radiation workers, while 8.3 were expected according to the health statistics of the USA. The study consequently arrived at the conclusion that there was no indication of an increased risk of leukemia among the shipyard workers and concluded that "the difference between our results and those of NAJARIAN and COLTON (1978) seem to reflect our more complete ascertainment of the PNS cohort, thus avoiding selection bias, and our more accurate classification of workers as to their radiation exposure."

3.6.6 All Workers in the Nuclear Industry Summed Up

BERAL et al. (1988) summarized the results of epidemiologic studies of radiation workers at Hanford (17 000), Sellafield (14 000), UKAEA (40 000), Portsmouth (25 000), and Oak Ridge (8000) as well as three other smaller studies [Rocky Flats (WILKINSON et al. 1983), 5000 workers; United Nuclear Corporation (HADJIMICHAEL et al. 1983), 4000 workers; and Pantex weapon industry (ACQUAVELLA et al. 1985), 4000 workers], thus amounting to a total of almost 120 000 workers in nuclear facilities. All were evaluated using the same criteria. The SMR for all observed 15 674 deaths was 82% (with fluctuations between 64% and 98%), for the observed 3477 cancer deaths, 86% (with fluctuations between 60% and 94%), and for the 131 leukemia cases, 90%, with fluctuations being higher due to the small numbers, ranging from 59% to 149%. Therefore, there is no proof of an increased mortality for workers in nuclear industry, either for all causes of death or for cancer or leukemia. In addition, no evidence could be provided for a significant increase in mortality for any specific type of cancer in any nuclear facility studied.

When calculating the trend for an increase in mortality for various types of cancer in relation to increasing radiation dose, only for one type of cancer was a significant trend found in more than one establishment: myeloma in Hanford and in Sellafield. Yet the overall number of deaths from myeloma was small, that is to say seven among more than 800 deaths among workers who had accumulated more than 50 mSv (3 of 293 in Hanford and 4 of 510 in Sellafield.) The absolute frequency of this rare disease in both plants is low, but statistically significant among the radiation workers. So far no specific cause, be it radiation or any chemical carcinogen, can be identified, as little is known about the possible causes of myeloma. Except for these seven observed deaths, compared to two to three expected deaths from myeloma, there is no strong evidence for an increased health risk to workers in the nuclear industry, either in any individual study or in the overall analysis of all published studies.

Therefore it was confirmed that if the established dose limits, as proposed by ICRP, are observed, the health risk to radiation workers is so small that it cannot be demonstrated, even in very comprehensive studies. The only scientifically tenable conclusion is that the risk factor suggested by the ICRP, even if not absolutely right, cannot be too low by more than 2000%. This was known before the studies began, so the costs of such epidemiologic studies seem disproportionately high.

The problems of such epidemiologic studies are very clearly represented in the report of the United States General Accounting Office to the Congress of the United States in January 1981 (pp. 63–79 in extracts):

Because of the large amount of work already done on radiation, further studies, in order to make useful scientific contributions, must be relatively precise. It is already known that radiation can cause cancer. What is needed now is an accurate quantitative description of that relationship ... *We do not believe that large-scale epidemiological studies of low-level exposures should be expected to provide reliable scientific data on the precise relationship between low-level ionizing radiation exposure and cancer* ...

Human epidemiological studies cannot be expected to determine the relationship between low-level ionizing radiation exposure and cancer.

A few epidemiological studies on low-level exposure groups may be warranted in an attempt to develop more accurate upper limits of risk from low-level radiation exposure. However, Federal agencies should carefully review each low-exposure study for its scientific merit.

The primary reason for initiating some studies of low-level exposure groups is to meet perceived legal or social responsibilities, rather than to develop scientific data.

We believe there is too much emphasis on epidemiological studies of low-level ionizing radiation exposure. Regardless of the reasons for conducting studies on low-level exposure, the inherent limitations on these types of studies make it highly unlikely that any reliable conclusions will be reached.

References

Acquavella JF, Wiggs LD, Waxweiler RJ, MacDonell DG, Tietjen GL, Wilkinson GS (1985) Mortality among workers at the Pantex weapons facility. Health Phys 48: 735-746

Anderson TW (1978) Radiation exposures of Hanford workers: a critique of the Mancuso, Stewart and Kneale report. Health Phys 35: 743-750

Aoyama T, Futamura A, Yamamoto Y, Kato H, Sugahara T (1983) Mortality study of Japanese radiological technologists. In: Biological effects of low level radiation. Vienna, IAEA-SM-266/43, 319-328

Aoyama T, Futamura A, Kato H, Nakamura M, Sugahara T (1987) Mortality study of Japanese radiological technologists. J Jpn Assoc Radiol Technol 58-63

Aurand K, Jacobi W, Schraub A (1955) Zur biologischen Strahlenwirkung des Radon und seiner Folgeprodukte. Sonderb. z. Strahlentherapie 35: 237-243

Bale WF, Shapiro JV (1955) Radiation dosage to lungs from radon and its daughter products. Proceedings of the 1st International Conference on peaceful uses of atomic energy. United Nations, Geneva, vol 13: 233-236

Baverstock KF, Papworth D, Vennart J (1981) Risk of radiation at low dose rates. Lancet I 430-433

Beral V, Inskip H, Fraser P, Booth M, Coleman D, Rose G (1985) Mortality of employees of the United Kingdom Atomic Energy Authority, 1946-1979. Br Med J 291: 440-447

Beral V, Fraser P, Booth M, Carpenter L (1988) Epidemiological studies of workers in the nuclear industry. In: Southwood R, Russel Jones R (eds) Radiation and health: The biological effects of low level exposure of ionizing radiations. John Wiley, Sussex

Checkoway H, Mathew RM, Shy CM et al. (1985) Radiation, work experience, and cause of specific mortality among workers at an energy research laboratory. Br J Ind Med 42: 525-533

Clough EA (1983) The BNFL radiation-mortality study. J Soc Radiol Protect 3: 24-27

Court Brown WM, Doll R (1958) Expectation of life and mortality from cancer among British radiologists. Br Med J II: 181-187

Frieben (1903) Cancroid des rechten Handrückens nach langdauernder Einwirkung von Röntgenstrahlen. Fortschr Röntgenstr 6: 106

Gilbert ES, Marks S (1979) An analysis of the mortality of workers in a nuclear facility. Radiat Res 79: 122-148

Gofman JW (1979) The question of radiation causation of cancer in Hanford workers. Health Phys 17: 617-639

Hadjimichael OC, Ostfeld AM, D'Atri DA, Brubaker RE (1983) Mortality and cancer incidence. Experience of employees in a nuclear fuels fabrication plant. J Occup Med 25: 48-61

Härting FM, Hesse W (1879) Der Lungenkrebs, die Bergkrankheit in den Schneeberger Gruben. Vjschr Ger Med 30: 296-309; 31: 102-132; 31: 312-337

Holthusen H, Meyer H, Molineus W (1959) Ehrenbuch der Röntgenologen und Radiologen aller Nationen. Sonderband zur Strahlentherapie 42. Urban & Schwarzenberg, Munich

ICRP (1928) International recommendations for X-ray and radium protection. Br J Radiol 1: 358-363

ICRP Publication 26 (1977) Recommendations of the International Commission on Radiological Protection. Annals of the ICRP 1, No 3

ICRP Publication 32 (1981) Limits for inhalation of radon daughters by workers. Annals of the ICRP 6, No 1

Jacobi W, Paretzke HG (1985) Risk assessment for indoor exposure to radon daughters. In: The science of the total environment 45. Elsevier, Amsterdam, pp 551-562

Kneale GW, Stewart AM, Mancuso TF (1979) Radiation exposure of Hanford workers dying from cancer and other causes. Health Phys 36: 87

Kunz E, Sevc J, Plcek V, Horacek J (1979) Lung cancer in man in relation to different time distribution of radiation exposure. Health Phys 36: 699-706

Mancuso TF, Stewart AM, Kneale GW (1977) Radiation exposure of Hanford workers dying from cancer and other causes. Health Phys 33: 369-384

Müller J, Wheeler WC, Gentleman JF, Suranyi G, Kusiak RA (1985) Study of mortality of Ontario miners. In: Occupational radiation safety in mining. Proc Intern Conf 1: 344-349

Najarian T, Colton T (1978) Mortality from leukaemia and cancer in shipyard nuclear workers. Lancet I: 1018-1028

NCRP Report 78 (1984) Evaluation of occupational and environmental exposures to radon and radon daughters in the United States. National Council on Radiation Protection and Measurements, Bethesda

Pott P (1775) Chirurgical observations relative to the cataract, the polypus of the nose, the cancer of the scrotum, the different kinds of ruptures and the mortification of the toes and feet. T. J. Carnegie, London

Rinsky RA, Zumwalde RD, Waxweiler RJ et al. (1981) Cancer mortality at a naval nuclear shipyard. Lancet I 231-235

Rowland RE, Stehney AF, Lucas HF (1978) Dose-response relationships for female radium dial workers. Radiat Res 76: 368-383

Seltser R, Sartwell PE (1965) The influence of occupational exposure to radiation on the mortality of American radiologists and other medical specialists. Am J Epidemiol 81: 2-22

Smith PJ, Doll R (1981) Mortality from cancer and all causes among British radiologists. Br J Radiol 54: 187-194

Smith PJ, Douglas AJ (1986) Mortality of workers at the Sellafield plant of British Nuclear Fuels. Br Med J 293: 845-854

Stehney AF, Lucas HF, Rowland RE (1979) Survival times of women radium dial workers first exposed before 1930. In: Late biological effects of ionizing radiation. Vienna, IAEA-SM-224

Tirmarche M, Brenot J, Piechowski J, Chamaud J, Pradel J (1985) The present state of an epidemiological study of uranium miners in France. In: Occupational radiation safety in mining. Proc Int Conf 1: 344-349

Tolley HD, Marks S, Buchanan JA, Gilbert ES (1983) A further update of the analysis of mortality of workers in a nuclear facility. Radiat Res 95: 211-213

United States Comptroller General (1981) Report to Congress: problems in assessing the cancer risks of low level ionizing radiation exposure. United States General Accounting Office

UNSCEAR (1982) Ionizing radiation: sources and biological effects. United Nations

Virchow R (1849) Mitteilungen über die in Oberschlesien herrschende Typhus-Epidemie. Arch Pathol Anat Physiol Klin Med 2: 143-322

Whittemore AS, McMillan A (1983) Lung cancer among U.S. uranium miners: a reappraisal. J Natl Cancer Inst 71: 489-499

Wilkinson GS, Voelz GL, Acquavella JF, Tietjen GL, Reyes M, Brackbill R, Wiggs ID (1983) Mortality among plutonium and other workers at a nuclear facility. Los Alamos National Laboratory Document LA-UR-83-266

4 Medical Aspects of Radiation Accidents

Otfried Messerschmidt

CONTENTS

4.1 Reactor Accidents and Nuclear Bomb Explosions - A Comparison

The discovery of nuclear fission by OTTO HAHN and FRIEDRICH STRASSMANN in 1938 not only provided a new means of producing large amounts of energy but also precipitated dangers which at first could not be foreseen but were suddenly forcibly brought to the awareness of mankind by the events in Hiroshima and Nagasaki.

Although nuclear power plants and nuclear weapons are different, in many ways a significant relation exists between them, as it is possible to produce the explosive material for a nuclear weapon in a reactor. Therefore a nation which possesses reactors as well as enrichment and reprocessing plants has the capability to become a nuclear power. This prospect is an extraordinary burden for the positive aspects of nuclear energy production, as the threats resulting from the nuclear arms race between the superpowers are transferred to the peaceful uses of nuclear energy. It is for this reason that the efforts to produce energy from nuclear power have developed into a first-grade political and psychological problem. People pay great attention even to small incidents in nuclear power plants, even if no significant amounts of radioactivity have been released. Thus the effects on the public were grave when a serious reactor accident with core melting actually happened on 26 April 1986 at Chernobyl. In some regions of Central Europe the fallout from the radioactive cloud of the reactor accident at Chernobyl was as high as the total activity deposited on the ground surface as fall out from all nuclear weapons testing by the nuclear powers in the years 1954-1966. Figure 4.1 shows that in Bavaria, for example, the amount of cesium 137 from nuclear weapons tests was significantly below the fallout from Chernobyl, while that of strontium 90 was higher.

Explosions of nuclear weapons and reactor accidents can lead to a comparable radiation burden to the general public over long distances. In both an explosion of nuclear weapons and a reactor accident, radioactive substances are released as a cloud emitting radiation and are then deposited on the ground as "fallout." The radioactive substances are decay products of the nuclear fission of uranium and plutonium. While in a nuclear weapon these

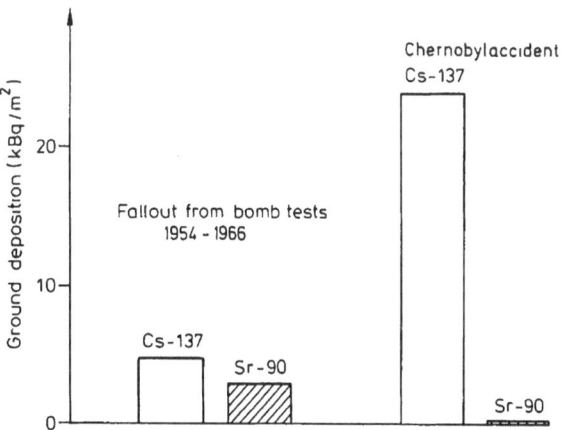

Fig. 4.1. Ground deposition of cesium and strontium from atomic bomb tests and the Chernobyl accident in Munich, Federal Republic of Germany (from GSF, Mensch und Umwelt, December 1986)

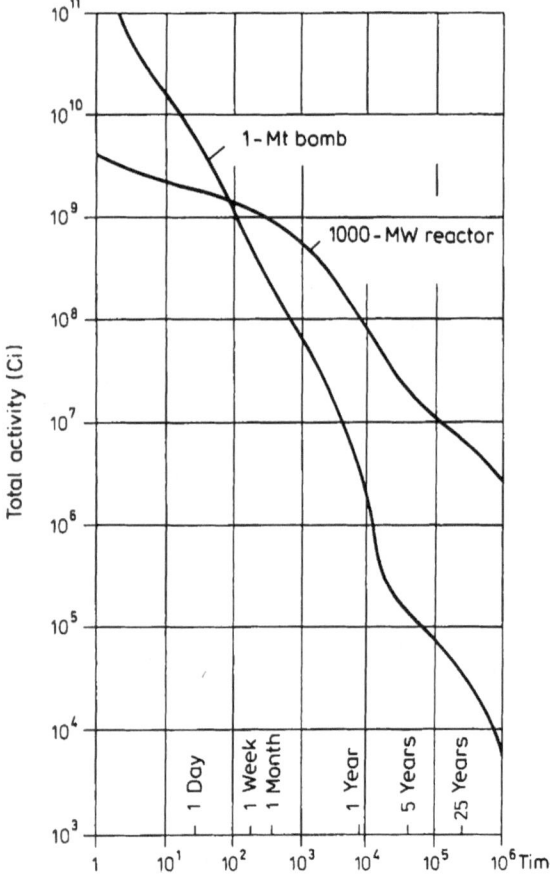

Fig. 4.2. Total radioactivity and decay of the fallout from a 1-Mt nuclear bomb and from the worst possible accident involving a 1000-MW nuclear reactor (after FETTER and TSIPIS 1981)

fission products are produced at the moment of explosion during a very short fission process, in a reactor they are produced in a delayed continuous fission process. The newly produced radioactive nuclei decay continuously into other elements, most of which are also radioactive. Therefore the core of a reactor contains fission products of different age (i. e., some which have already passed through a series of decay steps and some which have just been produced) while in the chain reaction of a nuclear weapon the fission products arise only at the moment of explosion. Therefore the decay curve of radioactive substances from a nuclear explosion, as shown in Fig. 4.2, is much steeper than the decay curve of the predominantly older fission products of a reactor core. Within 24 h, the radioactivity from a nuclear explosion decreases by a factor of 3000, in accordance with the equation

$$I = \frac{I_1}{t^{1.2}}$$

where I_1 is the dose rate 1 h after the explosion and I the dose rate at time $t + 1$ h. This means, in practice, that any increase in time by a factor of 7 decreases the dose rate by one order of magnitude, i. e., after 7 h to 10%, after 49 h to 1%, and after 343 h (2 weeks) to 0.1%.

One hour after the explosion the radioactivity of a nuclear weapon of 1 Mt energy yield is still considerably higher than the activity after the most serious reactor accident at a 1000-MW power plant. After about 4 days the radioactivity is about equal and after 1 year the radioactivity from the reactor accident is about 100 times higher than that from the nuclear weapon fallout. This correlates to the considerably higher amount of uranium contained in a 1000-MW power plant compared to a 1-Mt bomb (FETTER and TSIPIS 1981).

In a reactor accident the volatile fission products like the noble gases xenon and krypton as well as ruthenium, cesium, and tellurium and the various isotopes of iodine are released into the air, as the temperature of the melting reactor core is not high enough to vaporize heavier elements. Yet, the fireball of a 1-Mt bomb would totally destroy the reactor core during a nuclear weapon attack on a reactor. All fission products would be released and produce a radiation field which would result in a longer period of radioactive contamination than the fallout from the nuclear weapon itself.

A very serious reactor accident can lead to a disaster which in its dimensions is comparable to that involving a nuclear weapon explosion. Even the most serious reactor accident, however, cannot be compared with the consequences of a nuclear war, where several nuclear weapons are exploded and hundreds of thousands or millions may be killed.

4.2 Release of Radioactivity in an Accident

In contrast to the risks from all other parts of the fuel cycle (e. g., mining, fuel production, reprocessing, and waste storage), which are mainly long-term, a serious accident in a reactor can lead to a sudden release of large amounts of radioactivity and therefore to acute radiation exposure of reactor personnel and, under unfortunate conditions, even of the general public outside the reactor. The direct cause of the dramatic events in a serious reactor accident is the melting of the reactor core in which the production of energy takes place by nuclear fission.

To prevent melting of the core, extensive precautions are taken. For example, several emergency cooling pipes are provided in case a main cooling pipe ruptures. In the event of any accident the reactor is automatically shut down. To prevent the release of fission products and other radioactive substances, several barriers are created, e.g., in a pressurized water reactor (PWR) the fuel cladding, the reactor pressure vessel encasing the complete cooling system, and the gastight safety containment (the outer cover of reinforced concrete is not gastight, its main function being to protect the reactor from outside influences like an airplane crash). In addition there are further systems which are started in cases of faulty reactor operation and which are arranged in parallel (redundance) and work on different principles (diversity); they are spatially separated and they work automatically to exclude human error.

The asymmetric fission of the uranium nucleus produces elements of an atomic weight around 95 and 140. In a 1000-MW reactor the radioactivity of the fission products after 4 years of operation amounts to about 40 GCi (15×10^{20} Bq). Only 1 h after shutdown the activity falls to 10 GCi (3.7×10^{20} Bq) and 10 days later to about 1 GCi (3.7×10^{19} Bq).

In any accident only a limited amount of radioactivity would be able to escape from the reactor, even under the most unfortunate circumstances. Probably 100% of the noble gases would escape; the fraction of the other fission products escaping would depend on the time course and degree of the core melting. If the water in the containment is vaporized by the high temperature of the melting reactor core and pressure increases so rapidly that the containment is immediately fractured, about 50% of iodine and bromine may escape and up to 20% of tellurium. In a very delayed containment failure, the amount of radioactivity released would be less than 1%.

The actual danger to people close to the reactor depends on various factors, like:
- The amount of released radioactivity
- The distribution of radioactivity as a result of the weather at that time
- The population density
- The protection provided by people's whereabouts, i.e., whether they are outdoors, indoors, or in cellars
- Preventive measures like early warning, evacuation, provision of iodine tablets, and avoidance of contamination
- Provision of medical care

Depending on weather conditions, the ground contamination may reach up to 10^{13} Bq per m^2 at a distance of 1 km from the reactor. With increasing distance from the reactor, the ground contamination decreases, and at a distance of 10 km the activity per m^2 may be 10^{10}-10^{11} Bq (LÖSTER 1979). These activities decrease quickly due to the short half-life of many nuclides, but people who stay unprotected in the highly contaminated areas during the first few hours may receive considerable radiation doses. This can occur in different ways:
- External total body irradiation by γ-radiation from the passing cloud and the radioactivity deposited on the ground
- Internal radiation exposure as a result of incorporation by inhalation and ingestion and through open wounds
- Contamination of the body surface by deposition of radioactive substances on skin, mucous membranes, and clothing
- Combination of total body exposure with internal radiation, exposure and/or contamination of the body surface

Further complications may develop if, in addition to radiation exposure, conventional injuries occur as wounds or burns.

4.3 Results of Risk Studies of Possible Reactor Accidents

A nuclear power plant contains large amount of radioactivity and therefore represents a potential risk for the population living in the vicinity. Consequently running a reactor is only acceptable if the probability of a release of large amounts of radioactivity in the event of an accident is extremely low.

For this reason, "reactor safety studies" were performed to estimate the risk from a nuclear power plant. The first extensive reactor safety study, WASH 1400, was published in 1975 in the USA (RASMUSSEN 1975). The calculations for light water reactors were based on six location with different population densities. Weather data were collected at six real reactor locations. These data were used to describe the path of a radioactive cloud up to a distance of about 800 km in 16 sectors around the reactor. The biologic effects like acute death, acute radiation syndrome, thyroid nodules, cancer induction, and genetic damage were related to potential radiation doses.

After the publication of WASH 1400 it was necessary to study whether these results were transfer-

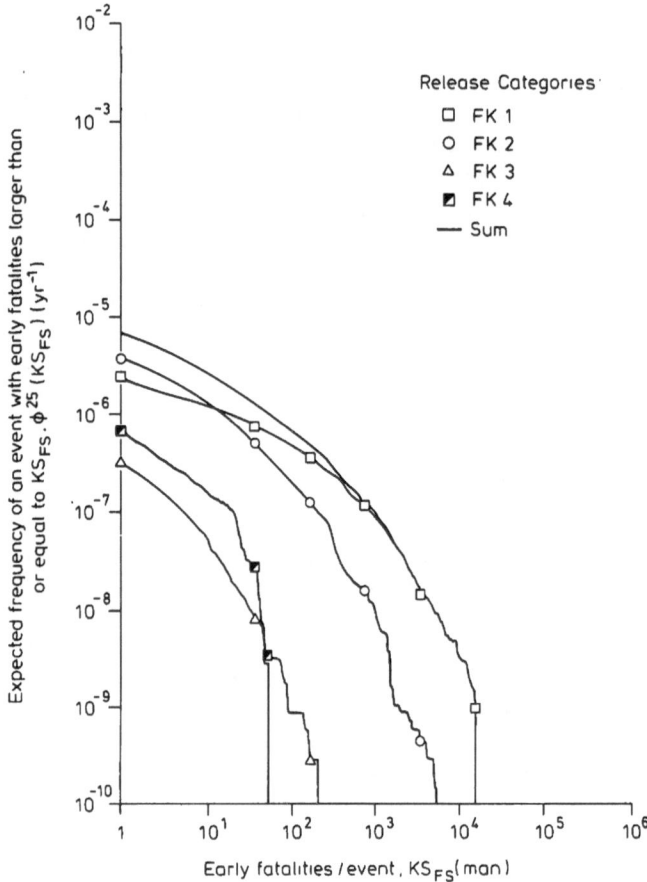

Fig. 4.3. Complementary cumulative frequency distribution functions of early fatalities, corresponding to 25 reactor units (Bayer and Heuser 1980)

able to other countries. In many countries similar studies were performed, usually following closely the general method of WASH 1400. The most extensive of these, the "German Reactor Safety Study," was published in nine volumes in 1979/1980. It was limited to the technology of PWRs. The technical details referred to the nuclear power plant Biblis-B, a 3750-MW (thermic) PWR (Deutsche Risikostudie Kernkraftwerke 1979). In this risk study 70 possible sequences of events were examined which could lead to a release of fission products. These were mainly small, medium, or larger leaks in the main cooling system; however, the highest relative contribution to the probability of a core melt by uncontrollable loss of cooling has been found to result from a small leak in the main cooling system. The various sequences of events could lead to different release mechanisms. To simplify the risk assessment, the various types of release were grouped into eight release categories.

Considering the various release frequencies, the German study estimated the risk of a reactor accident with core melting to be about 1:10000 (estimated range 1:2000 to 1:50000) per reactor year.

In the American WASH 1400 report a rate of 1: 20000 reactor years was given for core melting in light water reactors. But only 10% of these core-melting accidents would have led to a release of considerable amounts of radioactivity. The probabilitiy of unfortunate weather was estimated at 1:10 and the probability that a densely populated area would be affected by the radioactive cloud was calculated to be about 1:100. Therefore the probability of a catastrophic reactor accident with thousands of acute deaths is about 1:200 millions per reactor year, according to WASH 1400. This probability, even assuming a total number of nuclear power stations on earth of 1000, has to be regarded as so small that for the period over which man will be using nuclear fission reactors, such a gigantic reactor catastrophe is extremely unlikely. However, in neither risk study were events like sabotage and war considered.

Figure 4.3 shows the complementary frequency distribution of the acute deaths per year, in relation to 25 nuclear power stations representative for the Federal Republic of Germany. The frequency of fatalities, classified in release categories, is estimated

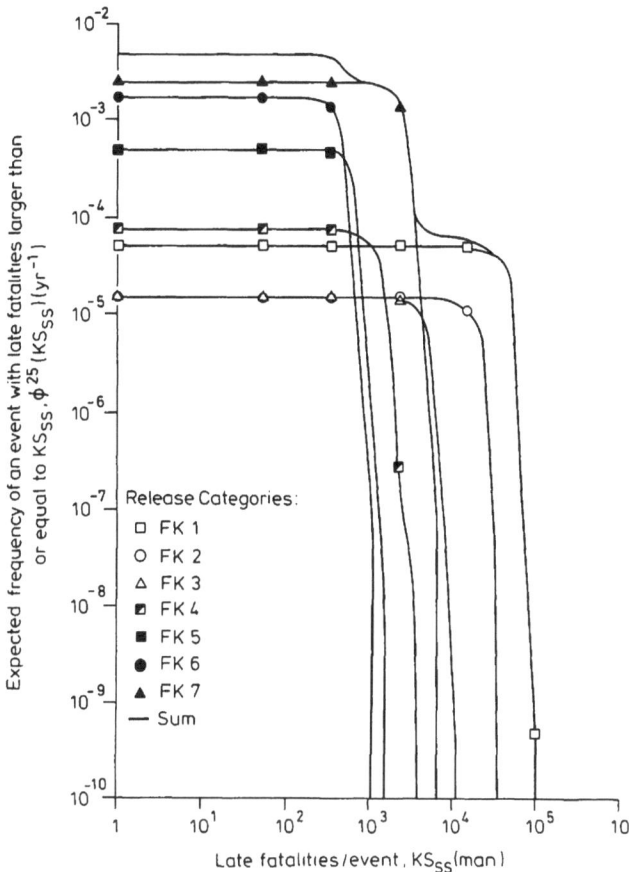

Fig. 4.4. Complementary cumulative frequency distribution functions of late fatalities, corresponding to 25 reactor units (BAYER and HEUSER 1980)

up to 1:100 000 per year. The probability of 1000 or more deaths would be about 1:10 millions. Figure 4.4 shows the complementary frequency distribution of the occurrence of lethal leukemia and cancer, using a proportional dose-effect relationship and a risk factor of 1.25×10^{-4} per 0.01 Sv, in accordance with ICRP 26. Radiation - induced cancer has to be expected even at distances very far from the reactor due to the drift of radioactivity. Therefore, as in the American WASH 1400 study, an area up to 800 km from the reactor is taken into consideration, covering a total population of more than 600 million people, i. e., the entire population of Europe. After a serious accident in one country the population of other countries may be affected by late injuries (cancer and genetic damage), but not by acute deaths.

After publication of the American and German risk studies, research into the mechanisms and the probability of reactor accidents was continued. More recent data suggest that core melting is at least one order of magnitude less likely than previously estimated. In addition, results of calculations on the "steam explosion" conclude that a failure in

the containment due to a steam explosion has to be expected in only 0.01% of core-melting accidents, compared to the 1% previously assumed. The most important results of the more recent calculations is, however, that the time of release due to containment failure would be 4.5 days after core melting and therefore much later than assumed in the safety studies. This delay reduces the release of radioactivity by a factor of 10^3 for tellurium, cesium, and rubidium, by a factor of 10^4 for barium and strontium, and by a factor of 10^5 for iodine, in comparison to the German risk study. Therefore it can be assumed that the scale of a reactor accident would be generally lower than estimated so far. This does, however, apply to PWRs only. Much less research has been done on the mechanisms of other reactor types, and the probability of accidents. Some comments will be made below on how the hypothetical accidents of WASH 1400 or the German study can be related to the real accidents of Three Mile Island and Chernobyl.

On Three Mile Island the accident happened because of failure of the reactor cooling system, due to a series of human errors which led to a dangerous increase in the reactor core temperature. In the further course of events contaminated cooling water flowed from the reactor containment into a separate building. In this way ^{133}Xe and a small amount of ^{131}I were able to escape into the environment (BELDA 1980). No members of the general population received a dose in excess of 1 mSv. The commission of the American Ministry of Health estimated the dose for 2 million people living in a radius of 50 miles around the reactor. These calculations were based on measurements of thermoluminescense dosimeters installed in the vicinity of the reactor. Extrapolated to the total "affected" population, the mean radiation exposure was about 0.015 mSv, which is about 1% of the normal annual radiation exposure from natural radiation sources.

The Three Mile Island accident is similar to model no. 9 of WASH 1400, i.e., that with the lowest rate of radioactivity release but the highest probability - 1:2500 with an estimated range between 1:250 and 1:25000 per reactor year. Considering an accumulated experience of about 400 light water reactor years in the USA, there was, according to RASHMUSSEN (1980), a definite chance that the accident at Three Mile Island would occur.

The catastrophic accident at Chernobyl cannot be compared with the results of the two risk studies. The characteristics of the accident reactor RBMK-1000, which are relevant to its safety, are very different from those of reactors analyzed in the safety studies. A direct transfer of the events in Chernobyl to West European or American nuclear power stations is therefore inappropriate.

According to the Soviet authorities the total nongaseous radioactivity released over a period of 10 days was 2×10^{18} Bq, which is equivalent to the most severe release categories of the German and American safety studies.

4.4 Planning Emergency Measures for Nuclear Power Plants

Although reactor catastrophes are very unlikely events, if they do happen they may have considerable effects on the health of the general population. Therefore government of countries running nuclear power plants set out guidelines for various precautions to reduce the consequences of such accidents (Rahmenempfehlungen in gem. Ministerialblättern No. 31, 1977, No. 13, 1981).

In planning emergency measures, a distinction has to be made between accidents in or close to the country concerned and those occurring in countries further away. If large amounts of radioactivity are released in the country concerned, preplanned, immediate measures including evacuation or relocation of a large part of the population might be necessary in regions close to the accident. Yet if a serious accident occurs far away, causing only a moderate amount of fallout, it is probably sufficient to take frequent measurements of radioactivity in the air, water, soil, plants, and food, and only then to recommend appropriate precautions, based on these measurements.

The size of the nuclear risk potential can be estimated from the number of existing or commissioned nuclear power plants. In 1986 there were over 370 nuclear power plants in operation worldwide, and another 151 under construction. In Europe they are concentrated in Central Europe, where 191 are in operation and 90 under construction.

The responsibility for organizing emergency measures varies considerably in the different countries of the world. In the Federal Republic of Germany it is the federal states that are responsible for protecting the population in times of peace. Consequently every state of the Federal Republic of Germany has its own legal and organizational system for protecting its population from injury in an emergency. The accident at Chernobyl showed that this privileged position of the states may be disadvantageous for the general population and resulted in totally different recommendations for permissible radioactivity levels in food in the different states.

In the following the situation in the Federal Republic of Germany is described as a model. Although there are significant differences in detail between the various countries, emergency planning is based on relatively uniform principles. However, everywhere the Chernobyl accident stirred up discussion on the adequacy of emergency planning.

The authority responsible for coordinating emergency measures is usually administrative district in which the accident took place. The authority makes the decisions on protective measures. However, experts, including doctors specialized in radiation protection, representatives of various rescuing organizations like fire brigades, and an experienced engineer from the nuclear power plant, are supposed to avise the authorities. Three levels of alarm may be considered by the authorities (STORNER 1981):

Emergency warning: will be announced in the event of a serious incident in a nuclear plant, if release is still below the criteria set for emergency alarm but the possibility of such criteria being met cannot be excluded with certainty.

Special alarm - water: will be announced if a considerable release of radioactivity into water has happened, but no significant release of radioactivity into the air. This is of special relevance for reactors situated upstream along rivers.

Emergency alarm: will be announced if during an accident in a nuclear plant a release of radioactivity into the air has happened which exceeds the set limits or there is concern that this is to be expected.

An emergency alarm has serious consequences. Monitoring groups have to enter the contaminated area, the general population will be warned by sirens and police cars and advised to stay indoors and if possible in basements, and potassium iodide tablets will be distributed. Evacuation of sectors of the population has to be considered, as do the closing of certain areas, limitations on traffic, confiscation of contaminated food and fodder, decontamination of people, animals, materials, houses, streets, and vehicles, etc. Medical care has to be provided.

To estimate the amount of radioactivity deposited in the environment, direct measurements and collections of samples of air, water, milk, grass, soil, and plants at various distances from the plant are necessary according to a preplanned schedule.

In the *first phase* of a serious accident, core melting may begin. Release of radioactivity has to be expected; however, its probability is perhaps hard to estimate. An emergency alarm should be announced.

In the *second phase,* i.e., the "early phase" according to the ICRP, release of radioactivity is about to take place or has just started. The emergency alarm has already been announced. During this period people are asked to stay in their houses; they might have to be evacuated at a later time. Potassium iodide tablets may be distributed.

During the *third phase,* i.e., the "intermediate phase" according to the ICRP, people may have to be decontaminated and provided with medical care. In the case of evacuation this happens mainly in "emergency stations." Serious cases will receive hospital treatment while others remain under supervision as outpatients.

The *fourth phase,* i.e., the "recovery phase" according to the ICRP, covers the period when the acute to subacute accidents are over. In this phase the main problems are monitoring of drinking water and food, control of contaminated or injured pets and livestock, decontamination of houses, streets, and ground, and finally the return of the evacuated population to their houses, if this appears safe according to the radioactivity measurements.

Table 4.1. Dose equivalent levels for early phase countermeasures

Countermeasure	Dose equivalent (mSv)	
	Whole body	Lung[a], thyroid, and any single organ preferentially irradiated
Sheltering and stable iodine administration		
Upper dose level	50	500
Lower dose level	5	50
Evacuation		
Upper dose level	500	5000
Lower dose level	50	500

[a] In the event of α-irradiation of the lung, the numerical values apply to the product of RBE and absorbed dose in mGy. This RBE is expected to be substantially less than 10 (ICRP 40)

4.5 Evacuation of the Population

If a large amount of radioactivity is released, evacuation might be necessary, especially for those people who live near a reactor. However, the risks associated with the process of evacuation should stand in an acceptable relation to the expected benefit. Psychological factors have to be considered, too.

Partial evacuation should also be considered. Before deciding to evacuate the entire population, pregnant women and children should be removed from the danger zone first. For seriously ill and bedridden patients, the strains of evacuation have to be balanced against the risks from radiation exposure, which may turn out to be rather low. However, it should be remembered that problems may arise for the remaining nursing personnel. Apart from these special cases there should be rigid rules governing the response of the general population, like those recommended by ICRP 40 (Table 4.1).

The question of evacuation also arose in the Three Mile Island accident. Although there was no massive release of radioactivity, there was apprehension that this could happen. For the government, the employees, and the management of the reactor, as well as for the public health system, a situation had developed which had not existed in the 25-year history of nuclear power plants in the USA. It became clear that there were emergency plans for possible accidents within the nuclear plant, e.g., a functioning medical service, but that these plans could not be applied to the medical problems arising outside the reactor. The protection of the public is the concern of the state and its authorities, and evidently the authorities in Harrisburg and Philadelphia were less competent to master the medical and logistic problems of such an accident than were the reactor personnel (LINNEMANN 1980).

Whereas evacuation turned out not to be necessary after the accident in Harrisburg, evacuation was indispensable following the Chernobyl accident (INSAG Report 1986). Monitoring groups were sent into the town of Pripyat, which is situated 3 km from the reactor, and demonstrated relatively high ground radioactivity only a few hours after the beginning of the accident. On the morning of April 26, dose rates of 0.14–1.4 mSv/h were measured in the streets. The following day the dose rate increased to values of 1.8–6 mSv/h, to reach 10 mSv/h in the afternoon of the same day, when the population had already been evacuated.

The population of the town of Pripyat was warned early. They were asked not to leave their houses and to keep the windows closed. Iodine tablets were handed out at an early stage. Thirty-four hours after the start of the accident the 45000 inhabitants of Pripyat were evacuated in an operation lasting only 2.5 h. After evacuation they were medically examined and monitored for external contamination. Incorporation of iodine was measured over the thyroid gland. Ten days after the start of the accident the remaining 90000 inhabitants in the 30-km zone around the reactor were evacuated, too, together with their pets and livestock. The total of 135000 evacuated people were estimated to have been exposed to a collective dose of about 16000 man-Sv from external γ-radiation. Those who lived between 3 and 15 km from Chernobyl were estimated to have received an average dose of 0.35–0.55 Sv. In individual cases exposure may have been several times this value, yet there is no report on serious radiation injury among members of the general population.

4.6 Prophylaxis of the Thyroid Gland

During a reactor accident, radioactive iodine is released in the form of ^{131}I, [132]I, [133]I, [134]I, and [135]I. At the high temperatures during the core melt, iodine is in gaseous form and contributes significantly to the radioactive cloud. Later it is deposited on ground, in water, and on plants. It may be incorporated by breathing air, by drinking water, by eating vegetables and fruit, and, most important for children, by drinking milk.

If iodine is incorporated by inhalation, it is totally resorbed by the lung into the blood circulation and, after transient accumulation in the salivary glands, it is stored in the thyroid gland. The lower the normal uptake of iodine with the diet in the particular region, the higher will be the proportion taken up into the thyroid gland. The iodine which is not stored in the thyroid gland will be excreted by the kidneys with a biologic half-life of about 6 h.

Accumulated radioiodine decays by emission of β- and γ-rays. The dose to the thyroid glands depends on the accumulated activity of radioiodine, on the composition of iodine isotopes (which differ in their half-lives and radiation quality), and not least on the functional state of the thyroid gland.

The dose estimates in the 86 inhabitants of the islands of Rongelap and Ailingnae from incorporation of radioactive iodine as a result of the fallout from the hydrogen bomb accident on the 1 March 1954 demonstrated remarkable high radiation exposures to the thyroid gland, especially for children. Those under 10 years had received 5–14 Gy, older children between 10 and 20 years about 5 Gy, while exposure of adults was considerably lower. This resulted in decreased function of the thyroid gland in about 15%, development of small nodules in the thyroid gland in about 36%, and development of thyroid carcinoma in four cases. Considering all inhabitants of this archipelago, some of them less affected by fallout, the number of thyroid carcinomas increased to seven among 243 exposed. No thyroid carcinoma developed in a comparable population of 504 inhabitants of other islands in the South Pacific (CONARD 1980). These observations contrast with the clinical experience with the use of ^{131}I in the diagnosis and therapy of thyroid gland disorders, where no significant increase was demonstrated in the rate of thyroid gland carcinoma among the investigated patient groups.

To prevent or to decrease the incorporation of radioactive iodine into the thyroid gland, authorities recommend provision of the population with potassium iodide tablets. As the intake of iodine may lead to side-effects, these tablets should be handed out only if a considerable release of radioiodine in a reactor accident is expected. Since it is assumed that in a serious accident at a reactor equipped with an adequate containment, release of radioactivity from the containment is delayed by many hours or even days, there should be sufficient time to distribute the iodide tablets after the accident has started.

The recommended dose of potassium iodide varies between different countries. In the Federal Republic of Germany an initial dose of 200 mg KI is recommended for adults. As iodine is eliminated by the kidneys with a half-life of about 6 h, a further 100 mg should be taken every 8 h, up to a total of ten 100-mg tablets over 3–4 days. The duration of iodine medication may be extended. This dosage also applies to pregnant women. Children (7–15 years) should start with a dose of 100 mg KI and then take 25 mg every 8 h up to a total of 500 mg. Infants (2–6 years) and babies should take a daily dose of 50 mg up to a total of 200 mg.

The Bureau of Radiobiological Health in the USA recommends a daily dose of 130 mg KI for adults and children, and a daily dose of 65 mg KI for children under 1 year, to block the thyroid gland in an accident. This medication can prevent up to 90% of iodine absorption, if taken before or immediately after incorporation. Fifty percent of uptake can still be blocked if KI is taken within 4 h after exposure.

Opinions differ more on the conditions under which prophylactic potassium iodide should be taken than on dosage. Due to the possible side-effects, the distribution of potassium iodide in order to protect the thyroid glands in a reactor accident is not beyond argument. There is a real risk of induction of hyperthyroidism and even thyrotoxicosis in healthy people living in iodine deficiency regions, like most mountainous countries. Possible causes of hyperthyroidism include the following preexisting conditions which often are not known to the individual affected:

- Graves' disease with diffuse autoimmune disorders
- Autonomous nodular adenoma
- Multiple autonomous microadenoma

The risk of induced hyperthyroidism in an iodine-deficient population may be estimated from an analysis of the experience in Tasmania. Iodine prophylaxis of the population against iodine deficiency was introduced by adding iodide to bread, and the prevalence of hyperthyroidism in the population increased by 5%. From these epidemiologic studies the conclusion may be drawn that supplying potassium iodide to the general population in an iodine-deficient region like Southern Germany may lead to clinical hyperthyroidism in between 0.1% and 1% of the population over 40, whereas the risk to younger people is considerably lower.

Balancing the risk of induction of thyroid cancer by incorporated radioiodines and the expected side-effects of iodine prophylaxis in the population, it is recommended that potassium iodide should not be provided unless radiation exposure of the thyroid gland to iodine of 1 Sv is likely to be exceeded.

4.7 Contamination and Decontamination of Body Surfaces

Calculations in the German reactor safety study showed that on average 6800 people would be residing within 8 km of the reactor in a sector of 30°, although in some locations the figure would be up to 20000. Assuming that 3% of the population will not go and remain indoors after the warning, up to 600 heavily contaminated people may have to be looked after with a surface contamination of skin and clothing up to 400 kBq/cm^2. For the affected person this means a total body exposure of about 30 mSv in 24 h and an additional β-radiation dose to the skin of about 10 Sv in 24 h. At this level of contamination early decontamination becomes an urgent task. First people should undress completely, as clothes are usually much more heavily contaminated than skin. Next, head, neck, and hands should be cleaned under running water at a washbasin. Taking a shower is usually not only unnecessary but also unwise, as it would spread the localized contamination to uncontaminated areas of the skin. Water is usually sufficient for decontamination. If not, slightly alkaline soaps may be used. To prevent injuries only soft brushes should be employed to clean the skin. Hard brushes may produce small injuries through which radioactivity enters the body. Generally, the intact skin is an impermeable barrier to radionuclides.

The decontamination process should be followed by measurements looking for any remaining contamination. In particular localized contaminations which are difficult to clean should be considered. If residual contamination is present, the process of decontamination has to be repeated; if complete decontamination is then still not achieved, one should not bother to do much more under the general conditions of a nuclear catastrophe. With a large number of affected people the time required for decontamination must be short. Persistent contamination on the skin will not spread easily.

Recommendations have been published for decontamination of the skin (Medizinische Maßnahmen bei Kernkraftwerksunfällen 1986). Experience with radiotherapy has shown that acute β-irradiation of limited skin areas with a dose of 5–10 Sv may produce transient reactions of the skin, like erythema. Serious, long-term injury has to be expected only after higher doses.

Table 4.2. Action levels for decontamination of skin (Medizinische Massnahmen bei Kernkraftwerksunfällen 1986)

Level	I	II	III	IV
Contamination (kBq/cm²)	0.4–4	4–40	40–400	>400
Decontamination	To be considered	Recommended	Necessary	Very necessary
Possible skin dose rate (Sv/24 h) (assuming typical nuclide mixture at light water reactors)	≤ 0.1	0.1–1	1–10	>10
Possible total body dose if contamination affects large parts of the body	≤ 0.3	0.3–3	3–30	>30
γ dose rate at 1 m distance (μSv/h)	≤ 1	1–10	10–100	>100

Contamination of the skin with about 40 kBq/cm² of a nuclide mixture typical for a light water reactor would lead to a β-radiation dose rate of about 50 mSv/h or about 1 Sv within 24 h. On this basis, a lower contamination limit for proper decontamination procedures of 40 kBq/cm² and an upper limit of 400 kBq/cm² have been recommended. A set of action levels for decontamination is given in Table 4.2.

For practical monitoring procedures the γ-radiation emitted together with the β-radiation by radionuclides is easier to measure. Therefore, β contamination may be estimated from the measurements of the γ dose rate. Contamination of 40 kBq/cm² of the expected mixture of nuclides is equivalent to a γ dose rate of 10 μSv/h at 1 m distance. As equipment for the helpers, portable proportional counters measuring an area of 70 cm² are recommended. They are capable of correct measurements of up to 10^6 counts per second, which is equivalent to a surface contamination of about 40 kBq/cm².

The contamined person is a source of radiation for people assisting with the decontamination procedures. Therefore contaminated clothes have to be stored and shielded appropriately. The risk from contamination for the helpers is unavoidable, but it is low if gloves are worn. The used water may, however, accumulate rather high activity levels and it may become necessary to collect it or to filter it through sand or soil.

Monitoring contaminated people may become a problem if the level of environmental radiation is higher than that emitted from lightly contaminated persons. Another problem may arise if there is so much radioactivity accumulated in the thyroid gland that the γ-radiation emitted from the localized source of the thyroid may wrongly suggest surface contamination.

4.8 Recommendations of the ICRP

In their publication 40 the International Commission on Radiological Protection (ICRP) made a series of recommendations for radiation protection of reactor staff and of the general population after a reactor accident. The introduction states that the primary contribution to the protection of the public lies in the safety standard of the reactor itself, the standard of technique, and the competence of the staff. As there are various types of reactor, planning for an accident cannot be the same in all cases. A variety of accident scenarios are always possible. Moreover, the more serious types have a lower probability of occurring. Thus the degree of detail as regards preparations may decrease with decreasing probability of an accident.

Although the source of radiation in the event of an accident is out of control, and dose limitations cannot be met, the principles recommended by the ICRP should be considered. These principles are as follows:

- Serious nonstochastic effects should be avoided by the introduction of countermeasures to limit individual doses to levels below the thresholds for these effects.
- The risk from stochastic effects should be limited by introducing countermeasures which achieve a positive net benefit to the individuals involved.
- The overall incidence of stoachastic effects should be limited, as far as reasonably practicable, by reducing the collective dose equivalent.

During a reactor accident the individual radiation doses to the rescue units will be particularly important: depending on circumstances, they might be higher than the recommended limits. Exposures of the population living outside the reactor environment are important, as they may occur at distances far from the accident and over long periods. Although individual doses might be lower than those to the rescue units, they affect a large group of people, leading to a high collective dose. It is also necessary to consider the level of radiation exposure for those people wishing to return to their homes after evacuation.

The expenses for decontamination of ground, roads, and buildings have to be placed in relation to the expenses arising from adverse health effects among the affected population. Difficult decisions have to be made if normal life is to be resumed in these regions. The measures which have to be considered for the welfare of the people include not just steps to combat immediate health risks, but also social and economic initiatives.

Whether there is exposure of rescue units or of the general population, implementation of countermeasures should follow certain principles, as defined by ICRP:

It should be possible to define, on radiation protection grounds, for each countermeasure, a lower level of dose below which introduction of the countermeasure is not warranted and an upper level of dose for which its implementation should almost certainly have been attempted. The dose range between these two levels is the one within which operational Intervention Levels should be set in the preparation of Emergency Plans.

In practice the level of dose at which action is taken should be that projected if a particular countermeasure is not introduced. Clearly, the dose actually averted may be less than the projected dose, because countermeasures may not be completely effective and some doses may be received during the implementation of the countermeasure.

The dose limits recommended by the ICRP are given in Table 4.1. They apply to the early phase, during and soon after the accident. The ICRP differentiates between three phases, the so-called early phase, the intermediate phase, and the recovery phase.

The Early Phase

(15) The early phase of an accident can be thought of as comprising two stages:

(i) while there is a threat of a serious release, i. e. that period from the time when the potential for off-site exposure of the public is recognized, to the time when significant amounts of radioactive material are released, or the plant is brought under control; and

(ii) the first few hours after the commencement of that release. For nuclear power plants the time between initiation of the accident sequence and the start of the release of radionuclides to atmosphere has been suggested to range between about half an hour and a day or more, although for other facilities shorter timescales could be applicable. The duration of the release may be from as little as half an hour to a few or more days depending on the particular circumstances.

(16) These two stages can be combined into one Early Phase because in both cases decisions designed to minimize doses to the public are likely to be made predominantly on the basis of sequences of occurrence and plant conditions which should have been identified in advance.

The Intermediate Phase

(17) The intermediate phase covers that period of time from the first few hours to a few days after the onset of the accident. It is presumed, generally, that the major part of the potential release from the installation to the atmosphere has already taken place and, unless the release consisted mainly of noble gases, there are likely to be significant amounts of radioactive material deposited on the ground. In this phase the most important routes of exposure will be:

(i) external exposure arising mainly from radioactive material deposited onto the ground; and

(ii) internal exposure arising mainly as a result of ingestion of water or foodstuffs contaminated directly.

(18) In the intermediate phase the results of environmental monitoring should be available for decision-making on the introduction of applicable countermeasures.

The Recovery Phase

(19) This phase is the one in which decisions are made concerning the return to normal living conditions and it may extend over a prolonged period. The objective during this phase will be to withdraw countermeasures which will have been introduced in the early and intermediate phases. This may require measures to reduce contamination of both built-up areas and agricultural land.

The basis for withdrawing countermeasures will be that radioactive contamination has been sufficiently reduced by the appropriate combination of radioactive decay, or weathering, and planned decontamination campaigns. There will need to be a number of social, economic and technical inputs to decision-making in the recovery phase.

4.9 Radiation Injury After Total Body Exposure

People suffering from radiation injury, in contrast to other physical injuries and burns, do not present immediately after exposure with any objective symptoms which might be indicative of the seriousness of damage, even leaving doubt as to whether any exposure has taken place at all.

Up to a total body dose of about 10 Gy, i. e., a dose at which prognosis can still be influenced by medical treatment, the critical phase develops only one or several weeks after radiation exposure. This radiation syndrome is primarily caused by disturbances in hemopoiesis in the damaged bone marrow. As described in detail by CRONKITE and FLIEDNER (1972), due to the hierarchic organization of hemopoiesis, with fixed turnover times and feedback mechanisms, a clinically significant drop in the white blood count does not occur until after a delay of 1 week. Only after granulocytes and platelets have decreased below a critical level does the radiation syndrome develop, with fever as a sign of a septic infection and bleeding.

Table 4.3. The incidence of initial symptoms after irradiation in relation to the severity of the radiation damage (acute whole body irradiation) (VOROBEV 1979)

Prospective severity of the radiation syndrome	Vomiting as main symptom; time of outset and duration	Indirect symptoms				
		General fatigue	Headache; awareness	Body temperature	Skin erythema; conjunctival reaction
Mild	Absent, or once after 3 h	Slight	Short-term headache; clear awareness	Normal	Mild conjunctival reaction possible
Moderate	1.5–3 h, twice or more	Moderate	Continuous headache; clear awareness	Normal or subfebrile	Erythema and conjunctival reaction
Severe	After 30 min to 1.5 h, several times	Serious	Continuous strong headache; clear awareness	Normal or subfebrile	Marked erythema and conjunctival reaction
Very severe	After 10–30 min, uncontrollable vomiting	Very pronounced	Strong headache; awareness can be disturbed	Subfebrile or febrile	Severe erythema; severe conjunctival reaction

Table 4.4. Clinical and hematologic signs and symptoms after neutron-gamma whole-body irradiation – therapeutic and prognostic conclusions (MESSERSCHMIDT 1979)

Signs and symptoms	Survival assured or probable (0.5–2 Gy)	Survival possible or questionable (2–6 Gy)	Survival improbable (more than 6 Gy)
Initial symptoms	Vomiting and nausea absent or slight within 2–6 h	In 70%–100%, severe vomiting within 1–2 h; weakness	In 100% severe retching and vomiting within 1 h; diarrhea; severe exhaustion
Early erythema	None	Slight reddening	Distinct reddening
Duration of latency period	More than 3 weeks	10–20 days	Up to 1 week
Lymphocytes	Slight to moderate decrease in 1–2 days to 1200/mm^3	Moderate to large decrease within first day to 1200–300/mm^3	Large decrease within hours down to 300/mm^3
Granulocytes	Decrease to 40%–50% of normal within 45 days	Initial granulocytosis; decrease in 7–20 days to about 20% of normal	Initial granulocytosis; decrease in 4–10 days; nadir 10% of normal
Platelets	Decrease to 50% of normal within 30 days	Decrease to 10%–30% of normal within 25 days	Decrease to almost 0% within 14 days
Reticulocytes	Unchanged	Slight decrease after several days	Marked decrease after a few days
Bone marrow	Occasional mitotic anomalies; slightly depressed cell count	Mitotic anomalies in 24–48 h; marrow hypoplasia	Within hours cellular decay; necroses; marrow aplasia
Clinical findings in the peak phase	None or slight	Epilation at 3 Gy; fever in about 14 days; hemorrhages after 2–3 weeks	Epilation in about 10 days; fever and hemorrhages within 8 days; abdominal pain
Result of examination	Treatment required within a few days	Treatment necessary with good expectations	Treatment necessary, but often in vain

A precise diagnosis is difficult before the onset of these symptoms, and problems will occur if, at an early stage, a large number of people have to be evaluated for future treatment need.

First signs of the degree of radiation exposure are early symptoms like dizziness, vomiting, exhaustion, cardiovascular symptoms, rise in temperature, and diarrhea, which start a few hours after exposure and often fade within 1 day. This is followed by a more or less symptom-free latent time of 1–3 weeks, until the climax of the radiation syndrome occurs. These early symptoms, some of which are hard to assess in an objective manner, may be exaggerated by psychological reactions. The relationship between the radiation dose and the intensity of these early symptoms has been doubted, yet the experience of Chernobyl has demonstrated their value. Especially sickness and vomiting are early reactions which give the doctor some initial diagnostic information before blood counts or bone marrow biopsies. The diagnosis of an early erythema lasting for 1 or 2 days indicates a radiation exposure of over 5 Gy. Table 4.3 shows a list of early symptoms from the Russian literature which might be useful for early diagnosis. As the lymphocytes are especially radiosensitive, the blood lymphocyte count decreases even at relatively low radiation exposures, reaching a plateau within the first 2 days after exposure. Therefore an earlier diagnosis is possible using the lymphocyte count than a count of any of the other blood cells. Table 4.4 gives a list of clinical and hematologic signs and symptoms which may be used for the evaluation of radiation damage. Figure 4.5 describes the decrease in lymphocytes and its relation to the degree of radiation damage.

If the total body radiation dose approaches 10 Gy, survival is only possible under optimal treatment and nursing conditions. In addition to bone marrow failure, persistent diarrhea as a sign of serious damage of the intestinal mucosa makes the prognosis worse and treatment difficult.

If the total body dose exceeds 10–30 Gy, there are effects on the central nervous system and the cardiovascular system, with serious cramps, loss of consciousness, and shock; the prognosis is then very poor and only symptomatic treatment is possible.

4.10 Medical Problems After a Reactor Accident

While the appropriate form of treatment of reactor accident victims outside the plant, including a significant proportion of the general population, is controversial, the principles for the medical treatment of workers affected by any accident in the plant are well established and have shown their value in various smaller industrial accidents.

Medical care for workers in industrial accidents can be divided into three levels. The first level is the first aid offered by nonprofessionals or first-aid attendants. As soon as possible a specialized doctor should be consulted. The first-aid measures aim at

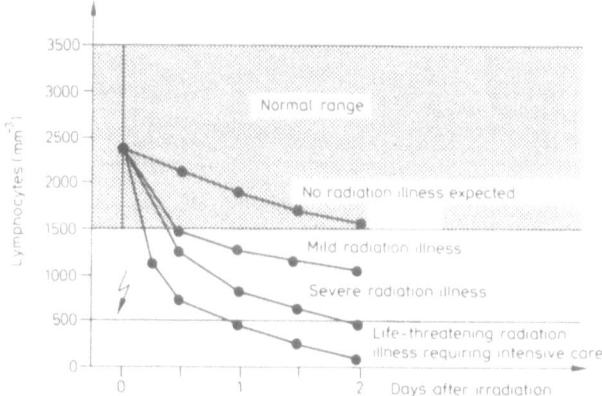

Fig. 4.5. Relationship between the time course of the fall in the lymphocyte count and severity of radiation illness (after SAENGER)

restoring or maintaining vital functions, since experience has shown that some serious reactor accidents result not only in radiation exposure but also in conventional injury, like burns or trauma. Another important step is to remove the injured person from the radiation zone or the contaminated area in order to avoid additional radiation exposure. For the same reason external decontamination should be carried out soon, to avoid transfer of radioactivity. Minor accidents like minor overexposures or contaminations can be dealt with in the first-aid station of the plant by internal personnel.

If radiation exposure exceeds a few millisievert, the help of regional radiation protection centers should be requested. These institutions should give advice and assist with the diagnosis (e. g., with total body counters, analyses of urine and feces or radioactivity, bone marrow biopsies, and blood counts). They should help to continue decontamination measures, to introduce decorporation measurements, and to estimate the individual radiation doses. The centers should cooperate with a larger hospital which has a well equipped department of nuclear medicine.

The third level of treatment consists of therapeutic measures which are performed in highly specialized hospitals for those patients who have received radiation exposure high enough to produce an acute radiation syndrome. The specialized hospital should be adequately equipped for intensive therapy for the acute radiation syndrome. For gnotobiotic treatment patients have to be isolated in a sterile environment. Besides intensive therapy with antibiotic drugs, substitution treatment with concentrated granulocytes and thrombocytes should be possible. Facilities for bone marrow transplantation should be available, too.

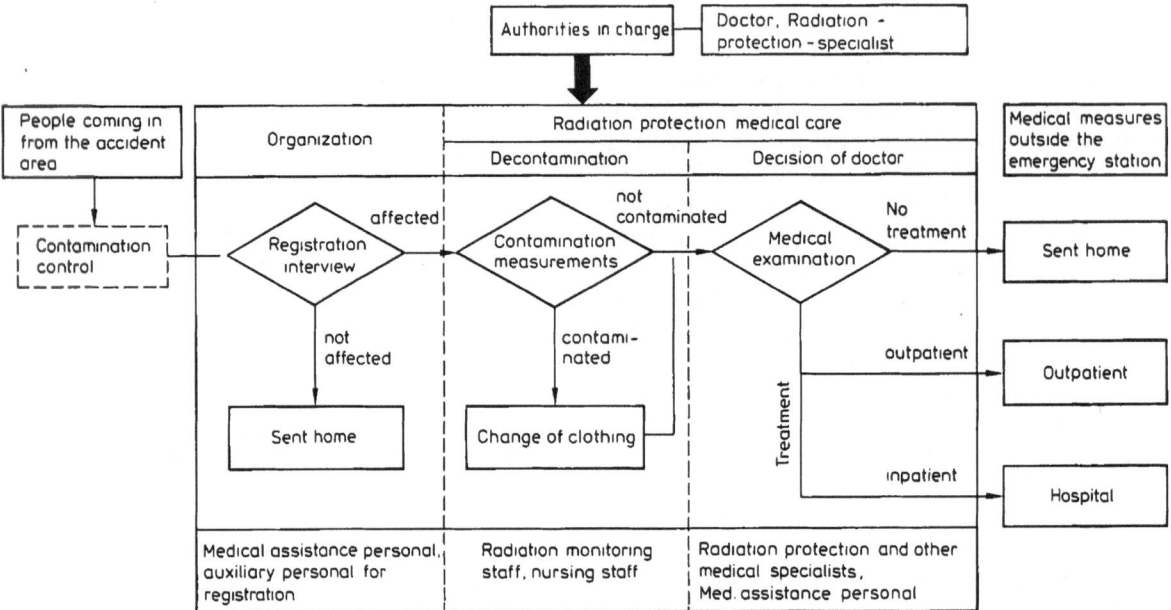

Fig. 4.6. Possible organizational structure for the medical management of the evacuated population in an emergency station (from Medizinische Maßnahmen bei Kernkraftwerksunfällen 1986)

This intensive care is only applicable to accidents involving a relatively small number of patients – up to a few hundred, as in Chernobyl. If, however, the release of large amounts of radioactivity leads to high potential radiation exposures of the population outside the nuclear power plant, very different problems arise. The planning for medical procedures in such a disaster is published in the book *Medizinische Massnahmen bei Kernkraftwerksunfällen* (1986).

Since people will be very worried after a reactor accident, a rush to the nearest hospital is likely to occur, mostly by people who have been only slightly or not at all affected by radioactivity. Present emergency plans therefore provide for emergency stations to filter those people who probably need proper care in hospitals. But even these emergency stations should, if possible, be attended only by those people who are suspected of having been exposed to radiation or contamination. Therefore people from areas not affected by the radioactive cloud or with only very low fallout (as determined by actual measurements) should be diverted away from the emergency station. Emergency stations should be established in buildings like schools and sports centers, where facilities for decontamination are available.

An example of the organizational structure of an emergency station is shown in Fig. 4.6. Three different areas should be established, one for registration, another for contamination measurements and decontamination, and a third for medical examination. The sequence of events in the emergency station might be as follows:

1. *Investigation of where the person stayed during the release of the radioactive cloud:* whether and how long he was outdoors or indoors, above ground, or in the basement. From this information a rough estimate of possible radiation exposure can be made if the ground contamination or local dose rate in the various places around the nuclear power plant has been monitored. The reliability of such information should, however, not be overestimated. This screening is more suited to the exclusion from further examination of those people who were not in an area which was significantly affected by the radioactive cloud.

2. *Monitoring of the body surface and clothes for the presence of radioactivity.* If contamination is found, there is also a very high probability that radiation exposure has taken place. The possibility of only secondary contamination, through contact with other people, and of rather low direct radiation exposure has to be considered. If contamination is found, the affected person has to be transferred to the decontamination group.

3. *Exploration of people for early symptoms of radiation illness,* like nausea, vomiting, dizziness, exhaustion, cardiovascular symptoms, increase of temperature, diarrhea, and early erythema.

Based on the result of these examinations those people should be selected who do not present with any of these symptoms and who were not exposed to life-threatening radiation doses. They will be much more numerous than those whose exposure was great enough to lead to early symptoms and who have to be examined more carefully, whether therapy or at least clinical supervision is necessary or not. In the latter case patients should be discharged and treated on an outpatient base by their family doctor, who should do regular blood tests during the next weeks.

The group of patients who need to be hospitalized is the smallest. For obvious reasons treatment of radiation patients cannot be done in the emergency station. Treatment of very serious forms of radiation injury requires hospitals with specialized and intensive care wards, where therapy under gnotobiotic conditions is possible, including facilities for bone marrow transplantation. The treatment of the acute radiation syndrome in reactor accidents was described by CRONKITE and FLIEDNER (1972).

In the emergency station it is not only radiation-exposed people who demand treatment; it is above all injured people and patients who were already sick before the accident who have to be treated. People injured during evacuation or by traffic accidents will need help. People who had been acutely or chronically ill are at great risk due to the separation from their family and their family doctor, and, in particular, as a consequence of the strain of a hasty evacuation. Patients suffering from diabetes or cardiovascular disorders may decompensate. The number of these injured and ill patients probably exceeds that of the radiation victims, especially if the release of radioactivity was small and the number of evacuated people large. General practitioners, internists, surgeons, and psychiatrists are more necessary for the medical care of the evacuated population than are radiation protection specialists.

Those patients who have probably been exposed to significant radiation doses but who do not need hospital treatment should receive outpatient care during the following weeks. In the first few days the most important diagnostic procedures is the determination of the concentration of lymphocytes and granulocytes in the peripheral blood. Furthermore, monitoring of the thyroid gland should be done, if a surface contamination of $40 \, kBq/cm^2$ or more was registered. This examination can be done with simple dose rate meters. An incorporated activity of 4×10^5 Bq iodine 131 in the thyroid gland produces an external dose rate (skin contact with the dose rate meter) of $10 \, \mu Sv/h$. This leads to a thyroid gland dose of about 150 mSv for adults and of about 1.2 Sv for infants.

The sequence of diagnostic procedures results either in further medical examinations or in hospitalization, if treatment is needed for radiation disease. The observation period should last about 5 weeks, as only then can the risk of a radiation syndrome needing treatment be excluded with certainty. By the 8th week people who did not develop severe radiation illness will have recovered completely.

4.11 Accidents Which Have Actually Occurred in Nuclear Power Plants

The first serious radiation accidents happened at critical piles testing plutonium for the development and construction of nuclear weapons in the USA.

During experiments with critical amounts of plutonium in Los Alamos, a release of radiation happened on 21 August 1945 which exposed one experimenter to a dose of about 3–6 Gy and caused his death within 26 days. A further very similar criticality accident happened 9 months later, in May 1946, again in Los Alamos, when the upper half sphere of a special containment for nuclear weapon plutonium fell down, forming a "critical mass." With the typical "blue glow," a large amount of radiation was released. One of the experimenters was exposed to a dose of about 11 Gy and died on the 9th day. Another person received a radiation exposure of about 2 Gy. She became ill but survived. The remaining six people were exposed to radiation doses of 0.1–0.6 Gy and did not develop any signs of radiation illness.

Further accidents happened during the following years, when nuclear explosives were handled, as in a laboratory for the recovery of uranium in Oak Ridge in June 1958. Unnoticed noncritical amounts of uranium nitrate accumulated in a containment, forming a critical mass. Eight people standing close by received radiation doses between 0.29 and 4.61 Gy. Several fell ill, but all survived.

On 30 December 1958 a technician in a plutonium recovery plant in Los Alamos received a dose of about 120 Gy, to date the highest radiation exposure of all registered accidents in nuclear power plants in the USA and Western Europe. The man suffered serious shock 15 min after exposure, became conscious again 6 h later, and died 35 h after the event showing the typical signs of the central nervous syndrome. Two other people received radiation exposures of 1.18 and 0.53 Gy. They developed slight changes in their blood count, but no symptoms of radiation disease.

Besides the criticality accidents in handling the fission materials for nuclear weapons, accidents happened in nuclear power plants after the war. When in 1961 the mechanics of the control rods of a transportable boiling water reactor of the US Army in Idaho Falls were about to be repaired, a sudden excursion happened with a subsequent steam explosion. Due to the pressure effects three technicians were killed. A considerable amount of radioactivity was released, but radiation was not the actual cause of death.

Special attention was paid to the accident in Vinca, Yugoslavia, on 15 October 1958. An experimental reactor became overcritical while the control system had been switched off. Only through the smell of ozone did those attending realize the danger. In four of the six affected people the dose was over 4 Gy. Bone marrow transplantations were carried out in the Curie Hospital in Paris. One victim died.

Detailed descriptions of all serious radiation accidents in reactors or critical assemblies cannot be given here; however, the reader is referred to the report on 89 accidents by Hübner and Fry (1980). This report also provides information on other radiation accidents in industry with various radiation sources, like ^{60}Co or ^{137}Cs. Altogether 82 serious accidents happened, and eight people died. In addition, 47 cases of localized irradiation to high doses are mentioned, mostly involving the arms and hands, as well as serious incorporation accidents as a result of intake of radium, transuranium elements, and fission products.

4.12 Medical Measures After the Chernobyl Accident

While the radiation accidents described above were confined to nuclear installations or industrial radiography and affected only a few people, the reactor accident at Chernobyl was a real catastrophe and people living outside the plant were affected as well.

The Soviet delegation reported to a meeting of the International Energy Agency in Vienna details about the diagnostic and therapeutic medical problems and about the organization of medical care (INSAG Report, 1986).

The accident began on the 26 April 1986 at 1.23 a.m. with an unexpected excursion followed by a gas explosion, and the reactor graphite started burning. About 15 min later, news of the accident reached the medical helpers on duty (three first-aid attendants), who, within the following 30-40 min, took care of 29 accident victims. All were highly contaminated. Initial decontamination consisted of taking off shoes and clothes. One victim was buried under the heap of rubble and could not be found; another died a few hours later due to serious thermal burns.

At 6 a.m. 108 patients were admitted for hospital treatment, and 24 more came the same day. Twelve hours after the accident, a specialized team of physicists, radiotherapists, hematologists, and laboratory assistants had arrived from Moscow. Within 36 h 350 people were examined and about 1000 blood tests were carried out. One hundred and twenty-nine patients, those who were most seriously ill, were transported to the specialized hospital in Moscow; other patients with lesser symptoms of radiation disease were taken to the general hospital in Kiev.

All patients were workers at the plant or firemen; members of the general population living nearby were not among them. During the first few days after the accident, 450 "medical brigades" were mobilized, consisting of 1240 doctors, 920 nurses, 720 medical students, 360 "doctors' assistants," and 2720 "assistants with secundary school education" to take care of the general population out side the power plant.

Two hundred and three workers developed an acute radiation syndrome. During the first 3 days, diagnostic interest was directed at the initial symptoms, like vomiting, erythema, and skin edema. The lymphocyte count was used for a gross estimate of total body dose. During the following 10-14 days the development of thrombocytopenia and the degree of granulocytopenia were recorded in each patient. Lymphocyte cultures were established to determine the number of dicentric chromosome aberrations for biologic dosimetry.

Based on hematologic data and early clinical symptoms, patients were classified into four groups. The fourth degree of injury, with lymphocyte counts below 100/µl, included the most seriously ill victims; 21 of the 22 patients in this group died within 28 days, all probably having received doses over 6 Gy. Out of 23 patients with the third degree of injury, who had been exposed to 4-6 Gy and had lymphocyte counts between 200 and 300/µl, seven died. Only one of 53 patients with the second degree of injury died. They had been exposed to doses between 2 and 4 Gy and had lymphocyte counts of 300–500/μl. No death occurred among the group of patients with the first degree of injury, exposed to less than 2 Gy and having lymphocyte counts between 600 and 1000/μl. The total death toll was 31.

In addition to their radiation injury, five patients suffered serious thermal burns. The majority of the heavily irradiated patients also developed severe radiodermatitis 2 weeks after the accident as a result of heavy skin contamination. These β-burns in some cases covered up to 90% of the body surface and were the most important factor determining prognosis. Many also suffered β-burns of the upper air passages.

The serious radiation damage to the skin caused the most dangerous complications, like brain edema, toxic encephalomyelopathy, liver failure, kidney failure, and toxic myocarditis. In some patients in the highest dose group the terminal phase was characterized by the development of pneumonitis and respiratory distress.

Bone marrow transplantation was carried out in 13 patients who had received total body doses over 6 Gy. Finding HLA-compatible donors proved a difficult organizational problem. However, bone marrow transplantations were not very effective, as in most transplanted patients life-threatening β-burns developed after transplantation and determined the outcome. Based on this experience, the place of bone marrow transplantation in the management of serious radiation catastrophes has to be regarded with some reservation.

Treatment of those patients who were exposed to lower doses and who had less radiation injury to their skin was considerably more successful. Treatment was purely symptomatic and many who would probably have died without treatment were saved. The main emphasis was on the prophylaxis and treatment of infections, as well as the substitution of platelets and erythrocytes. Detoxification therapy and total parenteral feeding were carried out in patients with extensive oropharyngeal radiation mucositis and with intestinal symptoms. Special care was taken to maintain the water and electrolyte balance of patients with intestinal syndrome and with radiation burns of the skin.

All patients with the second or a higher degree of radiation injury were treated in normal hospital wards, but under standard aseptic conditions: air sterilization with ultraviolet lamps, strict antiseptic procedures for the personnel, like special gowns and face masks, disinfection of shoes on a mat soaked with antiseptics, and daily changing of the patient's gowns. These simple aseptic conditions

were quickly established in the hospitals and the specially trained personnel cared for their patients adequately. Possible microbial contamination was checked regularly, and a low concentration of microorganisms was found in the room air of less than 500 colonies per m³. The diet was normal, but raw vegetables and fruits, as well as tinned food, were avoided. The effectiveness of these aseptic measures was demonstrated by the absence of exogenous bronchopulmonary infections. No patient with an acute radiation syndrome of the second or third degree developed infectious pneumonia.

In order to prevent endogenous infections, patients with second to fourth degree radiation injury were given biseptol and nystatin either 1 or 2–3 weeks before agranulocytosis to compare the effectiveness of these two methods of selective intestinal bacterial decontamination.

When fever occurred, two or three broad spectrum antibiotics were given intravenously (aminoglycosides, gentamicin, cephalosporin, and semisynthetic penicillins with activity against pyocyaneus bacteria). In half of the cases fever was promptly lowered after starting these antibiotics. If there was no effect within 24–48 h, intravenous γ-globulin was generously given at a dose of 6 g four to five times every 12 h. If the agranulocytotic fever did not settle within 1 week, amphotericin B was added. Acyclovir was used with good results during the initial stages of herpes infections of the facial skin, the lips, and the oral mucosa, from which about one-third of the patients suffered.

This empirical antibacterial, antifungal, and antiviral therapy was very effective and no patient succumbed to uncomplicated bone marrow syndrome.

Several patients with fourth degree radiation injury developed a diffuse, interstitial pneumonia, accompanied by rapid development of hypoxemia, which was fatal without exception.

Very important in the treatment strategy for patients with acute radiation disease was the transfusion of fresh platelets. The same donor was used four times. The indication for platelet transfusion was hemorrhage or decrease in the platelet count to below 20000 per μl. Usually about 300×10^9 platelets were given at each transfusion. Platelet transfusions were repeated after 1–3 days. Before transfusion, platelets and other blood components were exposed to 15 Gy ^{60}Co γ-rays to protect the patient from secondary disease. The platelet transfusions were very effective especially in patients with serious thrombocytopenia of less than 5000–10000 μl, which usually lasted about 2–4 weeks. As the amount of platelets required by dozens of patients during the period of most serious thrombocytopenia placed considerable stress on the blood donor service, allogeneic as well as autologous platelets were dry frozen during the first period and extensively used at the time of greatest demand.

There were no cases of incompatibility reactions after blood transfusions. On average three to five transfusions of platelets were needed for the treatment of a patient with third degree radiation injury. Granulocyte transfusions were not used for the treatment of infectious complications of agranulocytosis. Red blood cell transfusions were needed more often than expected, even for patients with second or third degree radiation injury, as radiation burns resulted in early and serious hemolytic anemia.

During the course of intensive clinical examinations and medical observations a large amount of data was collected, which now has to be evaluated. This is a unique and valuable experience, to which all doctors of the world should have access. This experience could help victims of any future radiation accident, which unfortunately cannot be excluded with certainty.

4.13 Combined Injuries After Nuclear Explosions

In a nuclear war, the scale of problems described in the previous chapters increase beyond human imagination, and it is very likely that the infrastructure of medical care would break down. Yet, the principles of medical management of radiation accidents remain valid and should be applied to the victims of nuclear warfare as far as the actual situation allows. There is, however, one major difference from catastrophic reactor accidents in that, especially if smaller, tactical nuclear weapons are deployed, a large proportion of victims will present with combined injuries, trauma, infections, wounds, burns, severe exhaustion, or stress, and the course and prognosis of the radiation syndrome will be markedly different from the uncomplicated radiation syndrome. The practical importance of this is well recognized, and it is estimated that nearly half of those fatally or severely injured by the nuclear explosions at Hiroshima and Nagasaki suffered from combined injuries. People outside their houses at the time of the explosion were hit directly by the flash and in addition to radiation doses received primary flash burns. People indoors were protected

from the flash but not from radiation and were wounded by collapsing houses.

GEIGER (1964) estimated that in any future nuclear war, 65%-70% of all the injured would have combined injuries; 5% would only have burns or wounds, 40% burns and high total body radiation doses, 5% wounds and radiation doses, and about 20% would suffer from all three injuries.

It appears that in patients suffering from combined injuries it is not just a question of the radiation syndrome being enhanced by the trauma (as if a higher radiation dose had been received) or the wound healing being delayed by the radiation; rather combined injuries lead to a clinical syndrome which differs significantly from those caused by radiation alone or by extensive burns alone, as has been stated explicitly by FEDEROV et al. (1958) on the basis of extensive animal experimentation. Experience gained with the treatment of the firemen at Chernobyl supports this statement. Reviews on the pathology and management of atomic bomb injuries have been published by MESSERSCHMIDT (1979, 1984).

The clinical picture of combined injury was first described by the Japanese doctors caring for the bomb victims in Hiroshima and Nagasaki. Mechanical wounds and burns initially showed the expected progress in healing, as in unirradiated patients. In the second week, however, the aspect of the wounds suddenly changed dramatically - they became superinfected and had a greasy appearance, granulation tissue broke down, wound margins became necrotic, and bleeding from the wounds started again. Histologically, massive necrosis and bacterial invasion was seen but the granulocytic wall which usually demarcates any necrotic tissue in the body was missing.

Animal experiments performed by various groups in the USA and USSR, summarized by MESSERSCHMIDT (1975), clearly demonstrate that it is especially the combination of burns with sublethal radiation doses which causes excessive lethality. Even burns covering only about 10% of the body surface combined with radiation doses well below the LD_{10} killed the majority of the animals since even radiation doses between 1 and 3 Gy decrease the local tissue response to infections sufficiently to leave the burns an open entry for all sorts of bacteria.

In addition to infection, the increased lethality in combined injury victims is also due to a higher susceptibility of the irradiated organism to shock. At the peak phase of the (nonlethal) radiation syndrome even a minor trauma can lead to fatal circu-

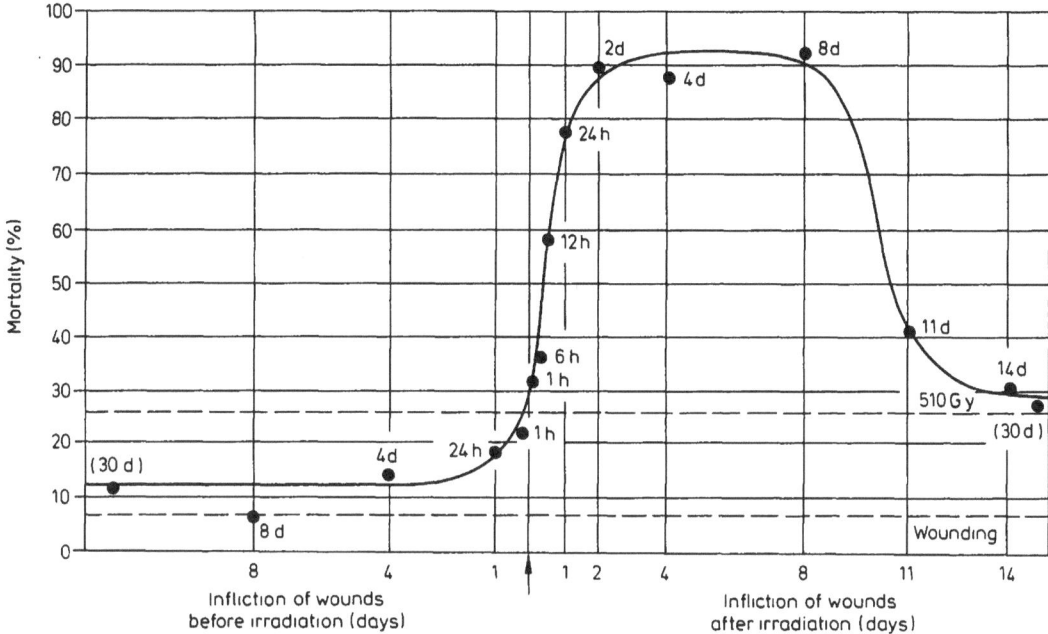

Fig. 4.7. Mortality of mice as a function of the interval between irradiation (5.1 Gy) and infliction of open skin wounds. *Dashed lines* indicate the mortality from pure irradiation (27%) and pure wounding (7%). (LANGENDORFF et al. 1964)

latory failure with a severe fall in systolic blood pressure which is unresponsive to the usual drugs (POLJAKOV 1964). In dogs, combined injuries resulted in deficient oxygen delivery to brain, liver, kidney, heart, and lungs. In mice, open skin wounds inflicted after total body irradiation led to disturbances in the microcirculation of liver, kidney, and lung and to ischemic myocardial degeneration (MESSERSCHMIDT and OEHLERT 1968). Also the consolidation of fractures is delayed by several weeks (RUBINSTEJN et al. 1957).

The sequence of total body irradiation and additional trauma and the interval between them are of crucial importance for the prognosis of combined injuries (Fig. 4.7). In experiments in mice, total body irradiation with 5.1 Gy resulted in about 25% lethality. If an open skin wound was inflicted at the time of irradiation, lethality increased to 30%; if the wound was made 2-8 days after total body irradiation, lethality was 90%, and it decreased to the control level only when the wound was made after an interval of 2 weeks. However, wounding a few days before total body irradiation increased radioresistance significantly, and lethality dropped from 25% to 10%, probably due to stimulated bone marrow proliferation (LANGENDORFF et al. 1964).

This detrimental effect of open skin wounds on the prognosis after total body irradiation, however, was not seen if a wound inflicted 2 days after irradiation was immediately closed again. A similar observation was made by RAZGOVOROV (1957) in dogs, rabbits, and guinea pigs inflicted with deep wounds 24 h after total body irradiation. By careful wound excision and primary closure immediately after wounding, lethality could be markedly decreased.

These experimental observations have important implications for the medical management of nuclear warfare victims (and, naturally, reactor accident victims) presenting with combined injuries. Most early symptoms of radiation injury which are used to estimate the degree of radiation exposure of the patient, e. g., nausea, vomiting, rise in body temperature, and even fall in lymphocyte count, are much less reliable than after uncomplicated radiation exposure. The additional trauma will usually increase these symptoms, suggesting a higher radiation dose than has actually been received. Based on extensive experience with animal experiments one may make the following recommendations for the doctor in charge of caring for patients with combined injuries (radiation syndrome plus trauma, e. g., burns and wounds):

1. Open wounds should be closed as soon as possible. However, blunt injuries, which are common in military surgery, should not be primary closed but treated by radical wound excision. Than in any case, secondary wound suturing can be indi-

cated. But during the phase of acute radiation syndrome surgery should be avoided at all costs.

2. If surgery is necessary it should be performed within 3-4 days after irradiation. Any surgical intervention after that time until recovery from the acute radiation syndrome 6-8 weeks after irradiation carries a grave prognosis.

3. Tolerance to various drugs may be reduced after total body irradiation; this is especially true for various narcotics such as chloroform, ether, and barbiturates. N_2O, halothane, cyclopropane, and diazepam appear to be less dangerous. If at all possible, local anesthesia or peridural anesthesia should be preferred.

Finally it should be stressed once again that, although it may save many patients if properly done, the medical care of victims of nuclear warfare suffering from combined injuries will not be very successful, mainly due to the expected breakdown of the medical infrastructure and the necessity of caring for large numbers of casualties at the same time, combined with heavy losses of doctors and, last but not least, completely inadequate training of doctors in the medical management of any form of catastrophe.

References

Bayer A, Heuser FW (1980) Übersicht über die deutsche Risikosicherheitsstudie. In: Messerschmidt O, Feinendegen LE, Hunzinger W (eds) Industrielle Störfälle und Strahlenexposition. Strahlenschutz in Forschung und Praxis, vol XXI. Thieme, Stuttgart, p 36

Belda W (1980) Analyse des Störfalls von Harrisburg. In: Messerschmidt O, Feinendegen LE, Hunzinger W (eds) Industrielle Störfälle und Strahlenexposition. Strahlenschutz in Forschung und Praxis, vol XXI. Thieme, Stuttgart, p 54

Cronkite EP, Fliedner TM (1972) The radiation syndromes. In: Hug O, Zuppinger A (eds) Strahlenbiologie 3. Springer, Berlin Heidelberg New York (Handbuch der medizinischen Radiologie, vol II/3, pp 299-340)

Conard RA, The 1954 Bikini Atoll Incident. An Update of the Findings in the Marshallese People. In: Hübner KF and Frey SA (eds) (1980) Medical basis for radiation accident prepared ness, REA/Ts International Conference 1979, OaK Ridge, Tenn. USA. p 55

Deutsche Risikostudie Kernkraftwerke (1979) Eine Untersuchung zu dem durch Störfälle in Kernkraftwerken verursachten Risiko. Eine Studie der Gesellschaft für Reaktorsicherheit, Verlag TÜV, Rheinland

Federov NA, Skurkovic SV, Samsina EV, Chochlova MP (1958) On the pathogenesis and treatment of the radiation-burn syndrome (in Russian). Wiss Allunionskonferenz für kombinierte Strahlenschäden, Moscow

Fetter S, Tsipis K (1981) Nukleare Katastrophen: ein Vergleich. In: Spektrum der Wissenschaft. Sci Am June 1981, p 130

Geiger K (1964) Grundlagen der Militärmedizin. Militärverlag der Deutschen Demokratischen Republik, Berlin

GSF Mensch und Umwelt, ein Magazin der Gesellschaft für Strahlen- und Umweltforschung. GmbH München, Dec 1986 p 35

Hübner KF, Fry SA (eds) (1980) Medical basis for radiation accident preparedness. REAC/TS International Conference 1979, Oak Ridge, Tenn/USA p 6

ICRP Publication 26 (1977) International Commission on Radiological Protection. Recommendations of the International Commission on Radiological Protection. Annals of the ICRP 1: No 3

ICRP Publication 40 (1984) Protection of the public in the event of major radiation accidents: principles for planning. A report of Committee 4 of the International Commission on Radiological Protection. Statement from the 1984 Stockholm Meeting of the ICRP. Pergamon, Oxford

INSAG Report (1986) International Atomic Energy Agency (IAEA) Safety Series No 75-INSAG-1. Summary report on the post-accident review meeting on the Chernobyl accident. Wien

Langendorff H, Messerschmidt O, Melching H-J (1964) Untersuchungen über Kombinationsschäden. Die Bedeutung des zeitlichen Abstandes zwischen Ganzkörperbestrahlung und Hautverletzung für die Überlebensrate von Mäusen. Strahlentherapie 125 (3): 322-340

Linnemann RE (1980) The Three Mile Island incident in 1979. The utility response. In: Hübner KF, Fry SA (eds) The medical basis for radiation accident preparedness. REAC/TS International Conference 1979, Oak Ridge, Tenn/USA p 501

Löster W (1979) Akute Gefährdungsmöglichkeiten der Bevölkerung bei Unfällen in Kernkraftwerken. In: Kirchhoff R, Linde H-J (eds) Reaktorunfälle und nukleare Katastrophen. Perimed, Erlangen, p 15

Medizinische Massnahmen bei Kernkraftwerksunfällen (1986) Leitfaden für ärztl. Berater der Katastrophenschutzleitung, Ärzte in der ambulanten Betreuung. Veröffentlichungen der Strahlenschutzkommission, vol 4. Fischer, Stuttgart

Messerschmidt O (1975) Kombinationsschäden als Folgen nuklearer Explosionen. In: Zenker R, Deuscher F, Schink W (eds) Chirurgie der Gegenwart, vol 4. Unfallchirurgie. Urban und Schwarzenberg, München

Messerschmidt O (1979) Medical procedures in a nuclear disaster, pathogenesis and therapy for nuclear weapons injuries. Thiemig, München

Messerschmidt O (1984) Biologische Folgen von Kernexplosionen, Pathogenese, Klinik, Therapie. Perimed Fachbuch Verlagsgesellschaft, Erlangen

Messerschmidt O, Oehlert W (1968) Untersuchungen über Kombinationsschäden. Histopathologische Untersuchungen an Mäusen nach Ganzkörperbestrahlung in Kombination mit offenen Hautwunden. Strahlentherapie 136: 229

Mitrofanov VB (1959) New findings on changes in the oxidation-reduction process of internal respiration and on some vitamins of the B complex in the event of traumatic shock, alone and in combination with radiation sickness (in Russian). Vestn Chir 7: 112

Poljakov VA (1955) Special features of the course and principles of treatment of combined radiation injuries (in Russian). Erweitertes Plenum der wiss. Räte der Traumatolog. Institute, Moscow

Rahmenempfehlungen für den Katastrophenschutz in der Umgebung kerntechnischer Anlagen. Gemeinsames Ministerialblatt 1977 No 31, p 638

Rahmenempfehlungen für den Katastrophenschutz in der Umgebung kerntechnischer Anlagen. Gemeinsames Ministerialblatt 1981 No 13, p 188

Rasmussen NC (1975) Reactor safety study - an assessment of accident risks in US commercial nuclear power plants, United States Regulatory Commission. WASH 1400 (NUREG 75/014)

Rasmussen NC (1980) A Review of the reactor safety study. In: Hübner KF, Fry SA (eds) The medical basis for radiation accident preparedness. REAC/TS International Conference 1979, Oak Ridge, Tenn/USA p 519

Razgovorov BL (1957) Primary wound suture in the presence of radiation sickness (in Russian). Exp Chir 2: 47

Rubinstejn JaG, Nemkin MN, Sisljannikova LI (1957) Heilung von Frakturen in verschiedenen Stadien der Strahlenkrankheit (in Russian). Voen med Journ 6: 33-37

Storner H (1981) Katastrophenschutz im Bereich kerntechnischer Anlagen in Bayern. In: Messerschmidt O, Betz B, Fliedner TM (eds) Medizinische Erstmassnahmen bei kerntechnischen Unfällen. Strahlenschutz in Forschung und Praxis, vol XXII. Thieme, Stuttgart, p 24

UNSCEAR Report (1977) Sources and effects of ionizing radiation. United Nations Scientific Committee on the Effects of Atomic Radiation

Vorobev AI (1979) In: Handbuch für medizinische Fragen des Strahlenschutzes. Militärverlag der Deutschen Demokratischen Republik, Berlin 1979 (first published in Russian in 1975)

Wald N (1980) The Three Mile Island incident in 1979: the state response. In: Hübner KF, Fry SA (eds) The medical basis for radiation accident preparedness. REAC/TS International Conference 1979, Oak Ridge, Tenn/USA p 491

5 Chemical Radioprotection in Mammals and in Man

Hans Mönig, Otfried Messerschmidt, and Christian Streffer

CONTENTS

5.1 Introduction

Experiments with enzymes in vitro (BARRON et al. 1949) and in microorganisms (LATARJET and EPH-RATI 1948) have shown that certain chemicals are

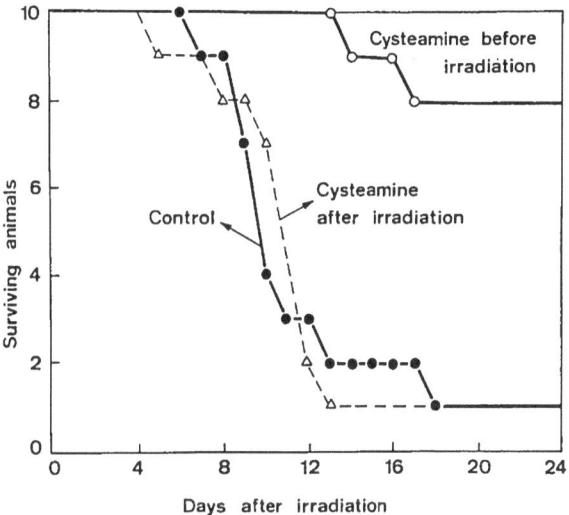

Fig. 5.1. Number of surviving mice as a function of time after X-irradiation (200 kV) with 7 Gy. One group received mercaptoethylamine (cysteamine) before irradiation (150 mg/kg b.w.), one group after, and the control group was irradiated only. (Redrawn from BACQ et al. 1951)

able to moderate the biologic effects of radiation. PATT et al. (1949) were the first to use a radioprotective agent, cysteine, in mammals exposed to whole-body X-irradiation. These authors observed a marked protective effect in rats that had been injected with cysteine just prior to irradiation (PATT et al. 1949, 1950). However, the agent did not confer protection when administered *after* the exposure. This phenomenon is illustrated by the experimental results of BACQ et al. (1951) with the agent mercaptoethylamine ($HS-CH_2-CH_2-NH_2$), or cysteamine, in mice. As Fig. 5.1 shows, a radioprotective effect does not occur when the drug is administered after irradiation. On the other hand, experiments in animals with reduced metabolism have shown that chemical radioprotection is possible even after the exposure has occurred. Thus, the whole-body irradiation of hibernating dormice *(Glis glis)* was found to have no injurious effects on animals injected intraperitoneally with cysteine at the time of awakening, which was 3 weeks postirradiation (KÜNKEL

Table 5.1. Interactions between chemicals and radiation (after PHILLIPS 1977)

Type of interaction	Example
Enhancement	
Synergistic	$2+1\rightarrow4$
Additive	$2+1\rightarrow3$
Subadditive	$2+1\rightarrow2.5$
Interference	$2+1\rightarrow1.5$
Antagonism	$2+1\rightarrow0.5$

and HECKMANN 1958). A comparable effect could not be demonstrated in hibernating ground squirrels *(Citellus tridecemlineatus),* however (SMITH 1959, 1960).

There are various ways in which chemicals and radiation can interact. The basic types of interaction are listed in Table 5.1. For applications in clinical therapeutics, enhancement of a radiation effect is usually referred to as "sensitization," and antagonism of the effect is called "radioprotection." LANGENDORFF (1965) states that the goal of chemical radioprotection is to "influence the radiosensitivity of a higher organism by the use of chemical agents in such a way that the organism either has no demonstrable response to radiation doses that would otherwise be harmful, or manifests a response that is markedly reduced." All measures that are taken to reduce injury after radiation exposure fall within the category of therapeutic procedures (BACQ 1965) and are outside the scope of this text.

5.2 Dose Reduction Factor and Radioprotective Efficacy

5.2.1 Dose Reduction Factor

To compare the effects of different radioprotective agents, we must have quantifiable biologic responses upon which to base the comparisons. The simplest radiation response, and the most extreme, is premature death. If several groups of animals are exposed to suitable radiation doses, the result can be represented as a dose-effect relationship using death or survival as the criterion. When we calculate the percentage of dead individuals for each group and plot these values against dose in a two-dimensional graph, we obtain an S-shaped dose-effect curve. By suitable transformation of the ordinate (probability grid), it is possible to represent many dose-effect relationships as straight lines. The left curve in Fig. 5.2 indicates the result in a proba-

bility grid for an irradiated but untreated population of mice. The percentage values (nonlinear) shown on the right-hand ordinate can be expressed in terms of "probits" (e. g., FINNEY 1964), which are shown on the left ordinate. The method of estimating specific parameters (e. g., the radiation dose for 50% lethality = LD_{50}) is based on a probit analysis of the experimental results (FINNEY 1964).

If we plot the results in a population treated with a protective agent before irradiation, we obtain a line that occupies a higher dose range in the graph (the right curve in Fig. 5.2). In many treatments, the tolerance variations are roughly the same as in the controls. This agreement is reflected in the parallel course of the probit regression lines (see Fig. 5.2). This makes it very easy to compare the results of different experimental series. The relative efficacy, or dose reduction factor (DRF), of the radioprotective chemical can be expressed in this case by the formula

$$DRF = D_2/D_1,$$

where D_1 and D_2 are the radiation doses for the control group and the treated group, respectively, that produce the same lethality. But, because the probit regression lines are not parallel in all cases, it is customary to express the relative effectiveness for a 50% lethality (LD_{50}) by the formula

$$DRF = \frac{LD_{50} \text{ for treated animals}}{LD_{50} \text{ for untreated animals}}.$$

The LD_{50} was a concept introduced in 1927 by TREVAN for measuring the toxicity of drugs (TREVAN 1927).

The DRF for the example in Fig. 5.2 would be calculated as follows:

$$DRF = \frac{8.21 \text{ Gy}}{5.94 \text{ Gy}} = 1.38.$$

In studies of cellular radiation effects, the DRF can also be determined from the slopes of the survival curves, which are exponential (see Figs. 5.4 and 5.5).

Some authors use "dose modification factor" (DMF) instead of "dose reduction factor." The terms are synonymous, but the DMF is most commonly used in reference to quantifiable tissue injury (see also ICRU Report 30, 1979).

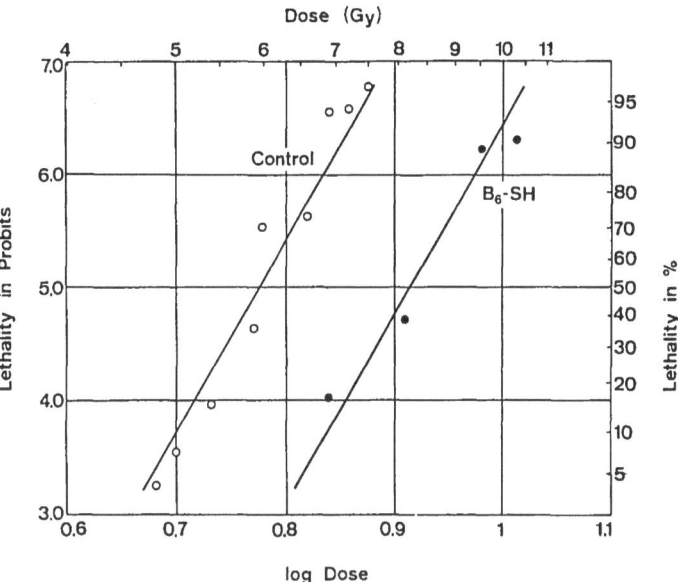

Fig. 5.2. Probit-transformed dose-effect curves for untreated mice (O) and mice treated with 5-mercaptopyridoxine (B_6-SH) (●). Lethality was studied for a period of 30 days after X-irradiation. The animals received 250 mg/kg B_6-SH by intraperitoneal injection 5-10 min before exposure. This gave a dose reduction factor of 1.38 with 95% confidence limits from 1.33 to 1.44. (From data of LANGENDORFF et al. 1958)

5.2.2 Toxicity of Radioprotective Agents

Like all drugs, radioprotective agents are toxic at sufficiently high concentrations. Table 5.2 shows the doses of selected radioprotectors in mg/kg body weight that cause a 50% lethality (LD_{50}). The LD_{50} for a particular agent depends not only on the species, age, and strain of the experimental animal but also on the mode of administration. It has been found, for example, that the toxicity of the radioprotector WR-2721 increases when it is administered in ten divided doses daily for 12 days. For a given mouse strain, the LD_{50} for WR-2721 decreased from 751 mg/kg with single dosing to 281 mg/kg with fractionated dosing (UTLEY et al. 1976b). A correlation appears to exist between the degree of protection afforded by the agent and its toxicity (DISTEFANO 1964). In addition, the radioprotective efficacy of many agents increases with the dosage level (see Fig. 5.3).

YUHAS (1970) showed that the DRF of the thiophosphate compound WR-2721 can be expressed as a power function of the administered dose *(X):* $DRF = K \cdot X^b$, where, for example, $K = 0.4$ and $b = 0.3$ and *X* is stated in mg/kg. Generally, in order to achieve a definite radioprotective effect in survival experiments, it is necessary to administer the drugs in high subtoxic doses (BRAUN et al. 1959). In discussing the practical value of chemical radioprotection, then, we must consider not just the magnitude of the DRF, but also the question of tolerance. Several parameters have been introduced which make it possible to express these relationships and compare them with one another. They are defined as follows (WESTLAND et al. 1968):

Protection Factor (PF)
PF = 1 + fraction of survivors

The PF is between 1 (0% survival) and 2 (100% survival).

Protective index (PI)
$$PI = PF \frac{LD_{50} \text{ (mg/kg) of the agent}}{\text{lowest effective dose (mg/kg)}}$$

The PI is a measure of the protection afforded by the smallest dose for which a radioresistance-enhancing effect can still be observed. Protective agents may be classified as follows:

Protective index	Classification
0-1	0 (ineffective)
2-5	+
6-10	+ +
11-15	+ + +
16-29	+ + + + (very effective)

Other classifications have also been proposed (e. g., BACQ 1965).

Effective dose is the dose of the agent in mg/kg that provides 50% survival at 30 days following radiation exposure. The PF in this case is 1.5.

Table 5.2. Doses of radioprotective drugs that cause 50% lethality (LD_{50}). Different authors use different observation periods for determining postinjection lethality

Agent	Animal species	Mode of administration	LD_{50} (mg/kg)	Authors
Mercaptoethylamine (cysteamine)	Mouse	i. p.	350 (base)	BACQ et al. (1951)
	Mouse	i. p.	270	DOHERTY et al. (1957)
	Mouse	i. p.	343 (323–364)[a]	YUHAS and STORER (1969 a)
	Rat	i. p.	143 ± 9[b]	MAISIN et al. (1964)
	Rat	oral	359 ± 20[b]	MAISIN et al. (1964)
	Rabbit	i. v.	150	BECCARI et al. (1955)
Cystamine (disulfide of cysteamine)	Mouse	i. p.	215	KOCH and SCHWARZE (1957)
	Rat	i. p.	126 ± 4[b]	MAISIN et al. (1964)
	Rat	oral	1035	MAISIN et al. (1964)
Aminoethylisothiuronium (AET)	Mouse	i. p.	690	DOHERTY and BURNETT (1955)
	Mouse	i. p.	600	SHAPIRA et al. (1957)
	Mouse			
	Adults	i. p.	475	ROUSANOV and NOVOSELOVA (1962)
	2–3 wks	i. p.	335	ROUSANOV and NOVOSELOVA (1962)
	Newborns	i. p.	520	ROUSANOV and NOVOSELOVA (1962)
	Mouse	i. p.	296 (284–310)[a]	YUHAS and STORER (1969 a)
	Rat	i. p.	410	BENSON et al. (1961)
	Rat	i. p.	378 ± 23[b]	MELVILLE and LEFFINGWELL (1962)
	Rat	i. p.	288	HANNA and COLCLOUGH (1963)
	Rabbit	i. v.	236	HANNA and COLCLOUGH (1963)
	Dog	i. v.	113	HANNA and COLCLOUGH (1963)
Para-aminopropiophenone (PAPP)	Mouse	i. p.	60 (56–64)[a]	YUHAS and STORER (1969 a)
5-Hydroxytryptamine creatinine sulfate (serotonin)	Mouse	i. p.	124 (117–131)[a]	YUHAS and STORER (1969 a)
S-(2-Aminoethyl) phosphorothioate (WR-638)[c]	Mouse	i. p.	777 (700–864)[a]	YUHAS and STORER (1969 a)
S-2-(2-Aminoethylamino) ethyl phosphorothioic acid (WR-2578)[c]	Mouse (C57BL/6J)[d]	i. p.	1300 (1233–1445)[a]	YUHAS et al. (1973)
S-2-(3-Aminopropylamino) ethyl phosphorothioic acid (WR-2721)[c]	Mouse (A/J)[d]	i. p.	554 (509–582)[a]	YUHAS (1970)
	(DBA/2J)[d]	i. p.	586 (549–626)[a]	YUHAS (1970)
	(C57BL/6J)[d]	i. p.	704 (665–745)[a]	YUHAS (1970)
	(BALB/cJ)[d]	i. p.	784 (744–825)[a]	YUHAS (1970)
S-2-(4-Aminobutylamino) ethyl phosphorothioic acid (WR-2822)[c]	Mouse (C57BL/6J)[d]	i. p.	292 (265–321)[a]	YUHAS et al. (1973)
S-2-(5-Aminopentylamino) ethyl phosphorothioic acid (WR-2823)[c]	Mouse (C57BL/6J)[d]	i. p.	345 (313–381)[a]	YUHAS et al. (1973)
S-2-(6-Aminohexylamino) ethyl phosphorothioic acid (WR-2824)[c]	Mouse (C57BL/6J)[d]	i. p.	343 (304–386)[a]	YUHAS et al. (1973)

Table 5.2. (Continued)

Agent	Animal species	Mode of administration	LD$_{50}$ (mg/kg)	Authors
S-2-[(3-Methylaminopropyl) amino]-ethyl phosphorothioic acid (WR-3689)[c]	Mouse (BALB/c)[d]	i. p.	1120 ± 35[e]	BROWN et al. (1982)[f]
N-(1-Adamantylmethyl)-2-mercaptoacetamidine (WR-109342)[c]	Mouse (C57BL/6J)[d]	i. p.	25 (22–28)[a]	CONNOR and SIGDESTAD (1982)

[a] 95% confidence limits
[b] Standard deviations
[c] Internal designation of Walter Reed Army Institute of Research, Washington, D. C.
[d] Designation of mouse strain
[e] Standard error of the mean (SE)
[f] Present a number of other compounds, some with high protective factors

The *therapeutic index* (TI) is defined by the formula

$$TI = \frac{LD_{50} \text{ (mg/kg) of the agent}}{\text{effective dose (mg/kg)}}$$

As with other drugs, the TI expresses the dose-response relationship, upon which the practical value of the drug is based. The larger the TI, the more successfully an agent can be used. The mode of administration of an agent has a major effect on its TI, as DAVIDSON et al. (1980) showed for various thiophosphate compounds in experimental animals.

Toxicity is a major factor in the clinical testing of radioprotective drugs. Applying the doses tolerated by small animals in mg/kg body weight directly to humans would lead to complications. Clinical trials of new drugs have shown that the results of small animal studies can be applied to humans more accurately in terms of the drug dose per unit body surface area than the dose per unit body weight (PAGET 1965). Studies on the distribution of the ^{35}S-labeled thiophosphate WR-2721 in various tissues of mice, rats, rabbits, and dogs confirm that the dose per unit surface area correlates with tissue distribution better than the dose per unit body weight (WASHBURN et al. 1976). These results led WASHBURN et al. (1976) to predict that a dose of 20 mg of WR-2721 per kg body weight would provide humans with the same amount of protection that a dose of 100 mg/kg provides in mice. Calculating the dose strictly by body weight would result in a 7 g dose of WR-2721 for a 70-kg man, which would certainly be much too toxic. Recently, therefore, the doses of radioprotective drugs for humans have been stated in mg/m^2 body surface area.

5.2.3 Effect of Route of Administration

Radioprotective agents can be administered orally, intravenously, intraperitoneally, intramuscularly, subcutaneously, or topically. The mode of administration has an effect on the absorption, distribution, transformation, storage, and/or excretion of the radioprotector. Efficacy is also affected by the solvent used; the pH of the solution may be a factor in this respect. While the large pH range of 2.6 to 10.0 for cysteamine used with phosphate buffers has not been found to alter its efficacy in mice to a significant degree, the agent S-(3-amino-2-hydroxypropyl) thiophosphate (Å-295; WR-77913) has shown substantial differences at pH values of 7 and 10 (ÅKERFELDT et al. 1967). The authors attribute this to sorption phenomena between the radioprotector and the blood proteins. Another thiophosphate, the diammonium salt of amidothiophosphoric acid (A-331), gave no protective effect at all at pH 3 (ÅKERFELDT et al. 1968).

In the animal studies of WR-2721, distilled water is frequently used as a solvent. But GLOVER et al. (1981) point out that this agent is transformed by acid hydrolysis into the more toxic and relatively inactive disulfide and thiol forms. Based on thin-layer chromatography and radiolabeling studies, the authors conclude that the best solvent for WR-2721 is a buffer solution with a pH of 7.20–7.25, which avoids undesired transformations. In experiments with mice, it was found that the use of a phosphate buffer at pH 7.4 did not alter the radioprotective effect of WR-2721 compared with distilled water solvent, but that the agent was less toxic when administered in the phosphate buffer (MÖNIG et al. 1986).

In the great majority of experiments on rodents, the radioprotector is given by intraperitoneal injection (see Table 5.2). Only a few studies have systematically compared the influence of different modes of administration on radioprotective effect. KUNA (1983) and KUNA et al. (1983a) investigated the effects of WR-2721 and mexamine administered by intraperitoneal, intramuscular, and subcutaneous injection. In the case of WR-2721, intramuscular injection proved most effective in mice based on the time interval between injection and irradiation (KUNA 1983). The intramuscular injection of mexamine also yielded the best results in rats. In this case, there was no apparent change in radioprotective effect when the drug was administered 15, 30, or 60 min before irradiation (KUNA et al. 1983a).

A number of studies have been done on the oral use of radioprotective agents. This route is of special importance for practical reasons, and most radioprotectors are active when given orally. Generally the values are higher for oral than for parenteral administration (see Table 5.2), but the protective effect is decreased.

As early as 1950, PATT et al. reported that cysteine is effective when administered orally in rats, but not glutathione. Cysteamine also protects rats on oral dosing of 450–600 mg/kg, with onset of effect occurring at 15 min and increasing by 30 min (BACQ 1956). Oral cysteamine in mice raised the $LD_{50/30}$ from 4.16 Gy in untreated controls to 5.97 Gy in the treated animals (DRF = 1.44) (MEWISSEN 1957). Aminoethylisothiuronium (AET) used orally in rats produces a significant increase in survival rates (LANGENDORFF and KOCH 1956). The same drug given orally in mice also gives significant protection when administered 2, 3, or 5 h before irradiation (DACQUISTO and BLACKBURN 1961).

AIRAPETYAN and ZHEREBCHENKO (1964) found that the sodium salt of 2-aminoethylthiophosphoric acid (WR-638) is a more potent radioprotector than cystamine when administered orally in mice. When injected, the compound does not incite a local inflammatory reaction, as can occur with cystamine. Experiments with cystaphos (WR-638) given intragastrically to rhesus monkeys showed that doses of 215 to 430 mg/kg are tolerated without adverse reactions (ZNAMENSKII et al. 1975). Higher doses of cystaphos lead to vomiting and depressed motor activity. We can estimate a DRF of approximately 1.3 based on the postirradiation survival rates reported by the authors.

TABACHNIK et al. (1982) describe the transformation of the thiophosphate WR-2721 after oral administration in rabbits and in man. One patient with cystic fibrosis received 5 mg/kg of WR-2721 four times daily in gelatine capsules for a period of 1 month. In the acid medium of the stomach, the WR-2721 is converted to the free thiol form N-2-mercaptoethyl-1,3-diaminopropane (MDP). MDP also has radioprotective properties (TABACHNIK et al. 1982).

The radioprotector N-(1-adamantylmethyl)-2-mercaptoacetamidine (WR-109342) appears to be highly effective with oral use. A DRF of 1.7 was achieved in mice after 75 mg/kg of the drug was given orally 30 min before irradiation (VOS 1980).

5.2.4 Effect of Time

The timing of administering of a radioprotective agent has a major bearing on its effectiveness. The protective effect is also influenced by the duration of radiation exposure. For most radioprotectors, there is an optimum period during which a maximal effect is obtained. Generally, this period is relatively brief. If irradiation occurs after the optimum period, the radioprotective effect is diminished or absent.

Table 5.3 shows the changes in the radioprotective efficacy of WR-2721 injected in mice between 15 and 120 min before exposure to X-rays. We see that the survival rate increases from 15 to 30 min and then decreases with time, showing an abrupt fall between 1 and 2 h.

In prolonged exposure tests in rats treated orally with AET, MELVILLE and LEFFINGWELL (1962) found a drastic reduction of efficacy when the ^{60}Co γ-radiation dose rate was decreased from 10.4 Gy/min to 4.67 Gy/min and to 0.93 Gy/min. Other studies showed that the daily intraperitoneal administration of cysteamine, serotonin, AET, or para-aminopropiophenone (PAPP) in chronically

Table 5.3. Thirty-day survival rates of NMRI mice given a single whole-body X-ray exposure versus time of intraperitoneal injection of 12.5 mg WR-2721 per animal before exposure (13.2 Gy at rate of 1 Gy/min). Only the 18% survival rate is statistically significant ($P < 0.001$) compared with the other values. The survival rate for unprotected animals is 0%. (After LANGENDORFF et al. 1974)

Time of injection of WR-2721	Number of irrad. animals	Number of surv. animals	Survival rate in %
15 min	100	54	54.0
30 min	79	49	62.0
60 min	49	30	61.2
120 min	50	9	18.0

irradiated mice did not prolong mean survival compared with untreated controls (DOULL et al. 1962). The same authors found no survival increase in mice chronically exposed to γ-radiation at 0.88 Gy/d even when cysteamine was added to the drinking water (1, 2, or 4 mg/ml). Similarly, neither cysteamine nor cystamine was found to prolong survival in mice chronically irradiated by the injection of 18.5 MBq or 37 MBq of radiogold (^{198}Au) (HERVÉ 1957). However, if the cysteamine injection was repeated within 3 days after administration of the ^{198}Au, a slight protective effect was achieved. Cysteamine also has proved less effective with fractionated irradiation than with single-dose irradiation (LANGENDORFF and CATSCH 1956).

Because of the short duration of action of most radioprotective chemicals, researchers have attempted to find agents that confer longer-lasting protection or to modify existing agents in ways that prolong their efficacy. So far, however, few cases are known in which a radioprotector can remain active for a prolonged period without substantial loss of effect. One such case involves the oral use of homocysteine thiolactone, whose protective effect in mice persists for 2–3 h (BRAUN et al. 1959). KOCH and SCHWARZE (1957) attribute this to a delayed release of the SH groups in the thiolactone.

Certain pharmacodynamically and hormonally active compounds are also known to confer long-lasting radioprotection. Examples are reserpine and the synthetic estrogen chlorotrianisene (Tace). Reserpine, the rauwolfia alkaloid, is effective in mice only when administered 12–14 h before irradiation (LANGENDORFF et al. 1957), in which case it gives DRFs of 1.2–1.4. Reserpine may derive its effect from the release of serotonin (VAN DEN BRENK and HASS 1961). The synthetic estrogen Tace has demonstrated a very long duration of action in experimental animals (FLEMMING and LANGENDORFF 1965). Male mice that received 1 mg Tace 30 days before exposure to 6.4 Gy showed a survival rate of 57%, as compared with 12% survival in controls. The effect was maximal (82% survival) when the drug was injected 10 days before exposure. This protracted effect may relate to the fact that Tace is initially stored in the body fat following injection, from which it is slowly released into the body.

Besides the agents already described, a number of other substances have been tested for delayed radioprotective effect. TIKHOMIROVA and ROGOZKIN (1979) administered adenosine diphosphate (ADP) to mice and guinea pigs 10–20 min before and also 10–20 min after prolonged exposure to 0.01 Gy/min with total doses of 16–17 Gy (mice) and

9–10 Gy (guinea pigs). The 30-day survival rate rose by 40% in the mice and by 25% in the guinea pigs. NIKOLOV et al. (1986) tested the protective effect of the Bulgarian drug Adeturone® and adenosine triphosphate in monkeys (Macaca mulatta) by administering the agents before and after long-term whole-body irradiation with 8.3 Gy (dose rate 9.2 mGy/min). Adeturone® is a salt of adenosine-5-triphosphate (ATP) and of S-(2-aminoethyl)-isothiourea. The radiation dose represented an LD$_{95/45}$ in unprotected animals, whereas 50% of the protected animals survived. Some symptoms of radiation sickness are moderated by Adeturone®. However, the authors obtained similar results with the injection of ATP alone. Tilorone, an inductor of interferon, confers moderate protection in mice when injected 6 or 18 h before irradiation (CHERTKOV et al. 1979a). The same authors obtained a protective effect with hydroxyurea injected in mice 1 or 2 days before long-term exposure to 0.01 Gy/min (CHERTKOV et al. 1979b). Positive aspects of hydroxyurea include its early onset of action and the high rate of repopulation of bone marrow stem cells. When mice were injected with ginseng extract 1 or 2 days prior to irradiation, 75% of the animals survived a dose of 7.2 Gy, while all the controls died within 30 days (TAKEDA et al. 1981). The extract did not afford radioprotection when given 3 or 5 days before the exposure.

A number of authors (SMITH et al. 1958; AINSWORTH and FORBES 1961; FLEMMING and FLEMMING 1966; LANGENDORFF et al. 1971b; BEHLING and NOWOTNY 1978) have achieved long-term radioprotective effects with bacterial lipopolysaccharides (endotoxins). Lipopolysaccharides from gram-negative bacteria (E. coli, S. minnesota) gave a DRF of approximately 1.2 in mice when administered 24 h before irradiation. By 48 h the radioprotective effect had dwindled. Studies with various structural components of lipopolysaccharides have shown that lipoid A is the effective component (LANGENDORFF et al. 1971c). BEHLING (1983) has published a comprehensive review of the radioprotective effects of bacterial endotoxins.

STREFFER and FLÜGEL (1973) were able to prolong the effect of serotonin by the prior injection of monoaminoxidase inhibitors. The authors showed that the maximum protective effect of this agent is not diminished, and that its duration of action can be substantially prolonged. Serotonin remained effective in mice for 40 min when administered alone, but its effect persisted for at least 1 h when monoaminoxidase inhibitors were injected first. ROMAN et al. (1982) were able to prolong the effect of cysteamine by encasing the drug in liposomes made of lecithin. This preparation provided a DRF of 1.30 when given orally to mice 3.5 h before irradiation. BENITA et al. (1984) reported on the prolonged release of cysteamine and cysteine from insoluble matrix tablets with different ethylcellulose/stearic acid ratios. The release rate of the aminothiol compounds could be controlled by varying the ethylcellulose concentration, and drug release was sustained for up to 3 h in kinetic experiments. So far these tablets have not been used in animal trials. The oral administration of liposome-entrapped cysteamine has been shown to provide a delayed release in rats (JASKIEROWICZ et al. 1985). This mode of administration also increases the concentration of exogenous sulfur compounds in the plasma, liver, and spleen compared with cysteamine administered in free form.

RINGSDORF (1967) discusses in some detail the possibility of using synthetic macromolecular compounds to achieve long-term radioprotection through a depot effect. This would result partly from a gradual release of fragments having a low molecular weight. HEISLER (1969) reports on the synthesis and investigation of N- and S-containing polymers as radioprotective compounds. A terpolymer composed of vinyl pyrrolidone (66 wt.%), acrylic acid (3 wt.%), and acryloylthiazolidone (31 wt.%) has proved especially promising. This agent remained active for periods of 6 h to 8 days while maintaining a relatively high DRF of 1.4 (HEISLER 1969; RINGSDORF 1978). BARNES et al. (1977) achieved radioprotection for 24 and 48 h in mice treated with N,N-diethyl-S-vinyldithiocarbamate and 1-vinyl-2-pyrrolidone copolymers. Following γ-irradiation with 9 or 9.5 Gy, which represented an $LD_{100/30}$ for unprotected mice, approximately half of the animals treated with the copolymer survived.

ROSS and PEEKE (1986) extended the effect of dextran sulfate (DS) by using the high-molecular-weight version D_{500} (500000 daltons). When DS_{500} is given to mice 1–3 days before irradiation, recovery of the hemopoietic system is markedly enhanced. DS_{500} is not effective when used postirradiation.

5.3 Chemical Radioprotection in Mammals

5.3.1 Acute Radiation Effects After Whole-Body Exposure

A single, brief whole-body irradiation can produce any of three symptom complexes in mammals, depending on the dose received (RAJEWSKY 1956). Each is characterized by a particular type of organ damage and length of survival. After exposure to very high radiation levels in excess of about 100 Gy, "central nervous system death" follows within a few minutes to 1–2 days, accompanied by severe CNS and cardiovascular manifestations such as convulsions, impaired consciousness, and circulatory collapse. With exposure to lower levels in the range 10–100 Gy, animal deaths are associated with severe gastrointestinal symptoms that include persistent diarrhea, water and electrolyte losses, and infections. This "gastrointestinal death" occurs within a few days, in contrast to "hematopoietic death," which takes several weeks to occur and is marked by symptoms of lymphatic and bone marrow disease. The associated leukopenia and thrombocytopenia give rise to lethal bacteremias and coagulation disturbances. The gastrointestinal mode of radiation death usually occurs within 5 days after exposure in mice, and hematopoietic death within 30 days. YUHAS and STORER (1969a) tested the ability of various drugs to protect against the three radiation-induced modes of death. Their results are summarized in Table 5.4. The authors evaluated gastrointestinal deaths in terms of the DRFs for 7-day mortality (DRF_7). As Table 5.4 indicates, the efficacy of six of the seven radioprotectors decreases with increasing dose. Only one agent, AET, affords greater protection against gastrointestinal death than it does against hematopoietic death. A decline in the effect of the drug WR-2721 on gastrointestinal death has also been demonstrated for observation periods of 5 days (LANGENDORFF et al. 1974) and 6 days (SIGDESTAD et al. 1975a). WR-77913, a thiophosphate of mercaptopropylamine, gave approximately equal protection for the bone marrow (1.91) and gastrointestinal tract (1.95) (MENDIONDO et al. 1982).

No agent, other than PAPP, shows an ability to protect against CNS death (see Table 5.4). This finding is supplemented by the results of WRIGHT and SHEWELL (1965) with mercaptoethylamine (cysteamine). In mice that received approximately 1000 Gy of radiation to the head, the authors found no difference in the mean survival time of about

Table 5.4. Protective effect of various radioprotective agents against hematopoietic death (DRF$_{30}$), gastrointestinal death (DRF$_7$), and CNS death (DRF$_0$) in mice. The agents were given by intraperitoneal injection 15 min before X-irradiation. (YUHAS and STORER 1969 a)

Agent	Code designation	Test dose (mg/kg)	DRF$_{30}$	DRF$_7$	DRF$_0$[a]
Para-aminopropiophenone	PAPP	40	1.7 ± 0.02	1.4 ± 0.04	1.28 ± 0.024
Aminoethylisothiuronium $\cdot$ Br $\cdot$ HBr	AET	200	1.4 ± 0.03	1.6 ± 0.04	1.02 ± 0.030
β-Mercaptoethylamine hydrochloride	MEA	200	1.6 ± 0.03	1.4 ± 0.03	1.00 ± 0.038
5-Hydroxytryptamine creatinine sulfate	Serotonin	100	1.5 ± 0.02	1.2 ± 0.04	1.07 ± 0.038
Sodium hydrogen S-(2-aminoethyl) phosphorothioic acid	WR-638[b]	500	2.0 ± 0.01	1.6 ± 0.03	0.98 ± 0.023
S-2-(3-Aminopropylamino)ethyl phosphorothioic acid	WR-2721[b]	500	2.7 ± 0.02	1.8 ± 0.04	0.49 ± 0.168
		250	2.3 ± 0.02	1.6 ± 0.03	1.02 ± 0.049
S-3-(3-Aminopropylamino)propyl phosphorothioic acid	WR-44923[b]	200	1.9 ± 0.02	1.4 ± 0.03	0.96 ± 0.030

[a] CNS death was determined by giving repeated exposures to 100 Gy at 5-min intervals until death occurred
[b] Internal designation of the Walter Reed Army Institute of Research, Washington, D. C.

20 min compared with untreated controls. The thiophosphate WR-2721 appears to sensitize animals to CNS radiation death when given in a dose of 500 mg/kg body weight (see Table 5.4). This may relate to a toxic drug reaction, as we see from the result for a smaller dose of the same agent. These results are consistent with those of experiments designed to test the effect of WR-2721 on motor impairment in rats induced by high radiation doses of 130 Gy (BOGO et al. 1985). A decline in the performance of conditioned motor sequences was observed in nonirradiated rats that had been treated with WR-2721. Moreover, this agent tended to produce a sensitizing rather than a protective effect in this experiment.

5.3.2 Studies in Organs and Tissues

As we saw in the previous section dealing with acute radiation effects, different tissues can manifest widely different responses to radioprotective agents. The thiophosphate WR-2721 provides low protection factors of 1.2-1.5 in the lung and kidneys (PHILLIPS et al. 1973) but can give factors as high as 3 for the bone marrow (YUHAS and STORER 1969a). Figure 5.3 shows the relationship between the administered dose of WR-2721 and the increase in the radiation resistance of the bone marrow and skin. The distribution of WR-2721 varies markedly from one tissue to the next (TANAKA and SUGAHARA 1980; RASEY et al. 1984). The agents S-2-[(3-methyl-aminopropyl)-amino]ethyl phosphorothioate (WR-3689, NSC-327729) and S-(3-amino-2-hydroxypropyl) phosphorothioate (WR-77913, NSC-318809) have exhibited distribution patterns in normal mouse tissues similar to WR-2721 (RASEY et al. 1986). However, the degree of radiation protection

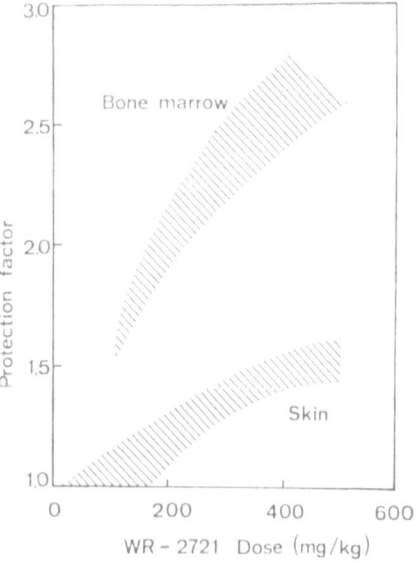

Fig. 5.3. Range of protective factors for skin and bone marrow as a function of the dose of WR-2721 according to published data (STEWART et al. 1983)

does not correlate directly with the local tissue concentration of the protective agent (TANAKA and SUGAHARA 1980) (Fig. 5.4). Different oxygen concentrations in the tissues may be responsible for these differences (STEWART et al. 1983; TRAVIS 1984). To be effective, the compound WR-2721 must be dephosphorylated. The phosphate group is split off by an enzymatic mechanism, and the rate of this reaction varies greatly from one tissue to the next (HARRIS and PHILLIPS 1971). The dephosphorylation process can be monitored in vivo by means of ^{31}P NMR spectroscopy (KNIZNER et al. 1986). SWYNNERTON et al. (1986) used high pressure liquid chromatography to study the behavior of WR-2721 (Ethiofos) and its metabolic products in the blood

Concentration of WR-2721

in μg per g or ml of tissue

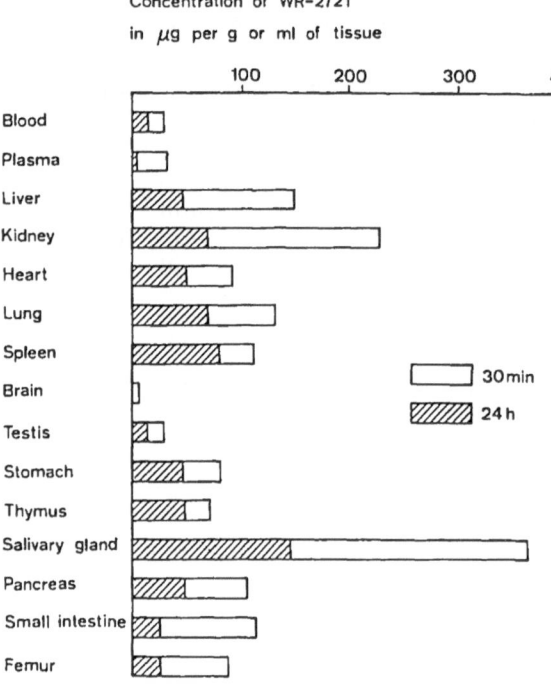

Fig. 5.4. Distribution of WR-2721 in various rat tissues 30 min and 24 h after intravenous injection of 100 mg/kg b. w. of the ^{14}C-labeled thiophosphate. Data on drug concentration pertain to the ^{14}C activity in the tissues. (TANAKA and SUGUHARA 1980)

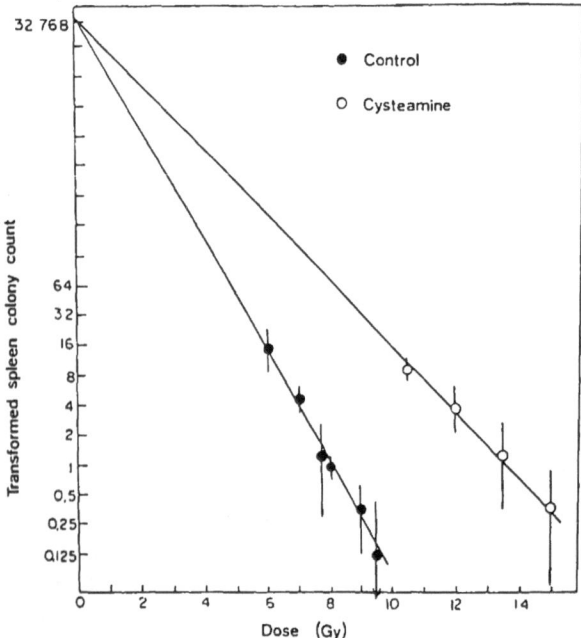

Fig. 5.5. Estimation of the dose reduction factor (DRF = 1.6) for cysteamine using the spleen colony technique in γ-irradiated mice. A modified logarithmic transformation was applied to the spleen colony counts. (After SMITH et al. 1966)

Table 5.5 a. Dose reduction factors as determined by 30-day survival and spleen colony reduction in γ-irradiated mice pretreated with various radioprotectors. The compounds are designated according to the nomenclature of the Walter Reed Army Institute of Research, Washington, D.C. (see Table 5.5 b). (After KINNAMON et al. 1980)

Agent	Dose reduction factor		Rank order	
	Sur-vival	Colo-nies	Sur-vival	Colo-nies
WR-2721	2.17	2.22	1	2
WR-3689	1.81	1.77	2	4
WR-2823	1.66	2.87	3	1
WR-638	1.60	1.75	4	5
WR-2529	1.59	1.56	5	6
WR-347 (cysteamine)	1.46	2.11	6	3
WR-151331	1.30	1.46	7	7
WR-108503	1.05	1.24	8	8

plasma of rhesus monkeys *(Macaca mulatta)* and beagle dogs. They observed a rapid clearance of the compound.

5.3.2.1 Hematopoietic Organs and Blood

The technique of spleen colony counts (colony-forming units, CFU) described by TILL and McCULLOCH (1961) is excellent for studying radiation damage to bone marrow stem cells. Figure 5.5 shows the results of CFU experiments on endogenous spleen colonies from gamma-irradiated mice. The slopes of both curves indicate a DRF of 1.6 for cysteamine administered before irradiation (SMITH et al. 1966). KINNAMON et al. (1980) made a comparison between the survival rate at 30 days and the spleen colony method in assessing the efficacy of eight radioprotective compounds (see Table 5.5). A comparison of the results indicates that both methods yield similar assessments. On that basis, KINNAMON et al. (1980) state that the CFU technique offers a means of testing the ability of compounds to protect irradiated bone marrow. However, a corresponding study of various lipopolysaccharides (endotoxins) from gram-negative bacteria showed no correlation between survival rate and spleen colony count (STEVENSON et al. 1981). SNYDER et al. (1986) obtained similar results with various nontoxic lipid A components of *S. minnesota*.

HARRIS and PHILLIPS (1971) used the CFU method to show that the thiophosphates WR-638 and WR-2721 confer only limited protection to hypoxic cells (Table 5.6). As a consequence, the oxygen enhancement ratio falls to 1.1 with WR-2721 and to 1.3 with WR-638. This finding is relevant to the radiotherapeutic application of radioprotective drugs,

Table 5.5 b. Structural formulas of the agents listed in Table 5.5 a. (After KINNAMON et al. 1980)

Designation	Structural formula
1. WR 2721	$H_2N(CH_2)_3NHCH_2CH_2SPO_3H_2 \cdot 1{,}28 \cdot H_2O$
2. WR 3689	$CH_3NH(CH_2)_3NHCH_2CH_2SPO_3H_2 \cdot H_2O$
3. WR 2823	$H_2N(CH_2)_5NHCH_2CH_2SPO_3H_2 \cdot 2\,H_2O$
4. WR 638	$NH_2CH_2CH_2SPO_3HNa$
5. WR 347	$NH_2CH_2CH_2SH \cdot HCl$

because solid tumors contain a significant number of anoxic cells.

Numerous studies have been done on the effect of radioprotectors on cells of the bone marrow and peripheral blood. BACQ et al. (1953 b) observed that cysteamine does not prevent leukopenia in mouse blood following irradiation, but that it does hasten regeneration of the leukocytes. Similarly, the intraperitoneal injection or oral use of AET in mice was

Table 5.6. D_0 values, dose modification factors (DMF), and oxygen enhancement ratios (OER) for bone marrow stem cells (CFUs) of mice exposed to whole-body 300 kV X-irradiation. The thiophosphates WR-638 and WR-2721 were injected intraperitoneally 20 min before irradiation (600 mg/kg). Hypoxia was produced by having the animals breath air with a low oxygen content. (HARRIS and PHILLIPS 1971)

Treatment	D_0 (Gy)	DMF	OER
Air	0.83		
			2.2
Hypoxia	1.83		
Air + WR-2721	2.53	3.0	
			1.1
Hypoxia + WR-2721	2.90	1.6	
Air + WR-638	1.91	2.3	
			1.3
Hypoxia + WR-638	2.35	1.2	

found to promote rapid recovery of bone marrow cell counts (DOHERTY 1960). A fall in the reticulocyte count of the peripheral blood in mice that received 0.75 Gy whole-body irradiation could be largely prevented by cysteamine and also by serotonin (ELTGEN et al. 1961). However, the authors did not observe a protective effect in analogous experiments on rats. KOCH (1965) used the radioiron test (^{59}Fe) in mice exposed to nonlethal radiation doses to show that a number of radioprotectors (cysteamine, histamine, serotonin, pyridoxal-5-phosphate) inhibit the depression of young red cells in the peripheral blood. The thiophosphate WR-2721, on the other hand, is active only above radiation doses of 0.5 Gy (MÖNIG et al. 1975). Mercaptopropionylglycine (MPG) given to mice exposed to 2.5 Gy or 5.0 Gy had no effect at all on the initial (approx. 24 h) fall of red blood cell count in the peripheral blood but did afford significant protection at longer intervals after irradiation (UMA DEVI and KUMAR 1981). No effect was apparent after exposure to 10 Gy. PANT and GHOSE (1981) made similar observations in mice irradiated with 7.6 Gy. However, AET was able to prevent completely the fall in red blood count in peripheral blood. MPG, on the other hand, reduces the early depletion or pronormoblasts and normoblasts from the bone marrow of irradiated mice and brings about an earlier and more rapid recovery of these cells (Saini and Uma Devi 1980).

5.3.2.2 Lymphatic System and Immune Responses

Some results of radioprotection studies of the lymphatic system are controversial. In general, it may be said that the regeneration of irradiated lymphatic tissues and organs is accelerated by sulfur-containing compounds such as cysteamine and AET (Bacq 1965). For example, the spleen weight of lethally irradiated mice began to recover on the 8th day in animals that were treated with mercaptoethylguanidine (MEG, a neutralization product of AET) prior to irradiation (Urso et al. 1958). Corresponding measurements of thymus weight indicated recovery after only 5 days (Urso et al. 1958). Kobayashi et al. (1965) observed a transient overshoot in the mitotic rate of thymus cells in (nonirradiated) mice for up to 12 h after treatment with cysteamine.

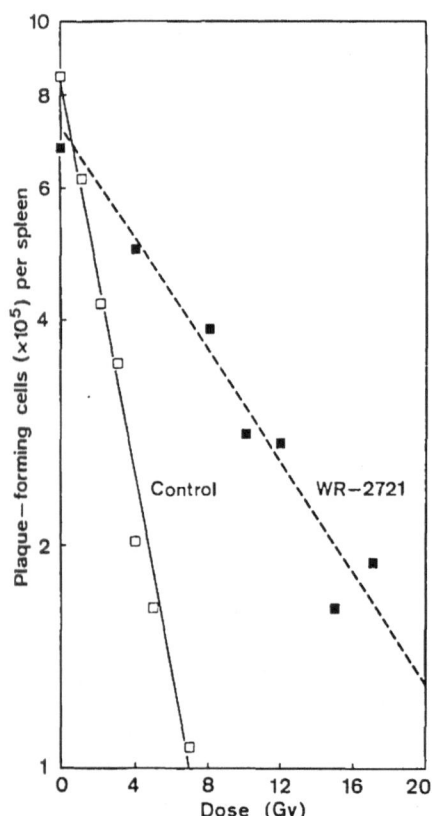

Fig. 5.6. Effect of WR-2721 on plaque-forming cells from the spleen of mice that were immunized with sheep red cells, as a function of radiation dose. The slopes of the curves indicate a DRF of 3.4. (After Yuhas 1972a)

On the other hand, Smith et al. (1965) found 20% lower lymphocyte counts for up to 48 h in mice treated with cysteamine than in untreated controls.

Suppression of immune responses following exposure to ionizing rays was first described in 1908 by Benjamin and Sluka and has since been confirmed by many authors. Doherty and Congdon (1959) noted prolonged survival times for skin allografts taken from mice treated with AET before irradiation. Similar experiments with the thiophosphate WR-638 did not confirm this result (Saltzstein et al. 1970). Cudkowicz (1962) likewise found no protective effect from AET or its derivative, aminopropylmethylisothiourea, for immunocompetent cells. On the other hand, Yuhas (1972a) obtained the high DRF of 3.4 with WR-2721 in studies of plaque formation in the mouse spleen after the injection of red blood cells from sheep (Fig. 5.6). Harris and Meneses (1978) found that the same agent gave a DRF of 1.6 for the recovery of viable lymphocytes. Cystamine showed an ability to reduce the change in the phago cytotic activity of mouse granulocytes caused by whole-body irradiation with doses up to 1 Gy (Hovestadt et al. 1983). Flemming (1977) reported on the protection of the reticuloendothelial system by cysteamine, serotonin, and lipopolysaccharides (endotoxin).

5.3.2.3 Skin

Numerous studies have been done on skin reactions following local irradiation of the limbs in mammals. Observable reactions include erythema, edema, hair loss, moist desquamation, and crusting. To obtain quantifiable results on the efficacy of certain radioprotectors, a grading system has been devised for assessing the severity of the skin damage. The dose-effect curves of the skin reactions of control mice and mice protected with WR-2721 are shown in Fig. 5.7 (Utley et al. 1976b). While this agent is highly effective in giving bone marrow protection, its dose reduction factors for the skin are not very high (see Fig. 5.3). Travis et al. (1982) obtained DRFs of 1.7 to 2.1 for skin desquamation in mice when this drug was administered 30–60 min before irradiation. Intravenous injection proved more effective than intraperitoneal injection.

As early as 1961, Bacq et al. observed that the concentration of ^{35}S-labeled cysteamine and cystamine in the skin did not correlate with protection against hair loss. While cystamine led to a high radiosulfur content in the skin with little protection against epilation, cysteamine showed a low skin

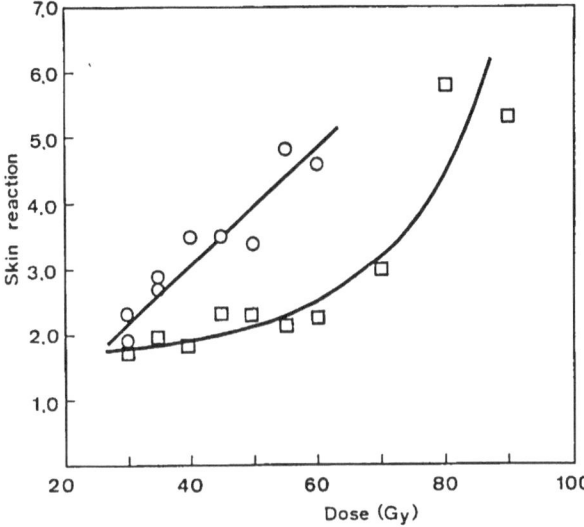

Fig. 5.7. Skin reaction in mice following a single local exposure of the foot and lower leg to ^{137}Cs γ-radiation. The dose-effect curves are shown for untreated animals (O) and for mice treated with WR-2721 before irradiation (□). The DMF equals 1.5-1.7 for doses up to 70 Gy (after Utley et al. 1976b). Skin reactions (erythema, edema, epilation, moist desquamation, crusting) were graded according to the system of Fowler et al. (1965) and Phillips et al. (1973)

concentration but was effective in preventing hair loss. Hofmann (1955) reported on an antierythema effect of cysteine on X-irradiated rat tails when applied in the form of a skin ointment. However, a subcutaneously injected cysteine solution was more effective in these studies. Cysteamine also gave gross and microscopic protection in rabbits when injected subcutaneously in 0.5% solution directly at the site of the contact irradiation (Bianchi and Gasparini 1955). The drug was not effective when injected intravenously.

While the foregoing results pertained to relatively early radiation-induced skin damage (up to about 40 days p. r.), 5-thio-D-glucose was able to afford skin protection in mice for 60-90 days after the irradiation (Schuman et al. 1983). The DRF in this case was 1.2 ± 0.1, as compared with 1.3 ± 0.1 for early skin reactions. The thioglucose is of interest by virtue of its cytotoxic and radiosensitizing action in hypoxic tumor cells (Song et al. 1977). A DRF of 1.5 was found for leg contractures in mice 182 days after irradiation in animals that had been injected with WR-2721 (Hunter and Milas 1983).

5.3.2.4 Digestive Tract

Besides protection against radiation-induced hematopoietic changes, a great deal of interest has focused on chemoprotection against radiation damage to the gastrointestinal tract. Experiments using various methods and drug combinations have demonstrated the ability of chemical agents to protect the digestive tract of radiation-exposed animals.

Sullivan et al. (1964) reported on the effect of cysteine as determined by histopathologic studies of the rat small and large intestine following fractionated whole-body X-irradiation. The authors found less intestinal damage in the treated animals than in the untreated controls. The cell-protecting effect of cysteine was also demonstrated by a study of the mitotic rate in the Lieberkühn crypts of the rat ileum (Beliles et al. 1959). Schwartz and Shapiro (1961) found that the intestinal absorption of ^{131}I-labeled oleic acid was improved by the pretreatment of X-irradiated mice with MEG. Gastric retention also was largely normalized by MEG. Orally administered MEG was somewhat more effective than parenteral doses. Maisin and Lambiet-Collier (1967) observed that cell renewal in the duodenum of X-irradiated mice was promoted more effectively by a mixture of AET, glutathione, and serotonin than by the drugs individually. This treatment reduced radiation damage to the stem cells of the small intestine but did not affect the delayed migration of crypt cells.

Dose-effect curves for crypt cells of the mouse small intestinal epithelium are shown in Fig. 5.8 (Utley et al. 1976b). Treatment with WR-2721 gave a DMF of 1.6 for 10% surviving crypt cells. This factor corresponds to the value obtained for observations of gastrointestinal radiation death (Langendorff et al. 1974; Sigdestad et al. 1975a). Ito et al. (1986) also report achieving a DRF of 1.6 for the mouse gastrointestinal tract by treatment with WR-2721. The small intestine and colon were evaluated by the microcolony method, and the esophagus by determining the LD_{50} in the 28- to 42-day range. Utley et al. (1978) calculated a DRF of 2.1 in their studies of "oral radiation syndrome" by determining the $LD_{50/8-10}$ following head irradiation in mice previously injected intraperitoneally with WR-2721.

Experiments by Sigdestad et al. (1975a) on the effect of WR-2721 on the survival of jejunal crypt cells in X-irradiated mice yielded shoulder curves, with "quasithreshold doses" D_q (Alper et al. 1962) equaling 6.49 Gy for the untreated mice and 14.06 Gy for the treated mice. A DMF of approxi-

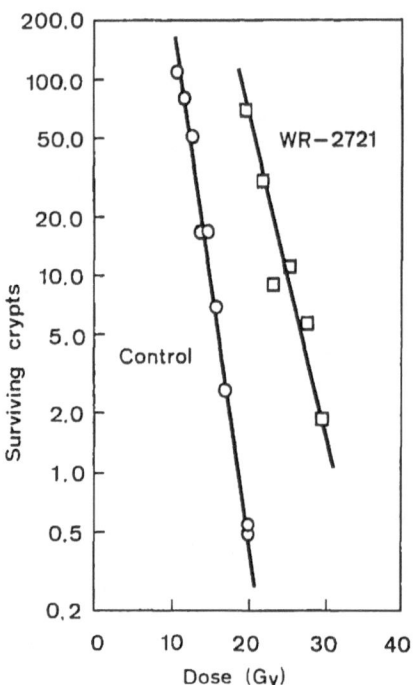

Fig. 5.8. Effect of radiation on the number of surviving clonogenic crypt cells from the small bowel epithelium of mice treated with WR-2721 (□) or untreated (○) before irradiation (after UTLEY et al. 1976b). The animals received whole-body irradiation from a ^{137}Cs γ-source, and the bowel was removed 3.5 days later and examined by the method of WITHERS and ELKIND (1970). The number of crypt cells per cross-section was 160 in unirradiated animals. The drug gave a DMF of 1.6 for the radiation-induced reduction of the surviving crypt cells to 10%

in mitochondria. In a study of the mouse jejunum, UMA DEVI et al. (1979) were able to reduce radiation damage with MPG, as determined from the total cell population, the number of mitoses, and the number of pyknotic nuclei and necrotic cells. The results showed that MPG protects the intestinal crypts by causing an early recovery from mitotic inhibition followed by an early regeneration of the crypt epithelium.

5.3.2.5 Respiratory Tract

Local irradiation of the rat thorax produces a "pulmonary radiation syndrome" culminating 30–90 days later in death from serosanginous pneumonia (DUNJIC et al. 1957). Intraperitoneal injection of cysteamine before irradiation gave a DRF of 1.35 for the observed $LD_{50/90}$. AET and serotonin also were able to prolong survival in tumor-bearing mice that received 75 Gy of local radiation to the thorax (COHEN and COHEN 1962). A long-term survival study of mice treated with a mixture of AET + glutathione + serotonin and then exposed to 15–20 Gy of whole-body X-irradiation showed that most of the mice died within 6 months after exposure (MAISIN and MATTELIN 1967). Histological examination of the tissues of these mice showed that they died of a pulmonary syndrome involving edema and congestion (stage 1 of lobar pneumonia). It should be noted that a study of this radiation syndrome would not have been possible without use of the drug combination, because the radiation-induced gastrointestinal syndrome would have been lethal at a much earlier stage. Most mice receiving 18 Gy of local thoracic irradiation likewise died within 6 monts from radiation pneumonia (MAISIN 1969). Even an expanded combination of radioprotectors (glutathione + AET + serotonin + cysteine + cysteamine) afforded only a weak protective effect.

An interesting attempt to achieve selective radioprotection of the lungs by inhalation of N-acetylcysteine (NAC) was undertaken in mice that received radiation to the upper body half (TARBELL et al. 1986). The LD_{50} values for 30 and 180 days' survival did not differ significantly from those in control mice not treated with NAC. Only the group exposed to 14 Gy showed a significant delay in mortality relative to the controls.

No radioprotection was achieved with vitamin E (α-tocopherol) in rats that received 15 or 20 Gy of whole thoracic irradiation (ROSTOCK et al. 1980). Vitamin E is a known antioxidant and is believed to activate free radical scavenger enzymes such as glu-

mately 1.7 was estimated for 50% surviving crypt cells from the dose-effect curves presented by SIGDESTAD et al. (1975a). A DMF of 1.3 was obtained in experiments with fission neutron irradiation (SIGDESTAD et al. 1976). The same authors tested the radioprotective efficacy of cysteamine and its thiophosphate derivative WR-638 on crypt cells of X-irradiated mice and obtained a DMF of approximately 1.5 for 50% cell survival (SIGDESTAD et al. 1975b). With fission neutron irradiation, cysteamine and WR-638 were far less effective than WR-2721. POULSEN and SZABO (1977) determined that the subcutaneous or oral administration of cysteamine (280 mg/kg) leads to duodenal ulceration or even perforation in nonirradiated rats. Electron microscopic studies of mitochondria from crypt cells of the upper small intestine in mice treated with protective sulfhydryl compounds (cysteine, 5-mercaptopyridoxine) demonstrated swelling of the cristae and external membrane (BRAUN and KOCH 1968). Nonprotective homologues of these compounds incited an opposite process (shrinkage) tathione peroxidase and superoxide dismutase. The rats received a vitamin E diet for 2 weeks prior to irradiation, and each was given an intraperitoneal injection of water-soluble α-tocopherol 4 h before irradiation. The 180-day survival data and histological examinations of heart and lung tissues showed no statistically significant differences from controls not treated with vitamin E.

5.3.2.6 Kidney

PHILLIPS et al. (1973) reported on renal damage in unilaterally nephrectomized adult mice and investigated the protective effect of the thiophosphate WR-2721. They observed the death of the animals from renal failure over a period of 1 year. The corresponding $LD_{50/365}$ was 17.9 Gy. The injection of WR-2721 increased the $LD_{50/365}$ to 26.1 Gy, corresponding to a DMF of 1.5. DONALDSON et al. (1984) achieved a protection factor of only 1.16–1.2 with WR-2721 for growth of the kidney irradiated with 10 Gy in young mice that had undergone unilateral nephrectomy.

5.3.2.7 Liver

ELDJARN (1954) showed that cystamine is converted to taurine in the liver of the rat, rabbit, and man. Cystamine inhibits oxygen consumption and phosphorylation in mitochondria taken from the liver (and spleen) of nonirradiated rats (FIRKET and LELIÈVRE 1966).

Histological studies of the irradiated rat liver have shown that the tissue returns to a normal state relatively quickly following pretreatment with cysteamine, but that radiation-induced fatty liver degeneration develops in animals not treated with cysteamine (VAN LANCKER and MAISIN 1953). In addition, histological studies of the mouse liver 6 h and 4 days after whole-body irradiation showed that cyste amine protects against spongy changes in the liver parenchyma (GEREBTZOFF and BACQ 1954). The reduction of radiation-induced glycogenesis by cysteamine has been followed histologically in the rat liver for up to 13 days after whole-body irradiation with 5 Gy (CHATTERJEE and BOSE 1959). According to MITZNEGG (1973 a), measurements of ^{3}H-thymidine incorporation into the DNA of mouse liver cells indicate that the radioprotective effect of cysteamine is based on a transient suspension of DNA synthesis. This is believed to accelerate and facilitate repair of the radiation-damaged DNA. The effect is age-dependent, as the process occurs only in the more mitotically active cells of younger animals.

5.3.2.8 Salivary Gland

In experiments with ^{35}S-labeled WR-2721 in BALB/c mice, UTLEY et al. (1976 a) found that the salivary gland contained the highest concentration of the radiolabeled drug of any organ in the body. However, a study of the protective effect of WR-2721 on the salivary gland of dogs that received 10–25 Gy of radiation to the head showed no significant change in the decreased salivary flow (UTLEY et al. 1978). SODICOFF et al. (1978 a), on the other hand, were able to demonstrate a protective effect of WR-2721 on the rat parotid gland. Over a 9-day period following X-irradiation with 16–64 Gy, the authors obtained DMFs of 2.5 for the gland weight, 1.7 for the amylase concentration, and 1.8 for the total amylase content in the gland. Protection was also demonstrated for the chronic phase of the radiation injury (60–90 days p. r.), with respective DMFs of 2.3, 3.2, and 2.0 (SODICOFF et al. 1978 b). Authors from the same group found in histological studies 60 days after irradiation that the thiophosphate is particularly effective in reducing the loss of the radiosensitive acinose cells of the parotid gland (PRATT et al. 1980). The synthetic β-adrenergic agonist isoproterenol also exerts a radioprotective effect on the rat parotid gland (SODICOFF and CONGER 1983). The β-receptor blocking drug propranolol abolishes this protective effect, but it does not alter the efficacy of WR-2721. On that basis the authors conclude that WR-2721 does not act via β-receptors, nor does it activate cAMP in the cell (see Sect. 5.5.3).

5.3.2.9 Thyroid Gland

UMA DEVI and JAGETIA (1981) showed in radioiodine experiments that external whole-body irradiation with 5 Gy in mice led to an increase in thyroid ^{131}I uptake and in plasma-bound iodine. Pretreatment with MPG did not alter this response significantly. Metabolic inhibition is suggested as the mechanism of action of the drug.

5.3.2.10 Gonads

Testes. Studies by various groups on the distribution of sulfur-containing compounds (usually ^{35}S-labeled) in various mammalian tissues have shown that the cystamine content in the mouse testis (GEN-SICKE et al. 1962) and rat testis (BETZ et al. 1962; MONDOVI et al. 1962) is relatively low compared with other tissues. The AET derivatives MEG and bis(2-guanidoethyl)-disulfide also were found to occur in the mouse testis in low concentration (SHAPIRO et al. 1963). The same is true of the thiophosphate WR-2721 in the rat (TANAKA and SUGAHARA 1980; see also Fig. 5.4).

While results on the radioprotection of the mouse testis have been positive, studies of the rat testis have produced conflicting findings. Histological studies by DESAIVE et al. (1953) in mice given cysteamine orally and exposed to 6 Gy whole-body irradiation showed a definite testicular regeneration of very rapid onset. WANG et al. (1959) tested the fertility of mice that were injected with cysteamine before whole-body irradiation (7 Gy) or local irradiation of the testis and found no difference relative to nonirradiated controls. SAVKOVIC et al. (1961) likewise demonstrated fertility protection 3-5 months after the local or whole-body irradiation (6 or 8 Gy) of infantile rats that had been treated with cysteamine. On the other hand, MAISIN et al. found no protection in rats injected with cysteamine and exposed to 7 Gy of whole-body radiation (VAN LANCKER and MAISIN 1953; MAISIN et al. 1955a). These studies were based on histological examinations and fertility tests. MANDL (1959a), however, observed a decrease in radiation damage for up to 8 days after the irradiation of rat testes with 2.3-4.6 Gy based on counts of surviving spermatogonia. ERSHOFF and BRAT (1960) found that AET had no protective effect on testicular damage in irradiated rats. AET was also found to decrease fertility slightly in nonirradiated mice (LEONARD and MAISIN 1964). A combination of protectors (glutathione + cysteine + cysteamine + AET + serotonin), however, was able to delay the disappearance of germ cells and hasten regeneration of the spermatogenic epithelium after local irradiation of the mouse testis with 13.5 Gy (LEONARD et al. 1969). The results indicate that the combination of different protectors is more effective than the isolated use of any single agent.

Comparisons of tissue drug concentrations with protective efficacy have led to various conclusions. While BETZ et al. (1962) believe that the cysteamine content does not necessarily correlate with the level of efficacy in the tissue, MONDOVI et al. (1962) reached an opposite conclusion on the basis of their findings.

MILAS et al. (1982a) were able to demonstrate a protective effect of the thiophosphate WR-2721 on

the mouse testes. The testes were exposed to local, single doses of 14 Gy, and 35 days later the spermatogenic epithelial stem cells were examined histologically using the method of WITHERS et al. (1974). Pretreatment of the animals with WR-2721 led to a DMF of 1.54 with 95% confidence limits between 1.49 and 1.62. The same agent also inhibits the reduction of testicular weight in mice exposed to 3-8 Gy of whole-body radiation (BHARTIYA and JAIMALA 1986).

Ovaries. Initial studies with very high radiation doses in mice and rats indicated that cysteamine and its disulfide cystamine afforded no radiation protection for the ovaries. In whole-body irradiation experiments by DESAIVE et al. (1952) using a dose of 9 Gy, mice protected with cystamine had negative fertility tests 1 month after exposure. Histological studies in rats exposed to 7 Gy of whole-body radiation also showed negative results in animals that had been treated with cysteamine (VAN LANCKER and MAISIN 1953). On the other hand, studies of the rabbit ovary following high dose irradiation with 17.6-30 Gy showed that cysteamine and sodium cyanide (NaCN) decreased ovarian radiosensitivity (DESAIVE 1954). It is likely that these agents protected mainly the primordial follicles against radiation injury.

Cysteamine and cystamine have shown protective value in mice and rats exposed to lower levels of radiation. In mice exposed to 1.5 Gy, a dose sufficient to cause 100% sterility within 30 weeks in the mouse strain tested, the total numbers of offspring during the 42 weeks of the experiment were higher in the mice pretreated with cystamine or cysteamine than in the untreated controls (RUGH and WOLFF 1957). The average period of fertility after irradiation was greater in the pretreated mice (cysteamine: 18 weeks, cystamine: 13 weeks) than in the controls (6.7 weeks). Histological studies of oocytes from rats that received local irradiation between 1 and 12.6 Gy also demonstrated radioprotective efficacy in animals injected with cysteamine prior to exposure (MANDL 1959b). Treatment of female mice with AET before whole-body irradiation also protected the reproductive capacity of the animals (EHLING and DOHERTY 1962). The number of offspring per animal was 10.6 in the untreated mice and 16.7 in the mice treated with AET. The period of fertility was 2.6 times longer in the AET-treated mice than in the unprotected animals. MPG was able to protect the graafian follicles of the irradiated mice (KUMAR and UMA DEVI 1983). The neuroleptic drugs hydergine, chlorpromazine, and promethazine and their combinations also conferred protection against sterility caused by the whole-body irradiation of mice with 0.25–1.5 Gy (RUPKEY et al. 1963).

MICHEL and FRITZ-NIGGLI (1973) reported a delay in radiation-induced sterilization in female mice fed a yeast preparation (Bio-Strath) before and after exposure to 0.25 or 0.5 Gy. The authors speculated on a possible radioprotective effect of the ribonucleic acid (RNA) contained in the preparation. Studies in which mice received RNA before or after irradiation have demonstrated a protective value against acute radiation death (EBEL et al. 1969).

5.3.2.11 Nervous System

To improve electivity in the radiation therapy of glioblastoma (malignant astrocytic gliomas), SPENCE et al. (1986) performed animal studies to investigate the degree to which the thiophosphate WR-2721 can protect normal CNS tissue. The authors exposed the cervical spinal cord of rats to γ-radiation doses of 20–38 Gy and observed the time course of radiation-induced paralysis of the forelimbs and hindlimbs. The radioprotector was administered by injection into the right lateral ventricle of the brain. A statistically significant delay of paralysis was observed in the animals treated with WR-2721. The mean DMF was 1.3 ± 0.01 for the forelimbs and 1.6 ± 0.15 for the hindlimbs. Histological studies showed no differences between the controls and the groups treated with WR-2721.

5.3.3 Genetic Studies

5.3.3.1 In Vitro Experiments on Human Lymphocytes

Lymphocytes from the peripheral blood were suspended in phosphate buffer with or without the addition of protective chemicals (various alcohols and mercaptoethanol, cysteine, cysteamine) and exposed to 3 Gy of γ-radiation (SASAKI and MATSUBARA 1977). It was found that the chemicals protect the chromosomes against the formation of exchange-type aberrations but not against terminal deletions. The protective mechanism may involve an interaction of the chemicals with OH radicals. Other authors have shown that the addition of L-cysteine prior to the X-irradiation of human lymphocytes in the G_0 phase in whole blood cultures

with doses up to 7 Gy leads to a transformation of the linear-quadratic dose-effect curve (VIRSIK and HARDER 1982). The protective effect is attributed to radical scavenging mechanisms or to binding of the cysteine molecule to DNA, which is thought to protect against "primary damage." It is interesting to note in connection with the results of in vitro experiments that MPG is known for its ability to protect the peripheral lymphocytes of radiotherapy patients against chromosomal damage (TANAKA and SUGAHARA 1970).

5.3.3.2 In Vivo Experiments in Mammals

KAPLAN and LYON (1953) tested the antimutagenic efficacy of cysteamine in male mice that received 5 Gy of local radiation to the testis. Immediately after the exposure, the male mice were mated with nonirradiated females, which were killed after 12–14 days' gestation to determine the number of live and dead embryos. In these experiments, which provide information on dominant lethal mutations, the ratio of live embryos to the total number of embryos showed no significant difference between the cysteamine-treated mice and the untreated mice. In similar experiments, however, LÜNING et al. (1961) were able to reduce the lethal mutation rate to 75% by treatment with cysteamine. The authors attributed the discrepant results to the fact that KAPLAN and LYON (1953) used a mouse strain with different radiosensitivity and administered the cysteamine at a different time (4–7 min before irradiation, as opposed to 15 min in the study of LÜNING et al.). LEONARD and MAISIN (1964) investigated the protective effect of AET against radiation-induced lethal mutations in male mice. The effects on spermatozoa, spermatids, and spermatogonia were studied separately by choosing different time intervals for the mating after irradiation of the testes with 4 and 12 Gy. On the 17th day after the start of the mating, the females were killed and counts were made of the corpora lutea and the number of live and dead embryos. After the exposure of spermatids to 12 Gy, AET decreased the radiation-induced preimplantation loss and slightly decreased the postimplantation loss after 4 Gy. The radiation-induced increase in the ratio of the number of dead embryos to the total number of implantations after irradiation of the spermatozoa and spermatids also was less pronounced after treatment with AET. The authors conclude that AET is able to protect against genetic damage.

Radioprotectors also have been tested for their ability to prevent chromosome damage in bone

marrow (DEVIK and LOTHE 1955) and in the duodenum (MAISIN and MOUTSCHEN 1960). Bone marrow
chromosome damage was evaluated 18 h after
whole-body irradiation with 2 Gy. Cysteine, cysteamine, and cystamine all were able to lower the rate
of chromosome damage, especially in cases where
an oxygen deficit existed during the irradiation
(8.5% O_2 in the ambient air; see also Sect. 5.5.2 on
the relationship between tissue oxygen tension and
radioprotective effect). The results of MAISIN and
MOUTSCHEN (1960) on the number of radiation-induced chromosome breaks in the duodenal cells of
whole-body-irradiated mice (2.25, 9.0, 15.0 Gy) indicate that the number of chromosome breaks depends
strongly on the radiation dose in the animals treated
with AET, with S-3-aminopropyl-N-methylisothiuro-
nium, and with S-2-aminobutylisothiuronium as well
as in the untreated controls. Following lethal irradiation of the treated animals, the number of chromosome
lesions was smaller than in the controls.

With lower level irradiation at 2.25 Gy, no differences were noted between the treated and untreated
animals.

In chromosome studies of bone marrow cells
from whole-body-irradiated mice, GUPTA and UMA
DEVI (1986) found that MPG and the thiophosphate WR-2721 were able to decrease the induction
of aberrations. These results imply that WR-2721
protects the chromosomes of bone marrow cells
both by reducing initial breaks and by stimulating
repair mechanisms. Studies based on thymidine incorporation into mammalian cells suggest that WR-
2721 aids repair processes through protection of the
repair enzymes (RIKLIS 1983; RIKLIS et al. 1983).

5.3.4 Radiation-Induced Developmental Anomalies

It is difficult to study the effect of chemical agents
on radiation-induced developmental anomalies, because the time at which the conceptus is exposed to
the radiation insult is of critical importance. The
greatest damage occurs to organs that are irradiated
during primary differentiation. Another difficulty
lies in the unacceptably high embryotoxicity of a
number of protective agents. Thus, for example, serotonin causes embryonic damage comparable to
that produced by moderate radiation doses, especially during the period of neurogenesis (KONER-
MANN 1972). WR-2721 also has shown a dose-de-
pendent embryotoxicity in rats, although it is not
teratogenic (SODICOFF et al. 1986). Experiments
with ^{14}C-labeled WR-2721 have shown that the thiophosphate passes rapidly through the placenta and
is absorbed by the fetus. A number of studies have
been done with sulfur-containing agents, which, apparently, are less toxic. Nonirradiated mouse fetuses that received cysteamine at 14.5 days' gestation by maternal injection showed no appreciable
differences relative to untreated controls when examined 4 weeks after birth (RUGH and CLUGSTON
1955).

The results of studies in mice are summarized in
Table 5.7. KONERMANN (1972) investigated a wide
range of radiation dose and phase-dependency during the course of mouse embryo development after
treatment with cystamine. He found that the effect
of cystamine was maximal at 3 and especially at
6 days' gestation and declined during organogene-
sis, indicating little effect on the development of

Table 5.7. Chemical protection against radiation-induced developmental anomalies in mice. Except for AET, all the radioprotective agents listed were effective. The animals received X- or γ-irradiation

Day of development[a] at time of irradiation	Dose or dose range (Gy)	Investigated radiation effect	Agent	Authors
3 to 15	0.5–5.5	Developmental anomalies	Cystamine	KONERMANN (1972)
8.5	2	Brain anomalies	e.g.: cysteamine, cystamine, AET, anoxia	RUGH and GRUPP (1960a)
8.5	2	Intrauterine death, fetal absorption, congenital anomalies		RUGH and GRUPP (1960b)
12	3	Weight and developmental anomalies	Cysteamine	WOOLLAM and MILLEN (1958)
13.5 to 19.5	3–12	Postnatal weight changes	Cysteamine	RUGH and CLUGSTON (1956)
14.5	3	Death	Cysteamine	RUGH and CLUGSTON (1955)
14¼ to 18¼	0.5	Postnatal weight changes	MPG[b]	DEV et al. (1982)
14¼ to 18¼	2.5	Postnatal weight changes	MPG[b]	DEV et al. (1981)

[a] Accurate to ±1 day. [b] MPG, mercaptopropionylglycine

teratogenic anomalies. DRF values as high as 2 were achieved in embryos irradiated on days 3 and 6. WOOLLAM and MILLEN (1958) investigated the effect of cysteamine on the teratogenicity of X-irradiation with 3 Gy on day 12 of embryonic development. The animals were killed on day 18, and the fetuses were examined for body weight and anomalies (syndactyly, micromelia, anophthalmia, hydrocephalus, meningocele, cleft palate). Except for cleft palate, fewer malformations were found in the cysteamine-treated group. Fetal weight loss also was lower in this group. RUGH and GRUPP (1960a, b) tested the efficacy of 15 agents in protecting against intrauterine death, fetal absorption, and congenital anomalies. Of all the agents tested, only cysteamine, cystamine, and hypoxia (6% O_2 + 94% N_2 in inspired air) exhibited a statistically significant radioprotective effect. AET, which confers protection in adult animals comparable to that of cysteamine, was not effective in these experiments. In contrast to these results, MAZUR (1985) observed a protective effect with AET in female mice exposed to 2 Gy immediately after the formation of a vaginal plug. The number of live and dead fetuses, fetal absorptions, and unimplanted embryos were determined on day 19 of gestation. Serotonin and cysteamine also were radioprotective in these experiments, but to a lesser degree than AET.

In rats irradiated with 3 Gy on day 15 or 18 of gestation, all newborn offspring died within 3 days. In rats pretreated with cysteamine, one-third of the offspring survived the first 4 weeks (MAISIN et al. 1955a). ROBERTS (1970) observed different effects of cysteamine in rats exposed to X-rays (150 kV) or to γ-rays (^{60}C) on day 16 of gestation. While cysteamine almost completely prevented neonatal and postnatal deaths in the X-irradiated offspring, no effect was apparent in the γ-irradiated animals. STARKIE (1961) studied the effect of cysteamine on the development of male gonads in the rat. Pregnant rats received 0.5–1.5 Gy of X-radiation to the abdomen between days 17 and 21 of gestation, and the testes of male offspring were examined histologically at 25 days after birth. It was found that testicular damage was reduced by maternal pretreatment with cysteamine.

GUPTA et al. (1981) produced chronic fetomaternal radiation exposure by the injection of ^{131}I (5.55 MBq/animal) and administered MPG daily from day 11.5 of pregnancy until parturition to study the effect on postnatal weight changes. While body weight differences between the controls and treated groups were minimal in the first 6 weeks after birth, the wet weights of some tissues, like testes,

spleen, thymus, and liver, were higher in the irradiated animals than in the controls, and they were higher still in the mice treated with MPG. This weight increase was attributed to compensatory reactions in the tissues during recovery. MPG apparently does not affect the suppression of brain development in fetal mice exposed to continuous β-irradiation (tritiated water) (BHATIA et al. 1986).

5.3.5 Late Effects

The results of animal studies on the ability of chemical agents to protect against late radiation injury, including alteration in long-term survival, are contradictory.

5.3.5.1 Tumors

A major difficulty in evaluating the protective effect of chemicals against radiation-induced tumor formation is that unprotected animals generally die before neoplasms have a chance to form. On the other hand, a predisposition to oncogenesis appears to exist in animals that survive acute radiation death after being treated with a radioprotective compound. For example, some surviving rats that had been treated with cysteamine developed neoplasms during the 7- to 19-month follow-up period, while nonirradiated animals of the same strain developed no spontaneous tumors (MAISIN et al. 1955b). The 203 surviving rats had a total of 16 tumors, including epitheliomas of the skin, stomach, intestine, prostate, and kidneys; myelomas; sarcomas of the pleura, testis, and stomach; and a Wilms' tumor of the kidney. Tumors did not form in areas that had been shielded from the radiation. The authors' conclusion was that cysteamine does not protect against neoplastic disease.

BRECHER et al. (1953) noted a high incidence of tumors in rats that had been treated with PAPP before whole-body irradiation with 7 Gy. Moreover, the high incidence of tumor types (e.g., intestinal adenocarcinoma) that rarely occur spontaneously in these animals indicated radiation-induced neoplasms. Because corresponding irradiated but untreated controls were not studied, it cannot be said whether the compound actually protected against late radiation effects. On the other hand, no tumors developed in a group treated with glutathione before irradiation. BOONE (1961) showed a definite protective effect of glutathione on the development of leukemia and lymphomas in mice exposed to 4 or 7 Gy of radiation. Studies on the effect of cysteine combined with hypoxia (6% O_2 + 94% He during irradiation with 9 Gy) on the formation of a number of tumors in whole-body-irradiated mice failed to demonstrate a protective effect (HOLLCROFT et al. 1957). The results are difficult to evaluate in this case, because the corresponding control groups were irradiated with the spleen shielded. MEWISSEN and BRUCER (1957) reported that the incidence of lymphosarcoma and lymphatic leukemia was significantly higher in γ-irradiated mice that had been pretreated with cysteamine (i. p. injection) or cystamine (gastric instillation). The surviving animals (also in the untreated control group) were observed for up to 300 days postirradiation. UPTON et al. (1959), on the other hand, found that MEG prevented the induction of granulocytic leukemia in mice exposed to levels of 1.5 and 3.0 Gy.

Irradiations of female mice with doses from 3.5 to 18 Gy showed that ovarian and mammary tumors developed more frequently than in nonirradiated animals (COSGROVE et al. 1964). The effect of AET on tumor incidence was questionable. Irradiation also increased the incidence of thymic lymphoma and myeloid leukemia, with AET preventing the induction of thymic lymphoma only when combined with the injection of isologous bone marrow. AET was unable to decrease mammary tumors in rats 11 months after whole-body irradiation with 2 or 4 Gy (CUDKOWICZ 1961).

In a study on the ability of combined radioprotectors (glutathione + cysteine + AET + cysteamine + serotonin) to decrease long-term radiation lethality in mice, MAISIN et al. (1978) investigated the causes of death. Animals that had received doses up to 20 Gy developed thymic lymphomas, nonthymic lymphomas, reticulosarcomas, myeloid leukemia, lung carcinomas, liver tumors, and sarcomas. A comprehensive analysis of the results showed that the mixture of the radioprotectors did protect against the formation of thymic lymphoma and possibly against liver tumors and leukemia. A definite overall protective effect was obtained with regard to all neoplasms. Thymic lymphoma also was the leading cause of death in mice given fractionated exposure to doses of 0.50–3.75 Gy four times at 1-week intervals (MAISIN et al. 1980). A mixture of the drugs listed above was effective in reducing this late radiation injury.

5.3.5.2 Cataract

In a study in which one rabbit eye was exposed to 15 Gy of X-irradiation (the opposite eye serving as a control), epilation and cataract were partially prevented by the intravenous administration of cysteine prior to irradiation (VON SALLMANN and MUNOZ 1952). After periods of 12–16 months, total cataract had developed in the control groups, while the lenses in the treated eyes showed some fine ring- or star-shaped opacities in the posterior cortex and delicate lesions on the anterior surface. Subconjunctival cysteine injections proved less effective in protecting against cataract. Glutathione and thiourea showed less effect than cysteine, and other chemicals such as dimercaprol, dl-α-tocopherol, and potassium cyanide produced no beneficial effect. As a follow-up to von Sallmann's experiments, PIRIE and LAJTHA (1959) performed similar experiments on rabbits' eyes to determine the possible mechanism of cysteine protection against radiation cataract. The authors found that cysteine injection produced mitotic inhibition in the lens epithelium of both the irradiated and nonirradiated eyes. The authors postulate that the protective effect of cysteine is based on its ability to arrest mitosis for a certain period of time.

Cysteamine also was able to retard cataract formation in X-irradiated rabbit eyes, as indicated by observations for up to 11 months after exposure to doses of 15–25 Gy (FRANCOIS and BEHEYT 1955). However, the local application of cysteamine in the form of eye drops had no effect on radiation cataract. The use of AET in rats that had received 24 Gy of γ-radiation to the eye gave significant protection against radiation damage to the lens epithelium, lenticular fibers, and eyelid (HANNA and O'BRIEN 1963). The protective effect was still apparent after 8 months. The studies were performed with tritium-labeled thymidine so that DNA synthesis could be observed in the phase of cell division. By contrast, COSGROVE et al. (1963, 1964) reported that AET did not prevent cataract formation in X-irradiated mice.

MENARD et al. (1986) achieved good results with S-(3-amino-2-hydroxypropyl) thiophosphate (WR-77913) in rats receiving 15.3 Gy of γ-radiation to the head. Unprotected animals developed cataracts 90–120 days postirradiation, but only minimal lens opacities formed in animals injected intraperitoneally with WR-77913 15–30 min before irradiation.

5.3.5.3 Other Late Effects

Radiation is capable of producing other long-term effects in tissues besides oncogenesis. A number of studies have dealt with the effects of radioprotective agents on damage to the kidneys and lung. HOLLCROFT et al. (1957) found in studies of whole-body-irradiated mice (9 Gy) that cysteine combined with hypoxia (6% O_2 + 94% He during exposure) lowered the incidence of glomerulosclerosis in females 14 months after irradiation. COSGROVE et al. (1964) reported that mice exposed to 7.5 Gy displayed a higher incidence of nephrosclerosis than of tumors. AET did not influence the renal damage, nor did it have a definite effect on radiation-induced life shortening. However, AET did reduce graying of the fur in irradiated mice.

A combination of the radioprotectors AET, cysteamine, cysteine, glutathione, and serotonin demonstrated a positive effect on glomerulosclerosis and noncancerous lung damage in mice (MAISIN et al. 1978).

UTLEY et al. (1981) tested the ability of the thiophosphate WR-2721 to protect against late radiation damage in skin, muscle, and vascular tissue over a period of 6 months in the rat. The animals received local radiation to one hind limb in doses of 20–80 Gy. While a DMF of 1.5 was determined for acute skin reactions, a DMF could not be established for skin changes past 6 months because of the diversity of the responses. A DMF of 1.5–2.0 was found for the skeletal muscle. Blood flow measurements showed no differences between the animals protected with WR-2721 and the nonirradiated controls.

5.3.6 Chemical Radioprotection in "Combined Injuries"

Radiation effects can be modified by exposure to other injurious agents or trauma, which tend to exacerbate the extent of the damage. This may occur in disaster situations, for example, where exposure to radiation is accompanied by the effects of explosions or fires in the form of mechanical injuries or burns.

The results of Hiroshima and Nagasaki illustrate this point very dramatically. But combined effects are seen in everyday hospital practice in cases where surgical operations are performed on patients who are receiving radiotherapy for neoplastic disease.

Experiments in animals have shown that addi-

Table 5.8. Combined effect of whole-body irradiation and burn on different experimental animals (after MESSERSCHMIDT 1977)

	Lethality
Dogs (Brooks, Evans, Ham, and Reid)	
20% burn	12%
1 Gy	0%
20% burn + 1 Gy	73%
Pigs (Bayter, Drummond, Stephens-Newsham, and Randall)	
10%–15% burn	0%
4 Gy	20%
10%–15% burn + 4 Gy	73%
Rats (Alpen and Sheline)	
31%–35% burn	50%
2.5 Gy	0%
5 Gy	20%
31%–35% burn + 1 Gy	65%
31%–35% burn + 2.5 Gy	95%
31%–35% burn + 5 Gy	100%
Guinea pigs (Korlof)	
1.5% burn	9%
2.5 Gy	11%
1.5% burn + 2.5 Gy	38%

tional insults can modify the response to *local irradiation,* but that the greatest potentiation occurs in *acute radiation syndromes following whole-body exposure.* Table 5.8 shows the effects of combined radiation and burns on the mortality of various mammalian species. Further studies have shown that the most striking increase in animal mortality occurs when the additional trauma takes place hours or days *after* the whole-body irradiation. This has important clinical implications in patients exposed to high levels of radiation and would imply a strict contraindication to surgical procedures during the leukopenic and thrombocytopenic stages of radiation sickness.

Given the traumatologic significance of combined injuries, it was natural to address the question of the potential benefit of proven radioprotective agents in cases where radiation damage coexists with other injuries. To explore this question, combined injuries were produced experimentally in animals (NMRI mice) by inflicting mechanical skin wounds or superficial burns at varying times after whole-body irradiation with various doses. The drug selected for pretreatment was WR-2721 owing to its high DRF. The animals were irradiated by exposure to 250-kV X-rays. Skin wounds were produced by excising a dime-size piece of dorsal skin, and burns were produced with a stamping tool water-heated to 95 °C and applied to the shaved dorsal

Table 5.9. Comparison of $LD_{50/30}$ values for irradiation and various combined injuries with standard deviations and 95% confidence limits. The slopes of the probit lines and corresponding dose reduction factors (DRF) are also shown. SW, skin wound; B, burn. (After SEDLMEIER and MESSERSCHMIDT 1980)

Type of insult	Control mice (without WR-2721)		Protected mice (with WR-2721)		DRF
	$LD_{50/30}$ in Gy	Slope	$LD_{50/30}$ in Gy	Slope	
Irradiation	6.29 ± 0.18 $(6.02 - 6.56)$	20.3 ± 3.6	15.21 ± 0.56 $(14.74 - 15.69)$	20.9 ± 2.9	2.42 ± 0.10
Irradiation + SW 10 min p.r.	6.47		14.37		2.22
Irradiation + SW 2 d p.r.	5.32 ± 0.21 $(5.19 - 5.45)$	17.2 ± 3.2	12.26 ± 0.44 $(11.64 - 12.82)$	12.0 ± 1.6	2.30 ± 0.12
Irradiation + SW 8 d p.r.	5.61 ± 0.21 $(5.49 - 5.72)$	21.4 ± 4.2	12.85 ± 0.45 $(12.46 - 13.29)$	12.2 ± 1.6	2.29 ± 0.11
Irradiation	6.06 ± 0.28 $(5.88 - 6.24)$	17.0 ± 3.1	12.88 ± 0.54 $(12.02 - 13.55)$	15.8 ± 2.4	2.12 ± 0.13
Irradiation + 2 B 10 min p.r.	5.37 ± 0.27 $(4.45 - 6.33)$	7.7 ± 1.5	10.79 ± 0.59 $(9.45 - 12.03)$	6.0 ± 1.0	2.01 ± 0.15
Irradiation + 2 B 2 d p.r.	5.27 ± 0.37 $(3.66 - 6.92)$	3.8 ± 0.9	10.06 ± 0.48 $(8.24 - 12.63)$	5.0 ± 0.9	1.91 ± 0.16

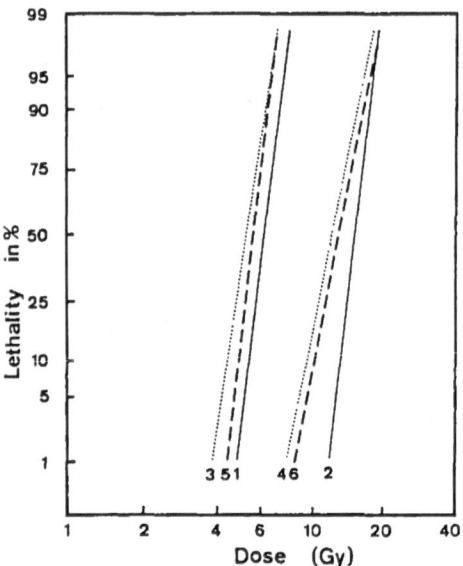

Fig. 5.9. Dose-effect curves for NMRI mice (Hannover strain) without WR-2721 (controls) and after treatment with 12.5 mg WR-2721 per mouse (protected animals). Combined injury: irradiation+skin wound (SW) (after SEDLMEIER and MESSERSCHMIDT 1980)

1. Irradiation (controls)
2. WR-2721 + irradiation } DRF 2.42

3. Irradiation + SW 2 d p.r. (controls)
4. WR-2721 + irradiation + SW 2 d p.r. } DRF 2.30

5. Irradiation + SW 8 d p.r. (controls)
6. WR-2721 + irradiation + SW 8 d p.r. } DRF 2.29

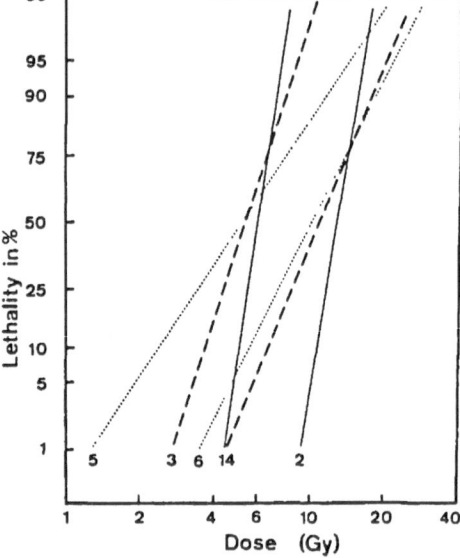

Fig. 5.10. Dose-effect curves for NMRI mice (Kisslegg strain) without WR-2721 (controls) and after treatment with 12.5 mg WR-2721 per mouse (protected animals). Combined injury: irradiation+burn (B) (after SEDLMEIER and MESSERSCHMIDT 1980)

1. Irradiation (controls)
2. WR-2721 + irradiation } DRF 2.12

3. Irradiation + 2 B 10 min p.r. (controls)
4. WR-2721 + irradiation + 2 B 10 min p.r. } DRF 2.01

5. Irradiation + 2 B 2 d p.r. (controls)
6. WR-2721 + irradiation + 2 B 2 d p.r. } DRF 1.91

skin (both were done under ether anesthesia). The wounds or burns were inflicted 10 min, 2 days, or 8 days postirradiation. The mice were injected intraperitoneally with 12.5 mg of WR-2721 (= approximately 500 mg/kg body wt) 15–20 min prior to irradiation.

The results of these experiments (Table 5.9) show that the DRFs achieved in mice with combined injuries were only slightly less than those obtained in mice with "pure" radiation injuries. This means that the protective agent reduced the radiation effect so much that the inflicted wounds were no longer able to cause a significant increase in mortality. Looking at the dose-effect curves in Figs. 5.9 and 5.10, we notice a marked flattening of the curves in cases where irradiation was compounded by a burn, i. e., the lethalities extend over a greater range of doses. These effects are also apparent in Table 5.9 and suggest that the combined injuries led to a burn shock which caused some deaths to occur even at low radiation levels.

SCHWEICKERT and LADNER (1964) showed that dogs treated with pyridoxal 5-phosphate before whole-body irradiation were able to tolerate major surgical operations. All unprotected animals exposed to 4 and 7 Gy of radiation died after the surgery. FARŠATOV (1986) reported on the favourable effects of cystamine in dogs. These animals were given combined injuries by irradiation and by inflicting mechanical wounds or large area burns. The finding is reported in this paper, that cystamine also influenced positively the healing of the wounds in those animals, which were not exposed to irradiation.

5.3.7 Effect of Radiation Quality

The great majority of studies on chemical radioprotection have been done with X-rays or γ-rays, i. e., radiation having a low linear energy transfer (LET). Because of the formidible equipment requirements, relatively little research has been done on chemical protection against radiation of high LET. Most experiments in mammals have employed neutron radiation.

Neutrons are uncharged particles that ionize indirectly. Collisions or nuclear reactions give rise to fast-moving secondary charged particles, which are capable of direct ionization. The main secondary particles produced by fast neutrons in tissues are fast protons as well as heavy recoil nuclei of carbon and oxygen (BEWLEY 1968). A major feature that distinguishes the heavy charged particles from the electrons that are released by γ- or X-rays is their

high LET. Radiobiologic studies have shown that this radiation possesses a higher relative biological effectiveness. In addition, tissue irradiated with fast neutrons shows less capacity for recovery.

Table 5.10 summarizes the results of various studies on acute radiation death in rodents. As the data indicate, the ability of protective agents to increase radioresistance in the setting of hematopoietic radiation syndrome (observation time: 30 days) is markedly less with neutron irradiation than with X- or γ-rays. Some experiments showed no protective effect at all following neutron irradiation. Even in mice irradiated with cyclotron neutrons ($\overline{E} \sim 9$ MeV), which has a lower LET than fission neutrons, the DMFs of WR-2721 for gastrointestinal and hemopoietic injury were markedly lower than with γ-rays (RASEY et al. 1984).

On the other hand, SIGDESTAD et al. (1975b, 1976) showed that cysteamine, WR-638, and WR-2721 protect the gastrointestinal tract against neutron irradiation at least as effectively as against X-irradiation (see Table 5.10). This result is noteworthy, given the fact that chemicals protect less well against gastrointestinal radiation syndrome than against hematopoietic radiation death (see Table 5.4). However, the same authors found that WR-2721 has its greatest protective effect in mice when it is administered 1 h prior to neutron irradiation, and 15 min prior to X-irradiation. The authors relate this discrepancy to different modes of restraint of the experimental animals during the neutron or X-irradiation, which may have affected the drug uptake for metabolism (SIGDESTAD et al. 1976). CONNOR and SIGDESTAD (1982) were also able to protect fission-neutron-irradiated mice against gastrointestinal death with the thiophosphates WR-2822 (S-2(4-aminobutylamino) ethylphosphorothioic acid) and WR-2823 (S-2(4-aminopentylamino) ethylphosphorothioic acid) and with an iminothiol derivative of 1-methylaminoadamatine (WR-109342). The DMFs were 1.51 for WR-2822, 1.21 for WR-2823, and 1.46 for WR-109342. In studies of intestinal crypt cells, the authors determined DMFs of 1.4 (WR-2822), 1.42 (WR-2823), and 1.16 (WR-109342) from the dose-survival curves. SIGDESTAD et al. (1986) found that the thiophosphate WR-151327 conferred even greater protection against gastrointestinal death in mice irradiated with fission neutrons (DMF = 2.2) than in mice exposed to ^{60}Co γ-rays. So far no interpretation has been offered for this result.

The reduced protective effect on hematopoietic radiation syndrome in neutron-irradiated animals can be explained in terms of the known differences

Table 5.10. Dose reduction factors (DRF) for acute lethality in rodents after neutron irradiation. The SULLIVAN study was done with rats, all others with mice

Agent	Observation period	Type of neutron radiation (γ component)	Comparative radiation	DRF		Authors
				Neu-trons	Comparative radiation	
Cysteine	30 d	Fission[a] (<10%)	γ-rays	1.1[b]	1.2[b]	PATT et al. (1953)
Cysteine	Intestinal death	Fission	250-kV X-rays	1[c]	–	SULLIVAN (1964)
Cysteamine	6 d	Fission: $\bar{E} = 1.2$ MeV (4%)	4-MeV X-rays	1.39[d]	1.26[d]	SIGDESTAD et al. (1975 b)
Cystamine	30 d	Fission: $\bar{E} \approx 2$ MeV ($\approx 8\%$)	–	1.20	–	SVERDLOV et al. (1969), SVERDLOV et al. (1973)
Cystamine	30 d	14.7 MeV (5%)	230-kV X-rays	1.15 ±0.02	1.83 ±0.06	LANGENDORFF et al. (1971 a)
WR-638 (cystaphos)	30 d	Fission: $\bar{E} \approx 2$ MeV ($\approx 8\%$)	–	1.35	–	SVERDLOV et al. (1969), SVERDLOV et al. (1973)
WR-638	6 d	Fission: $\bar{E} = 1.2$ MeV (4%)	4-MeV X-rays	1.42[d]	1.57[d]	SIGDESTAD et al. (1975 b)
WR-638	30 d	14.7 MeV (5%)	250-kV X-rays	1.18	1.72	MESSERSCHMIDT et al. (1978)
WR-2721	30 d	14.7 MeV (5%)	230-kV X-rays	1.17 ±0.02	2.20 ±0.07	LANGENDORFF et al. (1974)
WR-2721 (gammaphos)	30 d	Fission: $\bar{E} \approx 1.8$ MeV	X-rays	1.33	1.58	SVERDLOV (1974)
WR-2721	6 d	Fission: $\bar{E} = 1.2$ MeV (14%)	4-MeV X-rays	1.6	1.64	SIGDESTAD et al. (1975 a), SIGDESTAD et al. (1976)
WR-2721	30 d	Fission: $\bar{E} \approx 1.3$ MeV	250-kV X-rays	1.05 ±0.06	2.08 ±0.15	SEDLMEIER et al. (1981)
AET	30 d	Fission: $\bar{E} \approx 2$ MeV ($\approx 8\%$)	–	1.3	–	SVERDLOV et al. (1969), SVERDLOV et al. (1973)
Mixture of cysteamine + AET + serotonin	30 d	Fission (14%)	250-kV X-rays	<1.1	≈ 2	VOGEL et al. (1969)
5-Hydroxytryptamine (serotonin)	30 d	14.7 MeV (5%)	230-kV X-rays	1.06 ±0.02	1.57 ±0.07	LANGENDORFF et al. (1971 a)
5-Methoxytryptamine (mexamine)	30 d	Fission: $\bar{E} \approx 2$ MeV ($\approx 8\%$)	–	1	–	SVERDLOV et al. (1973)
Ethyron	30 d	Fission: $\bar{E} \approx 2$ MeV ($\approx 8\%$)	–	1	–	SVERDLOV et al. (1973)
PAPP	30 d	14.7 MeV (5%)	230-kV X-rays	1.09 ±0.02	1.36 ±0.05	LANGENDORFF et al. (1971 a)
Lipopolysaccharide (E. coli) (Endotoxin)	30 d	14.7 MeV (5%)	230-kV X-rays	1.06 ±0.02	1.22 ±0.02	LANGENDORFF et al. (1971 a)

[a] Slow neutrons were shielded
[b] Our evaluation based on the data of PATT et al.
[c] Studies of the passage of polyvinyl pyrrolidone from the blood into the bowel showed a DRF of 1.5 (SULLIVAN 1964)
[d] The stated DRF values pertain to studies of intestinal injury

between X-rays and γ-rays on the one hand and neutron radiation on the other. The LET of X- and γ-rays in tissues is considerably lower than that of neutron rays. As the LET of the radiation increases, the following processes take place in an aqueous milieu: (a) an increase in local energy deposition, (b) a decrease in indirect radiation effects and an increase in direct effects, (c) a decrease in the oxygen enhancement ratio (OER), and (d) a reduction of intracellular recovery, with an associated increase in irreversible radiation damage. Any of these processes can reduce the efficacy of a radioprotective chemical, depending on its mechanism of action. For example, the increased local energy deposition in the tissue cells can lead to greatly concentrated physicochemical changes, and the administered dose of the radioprotector may not be sufficient to deal with the excess. The second process listed above would tend to decrease the efficacy of radioprotectors that act by scavenging free radicals. With some agents (e. g., serotonin) that act by lowering tissue oxygen tension during irradiation, a decreased effect would be expected regardless of the administered dose, for the influence of oxygen content is less with neutron irradiation than with X- or γ-irradiation. The corresponding OERs in cell cultures are approximately 1.6 for 15-MeV neutrons and 2.5 for X- and γ-rays (BROERSE et al. 1968). In this case, the radioresistance-enhancing effect would stem from an anoxic state in the tissues. Some of the apparent protective effect against neutron irradiation is certainly based on a protective effect against the γ-radiation induced by the neutrons.

In mice pretreated with PAPP and exposed to fast protons at energies of 150–440 MeV, OLDFIELD et al. (1965a, b) obtained DRFs of 1.40 (150 MeV), 1.48 (440 MeV), and 1.58 (350 MeV) for the 30-day lethality, and a DRF of 1.56 for X-irradiation at 250 kV. Thus, the radioprotective effect of PAPP does not appear to depend critically on the proton energy or on the nature of the particle. The same authors achieved a higher DRF in mice pretreated with cysteamine and exposed to 440-MeV proton radiation (DRF = 1.61) than in mice exposed to 250-kV X-rays (DRF = 1.39) (OLDFIELD et al. 1965a). YARMONENKO et al. (1962) showed that the agents AET, cysteamine, cystamine, serotonin, and 5-methoxytryptamine exerted a protective action in mice exposed to whole-body irradiation with 660-MeV protons.

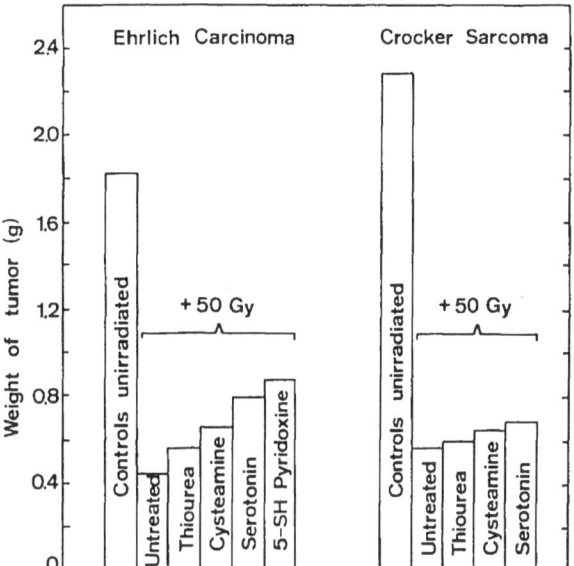

Fig. 5.11. Growth of solid tumors in mice irradiated locally with 50 Gy. The animal groups treated with the radioprotectors also received 50 Gy of local tumor radiation. All the drugs but thiourea gave significant protection of Ehrlich carcinoma compared with the untreated animals exposed to 50 Gy (cysteamine: $P < 0.05$; serotonin: $P < 0.01$; 5-SH-pyridoxine: $P < 0.001$). (After KOCH 1962)

5.4 Radioprotective Agents in Tumor Therapy

5.4.1 Animal Experiments

The basic goal of the radiotherapy of malignant tumors is to achieve maximum destruction of tumor with minimal damage to healthy tissues, although a "spillover" effect is often inevitable when ionizing radiation is used (STREFFER 1976). The radiotherapeutic control of a tumor depends not just on the radiosensitivity of the lesion but also on the sensitivity of the surrounding normal tissue. As a result, attention has been directed toward ways of enhancing the radiation resistance of normal tissues by chemical means. Studies are currently underway to find differences in the degree to which radioresistance is increased in normal and malignant tissues. Shortly after it was discovered that cysteine had radioprotective properties, that agent was utilized in animal experiments on Walker rat carcinoma (STRAUBE et al. 1950). The experiments showed that cysteine decreases the growth-inhibiting effect of X-rays. At the same time, mice with localized lymphosarcoma were found to tolerate higher radiation doses when treated with AET prior to whole-body irradiation. The reduction of tumor volume was greater in these mice than in untreated controls (ANDREWS and SNEIDER 1957).

Table 5.11. Results of animal studies on the protection of malignant tissues by radioprotective drugs

Agent	Tumor	Animal species	Protection	Authors
	Solid tumors			
Cysteine	Walker carcinoma	Rat	Yes	STRAUBE et al. (1950)
Cysteine	Lymphosarcoma	Rat	Yes	STORAASLI et al. (1953)
Cysteamine	Mammary carcinoma	Mouse	Slight	COHEN and COHEN (1959)
AET[u]	Yoshida sarcoma	Rat	No	HAAS and LORENZ (1960)
AET	Solid Ehrlich carcinoma	Rat	No	SHAPIRO et al. (1960)
AET	Crocker sarcoma	Rat	No	SHAPIRO et al. (1960)
Glutathione	Mammary adenocarcinoma	Mouse	Yes	MODLIN and MORRIS (1961)
AET, serotonin	Mammary carcinoma	Mouse	Yes	COHEN and COHEN (1962)
Cysteamine, serotonin	Solid Ehrlich carcinoma	Mouse	Slight[b]	KOCH (1962)
Cysteamine, serotonin	Crocker sarcoma	Mouse	No	KOCH (1962)
MEG[c]	Mammary adenocarcinoma	Mouse	No	SCHWARTZ et al. (1964)
Cysteamine	Mammary carcinoma	Mouse	No	KOCH (1967)
Cysteamine	Reticulum cell sarcoma	Mouse	No	KOCH (1967)
WR-2721	Mammary carcinoma	Mouse	Slight	YUHAS and STORER (1969 b)
WR-2721	Lung tumor (adenoma)	Mouse	No	YUHAS (1972 b, 1973)
WR-2721	Mammary carcinoma (EMT-6)	Mouse	Slight	PHILLIPS et al. (1973)
WR-2721	KHT sarcoma	Mouse	Slight	LOWY and BAKER (1973)
WR-2721	Lung adenoma	Mouse	No	ECHOLS (1973)
WR-2721	Mammary carcinoma	Mouse	No	ECHOLS (1973)
WR-2721	Mammary carcinoma (EMT-6) (euoxic)	Mouse	Yes	UTLEY et al. (1974)
WR-2721	Mammary carcinoma (EMT-6) (hypoxic)	Mouse	No	UTLEY et al. (1974)
WR-2721	Mammary carcinoma (3M2N)	Rat	No	YUHAS (1980 a)
WR-2721	Fibrosarcoma	Mouse	No	MILAS et al. (1982 b)
WR-2721	Fibrosarcoma	Mouse	Yes-no	STEWART et al. (1983)
	Ascites			
Cysteine	Ascites carcinoma	Mouse	Yes	BÄUMER et al. (1953)
Cysteamine	Ehrlich ascites carcinoma	Mouse	Yes	WENZ (1956)
AET	Ehrlich ascites carcinoma	Rat	No	SHAPIRO et al. (1960)
AET, ACT[d], Cysteamine	Ehrlich ascites carcinoma	Mouse	No	IRIE and YOSIHARA (1961)
Cysteamine, serotonin	Ehrlich ascites carcinoma	Mouse	Yes	KOCH (1962)
WR-638, WR-2721	Ehrlich ascites carcinoma (euoxic)	Mouse	Yes	HARRIS and PHILLIPS (1971)
WR-638, WR-2721	P-388 leukemia[e] (euoxic)	Mouse	Yes	HARRIS and PHILLIPS (1971)
WR-2721	P-388 leukemia[e] (euoxic)	Mouse	Yes	PHILLIPS et al. (1973)
	Leukemia			
MEG[c]	Myeloid leukemia	Mouse	No	SCHWARTZ et al. (1964)

[a] AET, aminoethylisothiuronium. [b] See Fig. 5.11. [c] MEG, mercaptoethylguanidine. [d] ACT, aminocyclohexanthiol. [e] Grown as ascites tumor

The effect of various radioprotectors on the growth of two locally irradiated solid tumors is shown in Fig. 5.11. It can be seen that radioresistance is increased only slightly, although all the agents but thiourea show a definitive protective effect relative to the untreated, irradiated control group.

A number of studies have examined the effect of radioprotectors on various experimental tumors following local or whole-body irradiation in animals (Table 5.11). As the table indicates, data are contradictory even for the same tumor type. It is possible that immunobiological tumor-host relationships have an influence on radioprotective effect (KOCH 1967). The oxygen content of the tumor tissue is, apparently, also a factor (HARRIS and PHILLIPS 1971; YUHAS et al. 1973; UTLEY et al. 1974; STEWART et al. 1983) (see also Sect. 5.5.2). Studies by HARRIS and PHILLIPS (1971) have shown that only euoxic tumor cells are protected by the thiophosphates WR-638 and WR-2721 (see Table 5.11). The results in Table 5.11 make it apparent that solid tumors receive minimal or no protection, while the resistance of ascites tumors is enhanced. Microaggre-

gates of solid tumors are protected better by
WR-2721 than larger tumor cell masses (MILAS et
al. 1984). MILAS (1984) stresses the need to investigate other factors which influence the radioprotection of solid tumors by WR-2721. YUHAS (1983) cit-
es the following reasons for the peculiar response
of ascites tumors to the thiophosphate WR-2721:
(a) the drug has ready access to ascites tumors when
given by intraperitoneal injection; (b) in their early
stages these tumors have a rich oxygen supply,
which enhances the effect of the drug; (c) the capacity to absorb WR-2721 apparently is not limited
in leukemic cells, as it is in most solid tumors; and
(d) the ascitic fluid can easily convert WR-2721 to a
radioprotective form.

Studies with labeled compounds of the thiophosphate WR-2721 have shown that this agent is distributed at different rates in normal and in solid tumor tissues (YUHAS 1980b). Figure 5.12 shows the
concentration of WR-2721 in the serum, in various
tissues, and in squamous cell carcinoma of the rat
as a function of time after injection. It is evident
that the rate of absorption of this thiophosphate is
considerably lower in the solid tumor than in normal tissues. The same observation was made for a
number of other solid tumor types (YUHAS 1980b).
The strongly hydrophilic WR-2721, not yet dephosphorylated, cannot surmount the barrier of the solid
tumor membrane (YUHAS et al. 1982). Initially it
was thought that differences in the absorption of
the thiophosphate related to the different vasculari-
zation of normal and neoplastic tissues (PHILLIPS
1977). But in vivo and in vitro analyses of the absorption kinetics of WR-2721 in normal tissues and
in solid tumors of mice, rats, and rabbits have
shown that normal tissues (except for CNS tissues)
can actively concentrate the thiophosphate against
a concentration gradient, whereas solid tumors absorb the compound passively (YUHAS 1980b).

Besides its ability to protect against ionizing rays,
WR-2721 also has demonstrated efficacy as a chemo-
protector (YUHAS 1979, 1980c; YUHAS et al.
1980b; GAUGAS 1982; GLICK et al. 1982; MILLAR et
al. 1982; TWENTYMAN 1983; GLOVER et al. 1986)
and as a chemotherapeutic agent for tumors (YU-
HAS et al. 1980a; IKEBUCHI et al. 1981). The literature also contains reports on the interaction of WR-
2721 with radiosensitizing drugs (YUHAS et al. 1977;
GRIGSBY and MARUYAMA 1981, 1982; ROJAS et al.
1982a, b, 1983; MENDIONDO et al. 1983).

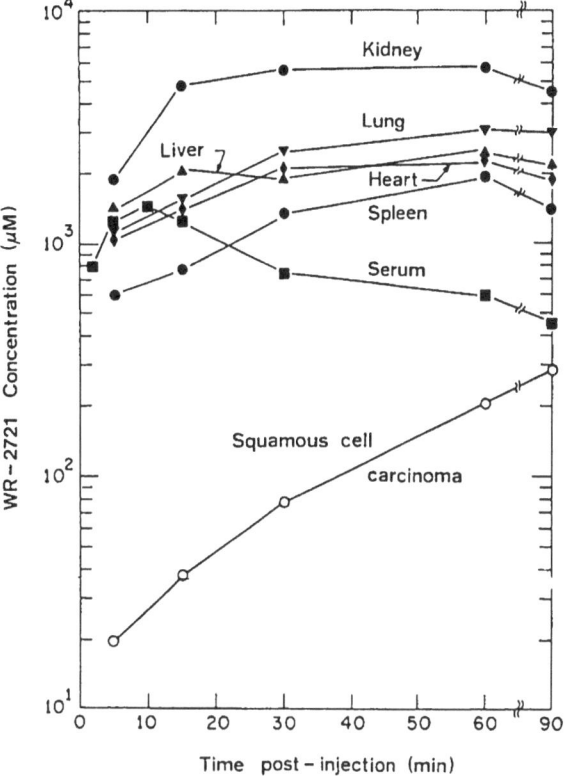

Fig. 5.12. Concentration of WR-2721 in the serum, in various
tissues, and in squamous cell carcinoma of the rat as a function of time after intraperitoneal injection of 200 mg/kg b. w.
of the drug (after YUHAS 1980b)

5.4.2 Experience with Radioprotective Agents in Man

Years before the discovery of chemical radioprotection, SHIRAI (1941) described the beneficial effect of
glutathione on complaints following radiotherapy.

The first clinical applications of cysteamine and
cystamine were described by BACQ and HERVÉ and
BACQ (1952a, b), BACQ et al. (1953a), and BACQ
(1954). A dose of 200–400 mg cysteamine or cystamine was well tolerated by patients when administered by slow intravenous injection (BACQ and
HERVÉ 1952). Cysteamine was administered in single or multiple doses before irradiation (HERVÉ and
BACQ 1952b) or immediately thereafter (HERVÉ and
BACQ 1952a). In nearly all cases this drug protected
against the symptoms of radiation sickness, especially gastrointestinal complaints such as anorexia,
nausea, vomiting, and diarrhea. Cystamine was in-
éffective. Excellent effects were obtained with cysteamine salicylate given orally in the form of gelatine capsules (300 mg t. i. d.) (BACQ et al. 1953a).
No toxic side-effects were observed on the blood

Table 5.12. Effect of cystamine on radiation sickness in X-irradiated patients (VACHTEL and SINENKO 1963)

Irradiated body region	Total number of patients	Single cystamine dose	Discontinuation of radiotherapy	With blood transfusion	Without symptoms of radiation sickness
Chest[a]	60 (control)	–	4	25	31
	130	0.8 g	0	25	105
Abdomen[b]	30 (control)	–	5	25	0
	72	0.8 g	0	36	36

[a] For neck and thoracic tumors. [b] For tumors of the abdomen and lesser pelvis

picture, blood pressure, or renal function. These initial applications were concerned more with investigating the curative effects of the drugs than their prophylactic effects. Like BACQ et al., HEUWIESER (1954) found that cysteamine had a salutary effect on radiation sickness when administered intravenously immediately before or after irradiation. BALDINI and FERRI (1957) reported on 35 cases in which intravenous cysteamine or oral cystamine was given to women undergoing radiotherapy after breast cancer surgery. Protective effect was assessed on the basis of white blood cell count, which was lower in the untreated control patients during the course of radiotherapy. In the treated patients, the protective effect was manifested chiefly in a more rapid recovery after the conclusion of therapy. Also radiation-induced gastrointestinal complaints were much less pronounced in the drug-treated patients. No difference was apparent between the women treated with cysteamine (i.v.) and cystamine (per os). DURKOVSKY and SIRACKA-VESELA (1958) also reported milder radiation sickness in 42 tumor patients given 0.2 g cysteamine by intravenous injection before or after irradiation. Results were somewhat better when the drug was given before the irradiation. VACHTEL and SINENKO (1963) did a comprehensive study of the effect of cystamine on radiation sickness in X-irradiated patients. As Table 5.12 indicates, radiotherapy had to be discontinued in 9 of 90 control patients who did not receive cystamine, but it could be continued to completion in all of the 202 patients treated with cystamine. The successful use of cystamine in radioherapy may have prompted Soviet officials to include cystamine tablets in the ABC self-protection kit for civil defense. According to a report by the International Civil Defense Organization (1983), six 200-mg tablets should be taken when there is danger of acute radiation exposure, with six more tablets at 4- to 6-h intervals in situations of chronic exposure. A dose of 600–800 mg cystamine taken twice daily at 6-h intervals will not impair the working ability of healthy individuals (ARNDT and RITTER 1979).

On the other hand, COURT-BROWN (1955) found that 160–300 mg cysteamine given intravenously either before or after therapeutic X-irradiation had no significant effect on symptoms, and he doubted the efficacy of the drug. A clinical trial of cystamine given to radiotherapy patients in the form of oral capsules also proved disappointing (HEALY 1960). The statistical conclusion from these studies (95% confidence limits) is that in a given series of 100 patients, radiation sickness is relieved in no more than 15 - a success rate comparable to treatment with lactose, which served as a control in these experiments.

Cysteine also has been tested clinically. LUDWIG (1955) administered 2 cm^3 cysteine hydrochloride (1.5%) intravenously to patients 30–60 min before their gynecological radiotherapy session. Clinical symptoms of radiation sickness were less frequent in the women so treated, and white blood counts were more stable. Follow-ups over a 14-month period showed no evidence of increased tumor resistance or of renal injury.

In clinical trials with AET, the rate of intravenous injection of the drug was limited by acute reactions consisting of nausea, retching, vomiting, coughing, burning of the eyes and face, and an intense flush involving the head, neck, and upper thorax (CONDIT et al. 1955). The maximum dose tolerated was 25 mg/min for 30 min. Patients receiving more than 500 mg AET orally developed dizziness and vomiting. At the doses employed, no modification in the response of the skin or hair to radiation was observed. ANDREWS and SNEIDER (1957) concluded that AET is too toxic for clinical use.

Clinical trials have been conducted in Japan with the drug MPG, known for its efficacy in experimental animals (SUGAHARA et al. 1977; TANAKA and SUGAHARA 1980). In patients receiving pelvic radiation for carcinoma of the cervix, peripheral blood cell counts and blood cell cultures for chromosome study were performed 1 or 2 weeks after the last exposure. The differences in the lymphocyte counts, leukocyte counts, and aberration yields between the treated and untreated groups were statistically significant. However, the data given by TANAKA and SUGAHARA (1980) on the effects of MPG on late injuries to the bowel (ileus symptoms and bleeding) do not withstand statistical testing (fourfold table). The same authors reported a beneficial effect of adrenochrome monoguanylhydrazone methanesulfonate in cervical cancer patients undergoing radiotherapy.

Recently thiophosphate compounds have received increased attention as protective agents in radiotherapy. TELIČENAS and KAROSENE (1973) published results on the clinical testing of cystaphos (WR-638) in patients with cancer of the breast or uterus. Their results indicate the following:

1. Cystaphos produces no toxic symptoms when given orally in a dose of 1 g before irradiation and is tolerated relatively well.
2. The drug does not alter the radiosensitivity of tumors.
3. Cystaphos significantly reduces systemic symptoms of radiation sickness such as nausea, headache, and insomnia.
4. The drug shows *no* protective effect on the development of radiation leukopenia and lymphopenia. It leads to a more rapid recovery of the leukocyte count, but not of the lymphocyte count.
5. Radiation thrombocytopenia does not occur following the use of cystaphos.

The thiophosphate WR-2721 has undergone initial clinical trials in the United States and in Japan (KLIGERMAN et al. 1980; TANAKA and SUGAHARA 1980). PHILLIPS (1980) reported on designs for phase I and phase II clinical trials of this agent. The pharmacokinetic properties of WR-2721 were investigated in 13 cancer patients who received a short-term infusion of 150 mg/m^2 of the thiophosphate (SHAW et al. 1986). High pressure liquid chromatography was used to measure the concentrations of WR-2721 and its metabolic products WR-1065 (dephosphorylated form of WR-2721) and WR-33278 (disulfide or WR-1065) in the blood and tissue. The results show that WR-2721 in man is absorbed very rapidly by the tissues and is rapidly metabolized.

In a double-blind experiment the drug was given orally to adult male volunteers in doses up to 5 g in 24 h. The principal side-effects were nausea, vomiting, seizures, diarrhea, fever, a transient rise of the serum creatinine, and a fall of the serum calcium and phosphate levels (CZERWINSKI et al. 1972). TA-NAKA and SUGAHARA (1980) found that 12 of 76 patients (15.7%) who received 2 mg WR-2721/kg i. v. daily exhibited side-effects. No toxicity was observed at doses less than 1 mg/kg body weight per day.

KLIGERMAN et al. (1980) investigated the effect of various intravenous doses of WR-2721 on gastrointestinal function and blood pressure in patients recovering from radiotherapy. The doses could be increased from 25 to 250 mg/m^2 (representing a total dose of 50–500 mg) without the occurrence of significant symptoms. In 65 patients receiving palliative radiotherapy, an attempt was made to determine the maximum tolerated dose of WR-2721 in a single dose and the highest dose of WR-2721 that could be tolerated daily in the greatest number of fractions (BLUMBERG et al. 1982). The single maximum tolerated dose was not reached, although 740 mg/m^2 was well tolerated, and one patient even tolerated 910 mg/m^2. In multiple dose trials, patients received 170 mg/m^2 four times a week and experienced toxic effects such as hypotension, hypertension, vomiting, and somnolence. In addition, three patients had allergic reactions in the multiple dose trial, including one which was life threatening. The main observation in another clinical trial of single doses of WR-2721 in 188 patients was the absence of significant toxicity in 95% of the patients (KLIGERMAN et al. 1984b). A maximum tolerated dose was not reached. No deaths occurred in either the single dose trial (188 patients) or the multiple dose trial (55 patients). Significant persistent hypotension was a dose-limiting factor in 5% of the patients. The frequency of side-effects after infusion of WR-2721 is summarized in Table 5.13. The trials established 740 mg/m^2 as an acceptable tolerated dose. In a final report of the phase I trial of single-dose WR-2721 application, results are presented for two infusion schedules (TURRISI et al. 1986). For future trials, the authors recommend a dose of 740 mg/m^2 infused in 15 minutes. On the basis of animal studies, CAIRNIE (1983) concluded that the side-effects of radiotherapy in humans are exacerbated by the use of WR-2721.

TANAKA and SUGAHARA (1980) used WR-2721 in the radiotherapy of patients with head and neck tumors. For its protective effect against radiation stomatitis, the authors determined DMFs of 1.7 for reddening of the oral mucosa and 1.3 for patchy mucosa. Nuclear medicine studies in patients receiving neck radiation have shown that treatment with WR-2721 (also known in Japan as amifostine or YM-08310) decreases chronic radiation damage to the salivary glands (TAKAHASHI et al. 1986). The

Table 5.13. Side-effects in patients after infusion of WR-2721 in 15 min (KLIGERMAN et al. 1984 b)

Dose (mg/m²)	Number of patients	Emesis	Hypotension	Somnolence	Sneezing
450	8	2	1	1	3
600	5	3	2	0	1
740	30	11	0	5	12
910	8	7	0	0	3
1100	4	2	1	2	1
Total	55	25 (45%)	4 (7%)	8 (15%)	20 (36%)

protective effect is roughly equivalent to a 10% reduction in the radiation dose (TAKAHASHI et al. 1984).

WR-2721 has been used in the hemibody irradiation of advanced cancer patients with multiple metastases (CONSTINE et al. 1986). Twenty-five patients received 600–900 mg/m² WR-2721 by intravenous injection 15–30 min before exposure to 6 or 7 Gy. The leukocyte and platelet counts for up to 4 weeks after the irradiation were compared with the blood counts of 20 patients who had not received the protective drug. Recovery of the leukocyte and platelet counts was more rapid in the treated patients, who experienced no life-threatening toxic effects from the radiation, in contrast to the unprotected group.

In tumor patients undergoing chemotherapy, GLICK et al. (1982) tested WR-2721 as a selective protectant against the toxic effect of alkylating agents (see p. 123), administering doses of 450–750 mg/m². Clinical experience with WR-2721 has been summarized by KLIGERMAN et al. (1984a).

5.5 Mechanism of Action of Radioprotective Agents

The results of previous studies on radioprotectors indicate that only a few of the many agents available exert a significant effect. Thus, we may limit our discussion of protective mechanisms to a few basic groups - most notably the sulfur-containing compounds (thiophosphates and compounds containing sulfhydryl groups) and the biogenic amines, as well as certain nucleotides and vitamins.

A great many studies have dealt with the mechanisms of action of radioprotective drugs. The processes have been discussed from a variety of perspectives, according to the area of interest of the particular author. Based on research to date, it seems clear that there is no single "mechanism of action," and that numerous processes play a role, depending on the biological system and the radioprotector. As with radiation effects, several mechanisms may act concurrently. We can divide these mechanisms into three broad categories: (a) physicochemical events, (b) pharmacodynamic processes, and (c) biochemical processes occurring chiefly on the cellular level (MELCHING and STREFFER 1966).

5.5.1 Physiochemical Mechanisms of Action

As mentioned in the Introduction, the concept of achieving chemical radioprotection in experimental animals was based on the results of irradiation experiments on biomolecules in dilute aqueous solution. Two possible mechanisms were considered for the effect of protective chemicals: (a) the agent reacts competitively with the radiolysis products of water (scavenger effect); (b) the agent reacts with the unstable intermediate product of the biomolecule produced by direct and indirect radiation effects and forms a stable compound. Whether or not radioprotectors require activation via radical intermediates should be considered according to more general pharmacologic principles (TRUSH et al. 1982). The "scavenger hypothesis" was derived from experiments in aqueous polymer systems (ALEXANDER et al. 1955). The observed breakdown of the polymers after exposure to radiation was attributed to HO_2 radicals (and to $\cdot OH$ radicals to a lesser degree). It could be shown that the protective agents reacted competitively with these radicals. The same authors assumed that HO_2 radicals were the major contributor to radiation-induced death in mammals, and that, therefore, the protective mechanism in vivo was similar to that in the in vitro polymer systems (ALEXANDER et al. 1955).

Physicochemical mechanisms of protection against indirect radiation effects caused by other radiolytic products of water have been demonstrated in numerous in vitro systems (survey in NAKKEN 1965). The pulse radiolysis technique has provided key insights into the kinetics of these processes (ADAMS 1967). BUXTON et al. (1988) compiled the reaction rate constants of various sulfur-containing agents with the radiolytic products of water. The values for the reactions of some of these agents with ·OH or e_{aq} are in the range of $10^{10} M^{-1} s^{-1}$. KOCH and SEITER (1964) stated that the increase in the survival rates of irradiated mice with increasing cysteamine concentration was attributable to radical scavenging mechanisms (see also p. 99); YUHAS 1970). A corresponding effect was not obtained with increasing histamine concentrations (KOCH and SEITER 1964). CHAPMAN et al. (1973) demonstrated a hydrogen donor mechanism for cysteamine in irradiated mammalian cells growing in vitro.

PETKAU was able to show that the enzyme superoxide dismutase (SOD) protects against radiation death in mice when administered before (PETKAU et al. 1975) or after (PETKAU et al. 1976) exposure. When injected before *and* after irradiation, the enzyme provides a DRF of 1.56 (PETKAU 1978), although ABE et al. (1981) were able to achieve a DMF of only 1.05 in comparable experiments with a different mouse strain. Because SOD is a specific scavenger for superoxide (O_2^-), it is reasonable to assume that this oxygen product is formed during and after the irradiation. Superoxide itself is not very reactive, but it can be transformed to H_2O_2 and ·OH (SAWYER and VALENTINE 1981). It has been shown in vitro that cysteamine and glutathione can increase the conversion of superoxide to hydroxyl radicals in the presence of iron salts (ROWLEY and HALLIWELL 1982). The protective value of SOD is further demonstrated by the observation that the D_0 values for leukocytes, lymphocytes, and platelets rise as the concentration of endogenous SOD increases (PETKAU et al. 1978). LIPECKA et al. (1982) were able to show that the $LD_{50/30}$ of whole-body-irradiated rats depended on the endogenous SOD activity in the red blood cells. On the other hand, BARTOSZ et al. (1980) dispute the benefit of an increased erythrocyte SOD content on radiation-induced hemolysis.

LOHMANN et al. pointed to the possible role of an electron donor mechanism in radioprotection. The results of studies using electron spin resonance spectroscopy indicate that SH-containing amino acids (LOHMANN et al. 1967) and the dephosphorylated thiophosphate WR-2721 (HAHN et al. 1975) form charge transfer complexes with metallic ions (Cu^{2+}, Fe^{3+}), which protect against radiation injury by serving as electron donors. In a study investigating the synergism of thiols with zinc aspartate in whole-body-irradiated mice, FLOERSHEIM and FLOERSHEIM (1986) found that zinc aspartate enhanced the radioprotective effect of AET.

ELDJARN and PIHL (1960) and ELDJARN et al.

(1956) proposed a special mechanism for the protection of target SH and target SS groups. Their hypothesis, intended mainly to explain the effect of radioprotectors in vivo, is based on the idea that certain agents containing sulfhydryl groups form mixed disulfides with the sulfhydryl or disulfide groups of biomolecules. This process is thought to confer protection against both the direct and indirect effects of radiation. Recent studies using the pulse radiolysis technique have shown that this mechanism can promote electron migration in proteins (PRÜTZ et al. 1982). Because of the high reduction potential for the disulfide cation with $E(RSS^+ R / RSSR) = +1.3$ V (BONIFAČIĆ and ASMUS 1986), it is necessary to modify the mixed disulfide hypothesis with regard to questions of radioprotective mechanisms (PRÜTZ and MÖNIG 1987). ELDJARN and PIHL (1956) showed that thiols and disulfides administered to mice can react quickly with the protein SH and protein SS in the body. The formation of mixed disulfides with human and rabbit plasma proteins has also been demonstrated for WR-2721 (TABACHNIK et al. 1982). After oral administration, this agent converts to the free thiol form in the acid medium of the stomach. WR-2721 has been utilized as a mucolytic agent in patients with cystic fibrosis (TABACHNIK et al. 1982).

RÉVÉSZ and MODIG (1965) proposed modifying the mixed disulfide model. They showed that the formation of mixed disulfides liberates glutathione from binding sites on the cellular protein, and this led them to conclude that glutathione is the real protector. In particular, the finding that cysteamine forms mixed disulfides with histones in Ehrlich ascites tumors, with glutathione liberated in very close proximity to the radiosensitive DNA, is believed to support this idea (MODIG 1973). A number of studies have described a positive correlation between radioresistance and the content of sulfhydryl groups that are not bound to proteins (RÉVÉSZ et al. 1963; SZUMIEL 1981). The major source of these sulfhydryl groups is endogenous glutathione. RÉVÉSZ (1985) has reviewed the importance of endogenous thiols for radioprotection. However, HARRIS and POWER (1973) showed that endogenous glutathione is of minimal importance for cellular radiosensitivity, and that the total content of SH groups correlates positively with radioresistance. This implies that the release of glutathione does not contribute significantly to radioprotection (POWER et al. 1974). It is noteworthy in this regard that mammalian cells with a genetic defect of glutathione biosynthesis are exceptionally sensitive to radiation (DESCHEVANNE et al. 1981). These data make it clear that the content of glutathione and SH groups generally does not determine the radiosensitivity of cells, but that it apparently can have an influence along with other endogenous factors.

Major importance is placed on the interaction of sulfur-containing compounds with the radiosensitive structures of the cell (see above, MODIG 1973). JELLUM (1965) showed that disulfides of various radioprotective drugs [cystamine, cystamine derivatives, the disulfide of MEG (di-(2-guanidoethyl)-disulfide)] can bind reversibly to DNA, RNA, and nucleoproteins, thereby increasing the thermostability of the DNA and RNA. The raised melting point of the DNA-drug complex has been confirmed by spectrophotometric studies (VASILESCU and RIX-MONTEL 1980; RIX-MONTEL et al. 1982). The aminothiols used in these experiments (cysteamine, cysteamine derivatives, and thiophosphates) undergo an electrostatic interaction with the phosphate groups of the DNA. Quantum mechanical simulations performed by the same group of authors support this assumption (BROCH et al. 1980; VASILESCU et al. 1986). Infrared and Raman spectroscopy helped LIQUIER et al. (1983) to develop a model for the mode of interaction of cysteamine with DNA. In this model the cysteamine molecule is bound at its two ends by electrostatic interaction with two consecutive phosphate groups of the same strand.

BACQ (1975) and CZAPSKI (1984) states that the foregoing physicochemical mechanisms are of minor importance in mammals, because the concentration of the protective agents in the cell is too low. The highest level ever reached in mammalian cells is 0.01 wt: % (BACQ 1975). This reasoning is supported by studies in irradiated hibernators showing that cysteine injected intraperitoneally at the time of awaking 3 weeks *after* radiation exposure exerts a total protective effect (KÜNKEL and HECKMANN 1958). It has been suggested, however, that the concentrations close to the cellular target sites are the same as the extracellular levels (CHAPMAN et al. 1973). In studies of fibroblasts in Chinese hamsters, EIDUS et al. (1982) showed that the radioprotective efficacy of cysteamine, thioglycolic acid, caffeine, and caffeine benzoate (weak electrolytes) correlated with their intracellular concentration, which in turn depends strongly on the pH of the medium.

5.5.2 Pharmacodynamic Mechanisms of Action

Oxygen and temperature are the prime determinants of the pharmacodynamic effects of radioprotective drugs. Both factors greatly influence the biological effects of radiation.

Hypoxia and anoxia lower the sensitivity of mammalian cells to radiation having a low LET (survey in VAN DEN BRENK 1969). HASEGAWA and LANDAHL (1967) demonstrated a quantitative link between the relative effectiveness of the radiation and the relative oxygen tension in the spleen and vena cava of mice (see Fig. 5.13). The quantity $f \cdot (1/DRF)$ is plotted on the ordinate in Fig. 5.13, where f is a factor that is valid for the entire range of oxygen tensions and which shifts the experimentally derived values onto the straight line for hypoxia. The factor f represents the component that is not related to oxygen tension. Thus, for example, serotonin gave a total DRF of $1.4 \cdot 1.26 = 1.77$ in the studies of HASEGAWA and LANDAHL (1967). The DRF of 1.26 with serotonin is due to the effect of hypoxia. The residual protective factor of 1.4 relates to mechanisms that are independent of hypoxia.

GRAY et al. (1952) conducted early studies on the protective efficacy of serotonin, which, as a vasoconstrictor, was thought to act by lowering the oxygen tension in the tissues. SUPEC et al. (1961) disputed the importance of the vasoconstrictive effect of

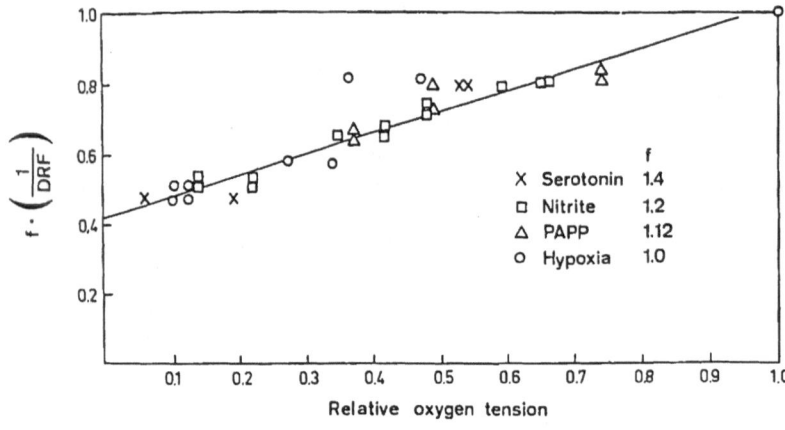

Fig. 5.13. Relations between relative effectiveness of radiation (1/DRF) and relative oxygen tension in the spleen and vena cava of mice after treatment with various protective drugs and under hypoxic conditions. Explanation in text. (After HASEGAWA and LANDAHL 1967)

serotonin, pointing out that psilocybin, described by GESSNER et al. (1960) as an even more potent vasoconstrictor, exerts no protective effect. BACQ (1963) also discounted the importance of vasoconstrictive properties in radioprotection by noting that tryptamine exerts a good protective effect, yet is a less potent vasoconstrictor than serotonin.

Vos et al. (1962) have shown that serotonin does not improve the survival rate of mammalian cells that have been cultured and irradiated in vitro. This finding clearly shows that the electron donor properties of serotonin are not sufficient to account for its radioprotective effect and implies the necessity of pharmacodynamic mechanisms, which may be mediated by the CNS. This concept is supported by a study of survival rates in X-irradiated mice that received serotonin by intracerebral injection before irradiation (STREFFER and FLÜGEL 1972). This mode of administration gave the same DRF as intraperitoneal injection. However, the intracerebral dose of serotonin could be decreased by a factor of 5 and still produce the same effect. Since another study has shown that serotonin given by intraperitoneal injection can readily enter brain areas like the hypothalamus and brain stem (KONERMAN and STREFFER 1974), we have strong evidence for CNS activity of this agent in radioprotection. Changes in liver metabolism also have been noted following the intraperitoneal and intracerebral injection of serotonin in mice (STREFFER 1977). This involved a substantial rise in the lactate-pyruvate ratio, a key index of intracellular redox equilibrium. These changes indicate that serotonin produces a condition of marked tissue anoxia, which apparently is mediated by central nervous mechanisms.

VAN DEN BRENK and HAAS (1961) assume a general, pharmacodynamically induced anoxia of the tissues and a lowering of the oxygen tension, while VAN DER MEER and VAN BEKKUM (1961) relate the protective action of serotonin in mice to a lowering of the spleen oxygen tension (see also Fig. 5.13). The notion of a general spleen effect is weakened by the finding of MELCHING et al. (1964) that the efficacy of serotonin is reduced in splenectomized animals, but not nearly to the degree caused by the administration of serotonin-specific antagonists or antimetabolites. The influence of serotonin on radiation effects in mammals, and the importance of pharmacodynamic effects, have been summarized by MELCHING (1965).

Changes in tissue oxygen tension also have been assumed or demonstrated for sulfur-containing agents like cysteine (MAYER and PATT 1953) and AET (DISTEFANO et al. 1960), with large differences

noted from one animal species to the next (DISTEFANO et al. 1962). But sodium thiophosphate, which is not a radioprotector, also leads to a fall in blood pressure and oxygen tension (CLEMEDSON et al. 1957). The general conclusion has been that the effect of sulfhydryl-containing radioprotectors is not based on a change of intracellular oxygen tension in the tissues (JAMIESON and VAN DEN BRENK 1966; BETZ et al. 1967). Studies with cysteamine (CRIBORN and RÖNNBÄCK 1979) and several thiophosphates (ÅKERFELDT et al. 1967) in mice have shown a decrease in oxygen consumption, respiration, and rectal temperature following injection of the compounds. Amphetamine given before the cysteamine injection compensates for the decrease in these parameters and significantly reduces the radioprotective effect.

The alteration of hemodynamic parameters in the rat such as arterial blood pressure, heart rate, and cardiac output following the use of cystamine and WR-2721 does not appear to correlate with radioprotective effect (KUNA et al. 1983b). Even so, the pharmacological actions of WR-2721 have been the object of detailed investigations (YUHAS et al. 1973). The injection of this agent in mice elicits a dramatic vasodilation that is most pronounced in the spleen. Tolerance to the toxic effects of WR-2721 is increased in splenectomized animals, but the protective effect is diminished. Thus at least part of the effect of WR-2721 is attributed to its ability to lower the peripheral oxygen tension (YUHAS et al. 1973). However, nonpharmacological studies in cell cultures (human renal cells) using cysteamine and the dephosphorylated form of WR-2721, known as WR-1065, have shown that both thiols rapidly consume all oxygen in the culture medium (PURDIE et al. 1983). This suggests strongly that the thiols can produce a local hypoxia in tissues, leaving few other molecules with SH groups to serve as hydrogen donors. This thesis is supported by the results of a study of the effect of WR-2721 on irradiated Chinese hamster cells grown as multicell spheroids (DURAND 1983).

The effect of the oxygen concentration in the tissue on the effectiveness of the thiophosphates makes it difficult to elucidate their mechanism of action. This property is characteristic not only of the thiophosphates, but of pharmacological agents in general (JONES 1981). Studies by HARRIS and PHILLIPS (1971) have shown that thiophosphates protect the bone marrow stem cells of mice less effectively under conditions of hypoxia than when ambient air is breathed (see Table 5.6). UTLEY et al. (1974) made the same observation in mouse mam-

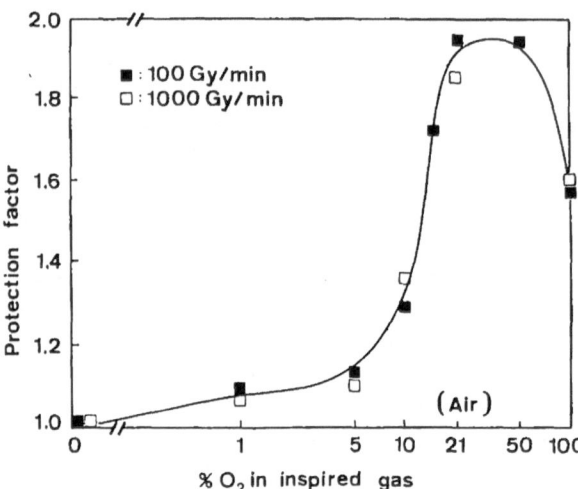

Fig. 5.14. Chemical radioprotection of the mouse skin by WR-2721 as a function of the oxygen content of the inspired air (after DENEKAMP et al. 1982)

mary tumors (EMT-6 carcinoma). However, the tumor showed increased radioresistance under conditions of hyperbaric oxygenation, which the authors attribute to the collapse of vascular channels. Clearly, these questions relating to the use of protective drugs are of extreme importance in the radiotherapy of tumors. Recently, very detailed studies have been done on oxygen as a factor influencing the effect of WR-2721 on the skin of irradiated mice (DE- NEKAMP et al. 1981, 1982; STEWART and ROJAS 1982). The experiments, which extended over a wide range of oxygen concentrations, have shown that the highest degree of radioprotection is achieved at a moderate oxygen concentration (see Fig. 5.14) corresponding to the radiobiological oxygen constant K. The value K represents the oxygen concentration at which radiosensitivity is exactly between the values for anoxic cells and cells that have a full oxygen supply (ALPER and HOWARD- FLANDERS 1956).

Since HOPE (1958) first described a parallel between radioprotective effect and hyperthermia in mice treated with AET and PAPP, there has been controversy concerning the link between temperature reduction and chemoprotection. Indeed, an externally induced hypothermia in rats, rectal-temperature 14-15 °C (HAJDUKOVIC et al. 1954), and in mice, rectal temperature below 20 °C (HORNSEY 1957), did lead to a increase in radioresistance. But BACQ (1965) points out that such low temperatures are never produced by the injection of radioprotective drugs. A mild hypothermia in mice can even cause radiosensitization (BLOOM and DAWSON 1961). On the other hand, increased radioresistance

has been observed after artificial hibernation in mice and rats (LANGENDORFF and KOCH 1955; CRI- BORN and RÖNNBÄCK 1979).

There have been many studies dealing with the effects of radioprotective agents on body temperature. In mice, treatment with AET (ASHWOOD- SMITH and SMITH 1959), cystamine (BETZ et al. 1962), cysteamine, diethyldithiocarbamate, or serotonin (LIÉBECQ-HUTTER and BACQ 1958) lowers the body temperature. Cysteine has a negligible effect on body temperature, even in large doses (LIÉBECQ- HUTTER and BACQ 1958). BACQ et al. (1965) deny that a connection exists between the protective efficacy of cysteamine, cystamine, serotonin, cyanide, and diethyldithiocarbamate and the fall in temperature produced by those agents. On the other hand, weak radioprotective effects are observed with fluoracetate, chlorpromazine, and reserpine when the body temperature is very low (BACQ et al. 1965). In a study on the efficacy of WR-2721 in mice, TRAVIS et al. (1982) noted a marked fall of rectal temperature between 30 min and 3 h following intraperitoneal or intravenous injection of the drug. This hypothermia does not appear to relate to the protective mechanism, because intravenous injection leads to higher DMFs than intraperitoneal injection, but the mode of administration does not affect the fall in temperature.

Polyethylene glycol is able to prevent a fall of rectal temperature in mice by about 4 °C following head and neck irradiation (SCHAEFFER et al. 1986). The temperature fall occurs 9 days after the exposure. When the drug is administered 20 min before whole-body irradiation, the $LD_{50/15}$ increases from 15.3 Gy to 19.9 Gy, which corresponds to a DMF of 1.24.

5.5.3 Biochemical Mechanisms of Action

The observation that a number of commonly used radioprotective chemicals occur naturally in mammals led to speculation as to whether exposure to high energy radiation might alter the endogenous content of these substances. This in turn raised the question of whether radiation-induced changes in endogenous chemicals might contribute to the protective effect of exogenous drugs. A number of authors have studied radiation-induced changes in the endogenous content of histamine, epinephrine, norepinephrine, serotonin and of compounds containing sulfhydryl groups. A detailed summary of these studies may be found in MELCHING and STREFFER (1966).

Studies on DNA synthesis in the spleen, thymus,

and liver of cysteamine-protected mice led HAGEN et al. (1958) to speculate that radioprotection by sulfhydryl compounds may be possible only when the cells are in the resting stage. In connection with the interaction of sulfur-containing radioprotective agents and DNA (see Sect. 5.5.1), KOVACS et al. (1972) observed in studies on the effect of cysteine on hematopoietic tissue that the bone marrow stem cells (CFUs) of mice were reduced 4 days after injection of the agent. They also noted a reduction of ^{14}C-thymidine incorporation into the DNA of the bone marrow cells 30 min after cysteine injection. The authors state that this finding probably results from an inhibitory effect of the cysteine on DNA synthesis during the S phase of the generative cycle of proliferating cells. A similar process is assumed to take place in mouse liver tissue following cysteamine use (MITZNEGG 1973b). Presumably this makes it possible for the radiation-damaged DNA to undergo repair before replication of the DNA occurs. This assumption has been confirmed by results on DNA synthesis in the mouse spleen after irradiation and pretreatment with serotonin (STREFFER 1974). Serotonin reduces DNA synthesis in the spleen. It cannot prevent the further radiation-induced depression that occurs several hours after the exposure; however, DNA synthesis recovers in subsequent days much faster in the animals that received serotonin before irradiation. Apparently a decrease in energy metabolism contributes to the observed effect (STREFFER 1974). VLADIMIROV et al. (1979) also describe an inhibition of DNA synthesis in the mouse spleen by cystamine, cystaphos (WR-638), gammaphos (WR-2721), mexamine, 5-amino-2,1,3-benzothiadiazole, and PAPP. A related study deals with the effect of AET, cysteamine, cysteine, glutathione, mercaptoethanol, mercaptopropionyl-glycine, and other compounds on the semiconservative and "unscheduled" incorporation of ^{3}H-thymidine into the DNA of rat thymocytes (TEMPEL et al. 1982). The results support the hypothesis that pro-

longation of the repair process with concurrent reversible inhibition of semiconservative DNA synthesis by aminothiols enhances the capacity for cellular repair. The mechanism observed in these studies is consistent with the hypothesis of BACQ and GOUTIER (1967) that the delay in the synthesis of DNA or in the mitotic phase produced by sulfur-containing agents increases the efficiency of the DNA repair process. This idea relates to another concept of BACQ and GOUTIER - that of "biochemical shock" as the cause of the resistance increase conferred by chemical agents.

The concept of "biochemical shock" further relates to a hypothesis of LANGENDORFF (1970) that a radioprotective chemical reacts with specific receptors on the surface of the plasma membrane. This triggers a stimulation of the adenyl cyclase system, leading to a normalization of various regulatory disturbances caused by the radiation. The effect of a β-receptor blocking drug on the efficacy of various radioprotective compounds is shown in Table 5.14. The data indicate that radioprotection is drastically reduced in some cases by treatment with the receptor inhibitor. The survival rate of irradiated cell cultures from Chinese hamsters has been increased by stimulating the intracellular content of cAMP with prostaglandins before irradiation or by the inhibition of phosphodiesterase, an enzyme that degrades cAMP (PRASAD 1972; LEHNERT 1975, 1979a,b). A rise of intracellular cAMP is considered as a possible mechanism for the radioprotection of the jejunal mucosal epithelium and hemopoietic stem cells observed after the injection of prostaglandins in mice (HANSON and AINSWORTH 1985). Increased radioresistance also has been found in human breast cancer cells when irradiation was preceded by phosphodiesterase inhibition (GRIEM et al. 1983). The authors point to the implications of these findings for radiotherapy.

MITZNEGG (1973b) showed in the study previously cited that cysteamine use in mice leads to a

Table 5.14. Effect of the β-adrenergic receptor blocker LB-46 (Visken) in mice treated with various radioprotectors before X-irradiation (LANGENDORFF and LANGENDORFF 1971)

Radioprotector	Radiation dose (Gy)	Surviving animals		Probability of error
		Without β blocker	With β blocker	
AET	11.2	50.0%	8.0%	<0.001
Cystamine	11.2	66.0%	37.0%	<0.01
Serotonin	10.5	56.0%	42.0%	>0.5
Tace[a]	8.1	54.0%	2.0%	<0.001
LPS[b]	8.1	59.6%	44.0%	<0.01

[a] Tace, 1,1,2-tri-n-anisyl-2-chlorethylene. [b] LPS, lipopolysaccharide of *E. coli*

rise of endogenous cyclic AMP in the liver cells. The author assumes that the protective effect of cysteamine is mediated by cyclic AMP as a second messenger. Cysteamine also has been shown to increase adenyl cyclase activity in the spleen, as has exposure to radiation alone (SOLTYSIAK-PAWLU-CZUK and BITNY-SZLACHTO 1976). No change in adenyl cyclase activity was found in animals treated with cysteamine and also exposed to high radiation doses. The action of physiological concentrations of catecholamines and dibutyryl cAMP on bone marrow stem cell solutions from mice increases the survivability of the stem cells (KULINSKII et al. 1977). The authors state that the radioprotection must be mediated by cyclic nucleotides in a process whereby the catecholamines activate adenylate cyclase via adrenoreceptors, and exogenous dibutyryl cAMP is able to enter the cell. The ability of catecholamines to raise the cAMP level in the bone marrow has been confirmed experimentally. VLADIMIROV et al. (1979) investigated the change in the cAMP content of the mouse spleen caused by various radioprotective drugs. They found that cystamine, mexamine, 5-amino-2,1,3-benzothiadiazole, and PAPP increased the cAMP content by 40%–110% of the control value, while cystaphos (WR-638) had no effect, and gammaphos (WR-2721) lowered the cAMP content in accordance with the dose administered. After administering WR-2721 in rats, TROCHA and CATRAVAS (1983) similarly found no rise of cAMP in the liver and spleen of nonirradiated animals, but the cAMP level was higher in irradiated rats that had been pretreated with WR-2721. The authors conclude that this factor and the other prostaglandin changes investigated may have an important bearing on the mechanism of radioprotection.

The influence of cAMP on the radiosensitivity of cells is further demonstrated by experiments in which the intraperitoneal injection of cAMP in mice 4 h or 3 days before irradiation led to a higher survival rate of the stem cells of the small intestine and the hair follicles (DUBRAVSKY et al. 1978). KIMURA et al. (1981) also report that the radioresistance of cultured mammalian cells is enhanced by dibutyryl cAMP. However, it was necessary to incubate the cells with this substance for several hours before and after irradiation to obtain a significant protective effect. Also, the concentration of the substance (1 mM) was relatively high. It is interesting to note in this context that the noncyclic mononucleotides of adenosine can also exert a good radioprotective effect when injected intraperitoneally in mice before irradiation. DRFs of approximately 1.5

have been reported (LANGENDORFF et al. 1962). These effects are increased when serotonin also is administered. This provides clear evidence that different mechanisms seem to be operative in the two classes of radioprotective drug.

Treatment with pyridoxal-5-phosphate, the coenzyme of decarboxylases and other enzymes, can also produce a radioprotective effect that is enhanced by the adenosine nucleotides (LANGENDORFF et al. 1960). Based on observations following the use of pyridoxal-5-phosphate and the determination of metabolites and enzyme activities that require pyridoxal-5-phosphate as a coenzyme, considerable importance has been ascribed to these metabolic processes in terms of the radiosensitivity of mammals and their modification after whole-body exposure (STREFFER 1971). LADNER et al. (1980) have shown that similar changes occur in tumor patients receiving therapeutic radiation. These data redirect our attention to the importance of endogenous substances and metabolic systems in chemical radioprotection. The mechanisms have been discussed by MELCHING and STREFFER (1966).

The diverse results of in vitro and in vivo investigations suggest that radioprotective agents can act through a great variety of mechanisms. While some agents contribute to radioprotection by modifying radiochemical reactions, others apparently exert their effect through biological phenomena, such as interaction with the receptors of biogenic amines. Observations that combinations of radioprotectors can exert a greatly enhanced effect give further evidence of the diversity of chemoprotective mechanisms. A major problem in human medicine is the need to use near-toxic amounts of the radioprotective drugs to achieve adequate protection. Another problem is the relatively short duration of action of most radioprotective drugs (30–60 min at most), and major improvements are needed before clinical use (e. g., in tumor therapy) becomes feasible.

For consistency in the reporting of data on radiation doses, we converted data expressed in roentgens (R) or rads to the SI unit gray (Gy), with 100 rads = 1 Gy. We converted the special unit of exposure R to Gy by assuming that 100 R ≙ 1 Gy, which may introduce a small error not exceeding 10%.

References

Abe M, Nishidai T, Yukawa Y, Takahashi M, Ono K, Hiraoka M, Ri N (1981) Studies on the radioprotective effects of superoxide dismutase in mice. Int J Radiat Oncol Biol Phys 7: 205–209

Adams GE (1967) The general application of pulse radiolysis to current problems in radiobiology. In: Ebert M, Howard A (eds) Current topics in radiation research, vol III. North Holland, Amsterdam, pp 35-93

Ainsworth JE, Forbes PD (1961) The effect of pseudomonas pyrogen on survival of irradiated mice. Radiat Res 14: 767-774

Airapetyan GM, Zherebchenko PG (1964) Some characteristics of the radioprotective properties of the monosodium salt of beta-aminoethylisothiophosphoric acid. Radiobiologiia 4: 259-265. Translat Ser Radiobiol (Washington) 4: 112-121

Åkerfeldt S, Rönnbäck C, Nelson A (1967) Radioprotective agents: results with S-(3-amino-2-hydroxypropyl)phosphorothioate, amidophosphorothioate and some related compounds. Radiat Res 31: 850-855

Åkerfeldt S, Rönnbäck C, Hellström M, Nelson A (1968) Radioprotective agents: further results with amidophosphorothioate and related compounds. Radiat Res 35: 61-67

Alexander P, Bacq ZM, Cousens SF, Fox M, Herve A, Lazar J (1955) Mode of action of some substances which protect against the lethal effects of x-rays. Radiat Res 2: 392-415

Alper T, Howard-Flanders P (1956) Role of oxygen in modifying the radiosensitivity of E. coli B. Nature 178: 978-979

Alper T, Fowler JF, Morgan RL, Vonberg DD, Ellis F, Oliver R (1962) The characterization of the "type C" survival curve. Br J Radiol 35: 722-723

Andrews JR, Sneider SE (1957) The modification of the radiation response. AJR 81: 485-497

Arndt D, Ritter M (eds) (1979) Handbuch für medizinische Fragen des Strahlenschutzes. Militärverlag der Deutschen Demokratischen Republik, Berlin

Ashwood-Smith MJ, Smith AD (1959) Radioprotective action of S-alkyliso-thiuronium salts. Int J Radiat Biol 1: 196-198

Bacq ZM (1954) The amines and particularly cysteamine as protectors against roentgen rays. Acta Radiol [Ther] (Stockh) 41: 47-55

Bacq ZM (1956) Efficacité et absence de toxicité de la cystamine en ingestion chez le rat. Bull Acad R Méd Belg VIth series 21: 121-129

Bacq ZM (1963) Radioprotection - the case of 5-hydroxytryptamine (5-HT). Strahlenschutz Forsch Prax 2: 172-175

Bacq ZM (1965) Chemical protection against ionizing radiation. Thomas, Springfield/Il

Bacq ZM (1975) Importance of pharmacological effects for radioprotective action. In: Bacq ZM (ed) Sulfur-containing radioprotective agents. International encyclopedia of pharmacology and therapeutics. Pergamon, Oxford, pp 319-323

Bacq ZM, Goutier R (1967) Mechanism of action of sulfur-containing radioprotectors. In: Recovery and repair mechanisms in radiobiology, no 20. Brookhaven Symposia in Biology, pp 241-262

Bacq ZM, Hervé A (1952) Sur un nouveau protecteur contre le rayonnement X. Schweiz Med Wochenschr 1952: 1018-1020

Bacq ZM, Hervé A, Lecomte J et al. (1951) Protection contre rayonnement X par la β-mercaptoéthylamine. Arch Internat Physiol 59: 442-447

Bacq ZM, Deschamps G, Fischer P et al. (1953a) Protection against x-rays and therapy of radiation sickness with β-mercaptoethylamine. Science 117: 633-636

Bacq ZM, Hervé A, Scherber F (1953b) Action de la mercaptoéthylamine sur la régénération des leucocytes chez la souris après irradiation aux rayons x. Arch Int Pharmacodyn 94: 93-102

Bacq ZM, Beaumariage ML, Radivojevitch DV (1961) Protection chimique locals et générale contre l'épilation par le rayonnement X. Bull Acad Roy Méd Belg, VIIth series 1: 519-550

Bacq ZM, Beaumariage ML, Liébecq-Hutter S (1965) Relation entre la radio/protection et l'hypothérmie induite par certaines substances chimiques. Int J Radiat Biol 9: 175-178

Baldini G, Ferri L (1957) Experimental and clinical research on the radioprotective action of cysteamine and cystamine. III. Clinical research. Br J Radiol 30: 271-273

Barnes JH, Fatome M, Esslemont GF, Andrieu L (1977) Prolonged radioprotective action of water-soluble polymers with labile sulphur-containing side-chains. Eur J Med Chem-Chim Therap 12: 467-470

Barron ESG, Dickman S, Muntz JA, Singer TP (1949) Studies on the mechanism of action of ionizing radiation. I. Inhibition of enzymes by X rays. J Gen Physiol 32: 537-552

Bartosz G, Leyko W, Kedziora J, Jeske J (1980) Superoxide dismutase and radiation-induced haemolysis: no benefit of its increased content in red cells. Int J Radiat Biol 38: 187-192

Bäumer J, Hofmann D, Kepp RK (1953) Die Strahlenschutzwirkung des Cysteins bei dem Asziteskarzinom der Maus. Strahlentherapie 92: 25-32

Beccari E, Bianchi C, Felder E (1955) Chemisch-physikalische, pharmakologische und klinische Untersuchungen über β-Mercaptoäthylamin, besonders im Hinblick auf die Bleivergiftung. Arzneimittelforschung 5: 421-428

Behling UH (1983) The radioprotective effect of bacterial endotoxin. In: Nowotny A (ed) Beneficial effects of endotoxins. Plenum, New York, pp 127-148

Behling UH, Nowotny A (1978) Long-term adjuvant effect of bacterial endotoxin in prevention and restoration of radiation-caused immunosuppression. Proc Soc Exp Biol Med 157: 348-353

Beliles RP, Kereiakes JG, Krebs AT (1959) Influence of cystein on the intestinal epithelium of x-irradiated rats. J Natl Cancer Inst 22: 1045-1057

Benita S, Shani J, Abdulrazik M, Samuni A (1984) Controlled release of radioprotective agents from matrix tablets - effect of preparative conditions on release rates. J Pharm Pharmacol 36: 222-228

Benjamin E, Sluka E (1908) Antikörperbildung nach experimenteller Schädigung des haematopoetischen Systems durch Röntgenstrahlen. Wien Klin Wochenschr 21: 311-312

Benson RE, Michaelson SM, Downs WL, Maynard EA, Scott JK, Hodge HC, Howland JW (1961) Toxicological and radioprotection studies on S,β-aminoethylisothiuronium bromide (AET). Radiat Res 15: 561-572

Betz EH, Mewissen DJ, Lelièvre P (1962) Protective effectiveness of cystamine versus delay of exposure, body temperature and protein linkage. Int J Radiat Biol 4: 231-238

Betz EH, Lelièvre P, Smoliar V (1967) Protective effectiveness of some sulphur-containing substances and oxygen uptake in the rat. Int J Radiat Biol 12: 163-168

Bewley DK (1968) Calculated LET distributions of fast neutrons. Radiat Res 34: 437-445

Bhartiya HC, Jaimala (1986) Inhibition of reduction in the testicular weight by WR-2721 in relation to the body weight after whole-body gamma irradiation. Strahlenth Onkol 162: 68-70

Bhatia AL, Sisodia R, Saraswat A (1986) Role of 2-mercaptopropionylglycine against ^{3}H β-rays in the development of mice brain. Curr Sci 55: 163-165

Bianchi E, Gasparini S (1955) L'aumentata resistenza tissulare cutanca alle radiazone X in rapporto alla somministrazione di beta mercaptoetilamina al 0,5% per infiltrazione sottocutanea nel tratto da irradiare. Radiologia (Roma) 11: 1137-1153

Bloom HJG, Dawson KB (1961) Enhanced effect of total-body x-irradiation in mice under mild hypothermia. Nature 192: 232-233

Blumberg AL, Nelson DF, Gramkowski M, Glover D, Glick JH, Yuhas JM, Kligerman MM (1982) Clinical trials of WR-2721 with radiation therapy. Int J Radiat Oncol Biol Phys 8: 561-563

Bogo V, Jacobs AJ, Weiss JF (1985) Behavioral toxicity and efficacy of WR-2721 as a radioprotectant. Radiat Res 104: 182-190

Bonifačić M, Asmus K-D (1986) One-electron redox potentials of RSSR$^+$-RSSR couples from dimethyl disulphide and lipoic acid. J Chem Soc, Perkin Trans 2 (11): 1805-1809

Boone IU (1961) Effect of preirradiation treatment with glutathione on lifespan and tumor incidence in CF$_1$ mice. Radiat Res 14: 453

Braun H, Koch R (1968) Untersuchungen über einen biologischen Strahlenschutz. 86. Mitt Veränderungen der Mitochondrien nach strahlenschützenden Sulfhydrylkörpern bzw nicht schützenden Homologen. Strahlentherapie 135: 628-631

Braun W, Kirnberger E-J, Stille G, Wolf V (1959) Vergleich einiger Sulfhydrylverbindungen im Strahlenschutz nach Wirkungsgrad und Zeitabhängigkeit. Strahlentherapie 108: 262-268

Brecher G, Cronkite EP, Peers JH (1953) Neoplasms in rats protected against lethal doses of irradiation by parabiosis or para-aminopropiophenone. J Natl Cancer Inst 14: 159-175

Broch H, Cabrol D, Vasilescu D (1980) Quantum mechanical simulation of the interaction between the radioprotector cysteamine and DNA. Int J Quantum Chem: Quantum Biol Symposium 7: 283-295

Broerse JJ, Barendsen GW, Keesen GR van (1968) Survival of cultured human cells after irradiation with fast neutrons of different energies in hypoxic and oxygenated conditions. Int J Radiat Biol 13: 559-572

Brown DQ, Pittock JW, Rubinstein JS (1982) Early results of the screening program for radioprotectors. Int J Radiat Oncol Biol Phys 8: 565-570

Buxton GV, Greenstock CL, Helman WP, Ross AB (1988) Critical Review of rate constants for reactions of hydrated electrons, hydrogen atoms and hydroxyl radicals ($\cdot$OH/ $\cdot$O$^-$) in aqueous solution. J Phys Chem Ref Data 17: 513-886

Cairnie AB (1983) Adverse effects of the radioprotector WR2721. Radiat Res 94: 221-226

Chapman JD, Reuvers AP, Borsa J, Greenstock CL (1973) Chemical radioprotection and radiosensitization of mammalian cells growing in vitro. Radiat Res 56: 291-306

Chatterjee RA, Bose P de (1959) Evaluation of radioprotective efficacy of cysteamine (with rat-liver-cell glycogen as reference system). Int J Radiat Biol 4: 420-424

Chertkov KS, Talosh M, Mosina ZM, Preobrazhenskii Yu Yu (1979a) Radioprotective efficacy of tilorone. Radiobiologiia 19: 455-458. Translat Ser Radiobiol (Washington) 19: 163-166

Chertkov KS, Rogozkin VD, Dikovenko EA, Mosina ZM (1979b) Kinetics of hemopoietic stem cells and survival of mice treated with hydroxyurea and exposed to prolonged γ-radiation. Radiobiologiia 19: 863-867

Clemedson C-J, Frederikson T, Sörbo B (1957) On the biochemistry and pharmacology of sodium monothiophosphate. Acta Physiol Scand 41: 269-276

Cohen A, Cohen L (1962) Effects of aminoethylisothiouronium bromide and 5-hydroxytryptamine on the response of C$_3$H mammary tumor isografts to irradiation in vivo. Br J Radiol 35: 200-204

Cohen L, Cohen A (1959) Experimental evaluation of systemic medication (cysteamine, menadione, flavonoids and corticoids) modifying reactions to radiotherapy. Br J Radiol 32: 18-21

Condit PT, Levy AH, Scott EJ van, Andrews JR (1955) Some effects of β-amino-ethylisothiuronium bromide (AET) in man. J Pharmacol Exp Ther 122: 13 A-14 A

Connor AM, Sigdestad CP (1982) Chemical protection against gastrointestinal radiation injury in mice by WR 2822, WR 2823, or WR 109342 after 4 MeV x ray or fission neutron irradiation. Int J Radiat Oncol Biol Phys 8: 547-551

Constine LS, Zagars G, Rubin P, Kligerman M (1986) Protection by WR-2721 of human bone marrow function following irradiation. Int J Radiat Oncol Biol Phys 12: 1505-1508

Cosgrove GE, Upton AC, Congdon CC, Doherty DG, Gosslee DG (1963) Effects of AET and bone marrow on delayed somatic effects of radiation in mice. Radiat Res 19: 231

Cosgrove GE, Upton AC, Congdon CC, Doherty DG, Christenberry KW, Gosslee DG (1964) Late somatic effects of X-radiation in mice treated with AET and isologous bone marrow. Radiat Res 21: 550-574

Court-Brown WM (1955) A clinical trial of cysteinamine (beta-mercaptoethylamine) in radiation sickness. Br J Radiol 28: 325-326

Criborn C-O, Rönnbäck C (1979) Pharmacologic effects of radiation protective compounds related to their protective effect in mice. Acta Radiol Oncol 18: 31-44

Cudkowicz G (1961) Mammary gland neoplasia in irradiated rats given the radioprotective drug AET. Proc Am Assoc Cancer Res 3: 217

Cudkowicz G (1962) Relative inability of AET and APMT to protect immunologically competent cells against radiation injury. Transplant Bull 29: 109-112

Czapski G (1984) On the use of OH$^\cdot$ scavengers in biological systems. Israel J Chem 24: 29-32

Czerwinski AW, Czerwinski AB, Clark MC, Whitsett TL (1972) A double blind comparison of placebo and WR-2721 AE in normal adult volunteers. Report MCA 1-33 to US Army Medical Research and Development Command

Dacquisto MP, Blackburn EW (1961) Protective effect of orally administered S,β-aminoethylisothiuronium $\cdot$ Br$\cdot$HBr against x-radiation death in mice. Nature 190: 270 only

Davidson DE, Grenan MM, Sweeney TR (1980) Biological characteristics of some improved radioprotectors. In: Brady LW (ed) Radiation sensitizers. Masson, New York, pp 309-320

Denekamp J, Michael BD, Rojas A, Stewart FA (1981) Thiol radioprotection in vivo: the critical role of tissue oxygen concentration. Br J Radiol 54: 1112-1114

Denekamp J, Michael BD, Rojas A, Stewart FA (1982) Radioprotection of mouse skin by WR-2721: The critical influence of oxygen tension. Int J Radiat Oncol Biol Phys 8: 531-534

Desaive P (1954) Influences du monde d'irradiation, de l'hypophysectomie, des hormones gonadotropes et des radioprotecteurs chimiques sur la réponse de l'ovair le lapine aux rayons röntgen. Acta Radiol [Ther] (Stockh) 41: 545-557

Desaive P, Bacq Z, Herve A (1952) Causes de la stérilité des souris femelles irradiées in toto et protégées par la cystinamine. Experientia 8: 436-437

Desaive P, Bacq Z, Hervé A (1953) Du comportement des testicules chez des souris irradiées et protégées par un sel bécaptan. J Belge Radiol 36: 504-525

Deschevanne PJ, Midander J, Edgren M, Larsson A, Malaise E, Revesz R (1981) Oxygen enhancement of radiation induced lethality in glutathione deficient fibroblasts. Biomedicine 35: 35-37

Dev PK, Gupta SM, Goyal PK, Mehta G, Pareek BP (1981) Radioprotective effect of MPG (2-mercaptopropionylglycine) on the postnatal growth of mice irradiated in utero. Strahlentherapie 157: 553-555

Dev PK, Gupta SM, Goyal PK, Mehta G, Pareek BP (1982) Protection from body weight loss by 2-mercaptopropionylglycine (MPG) in growing mice irradiated in utero with gamma radiation. Experientia 38: 962-963

Devik F, Lothe F (1955) The effect of cysteamine, cystamine and hypoxia on mortality and bone marrow chromosome aberrations in mice after total body roentgen irradiation. Acta Radiol [Ther] (Stockh) 44: 243-248

Distefano V (1964) Some remarks on the pharmacology of radioprotectant agents. Ann NY Acad Sci 114: 588-596

Distefano V, Korn PS, Leary DE (1960) The blood pressure effects of 3-aminopropyl-N'-methylisothiuronium bromide hydrobromide (APMT) in the cat. Fed Proc 19: 356

Distefano V, Klahn JJ, Leary DE (1962) The pharmacological effects of some radioprotective agents in mice. Radiat Res 17: 792-800

Doherty DG (1960) Chemical protection to mammals against ionizing radiation. In: Hollaender A (ed) Radiation protection and recovery. Pergamon, Oxford, pp 45-86

Doherty DG, Burnett WT (1955) Protective effect of S,β-aminoethylisothiuronium-Br-HBr and related compounds against X-radiation death in mice. Proc Soc Exp Biol Med 89: 312-314

Doherty D, Congdon CC (1959) Prolongation of homograft survival in AET-protected and isologous bone marrow-treated irradiated mice. Fed Proc 18: 216

Doherty DG, Burnett WT, Shapira R (1957) Chemical protection against ionizing radiation. II. Mercaptoalkylamines and related compounds with protective activity. Radiat Res 7: 13-21

Donaldson SS, Moskowitz PS, Evans JW, Fajardo LF (1984) Protection from radiation nephropathy by WR-2721. Radiat Res 97: 414-423

Doull J, Plzak V, Root M (1962) Protection against chronic radiation lethality in mice. Radiat Res 16: 578-579

Dubravsky NB, Hunter N, Mason K, Withers HR (1978) Dibutyryl cyclic adenosine monophosphate: effect on radiosensitivity of tumors and normal tissues in mice. Radiol 126: 799-802

Dunjic A, Maisin H, Maldague P (1957) Protection afforded by cysteamine against acute pulmonary syndrome in rats after x-irradiation. Arch Int Physiol Biochim 66: 22-28

Durand RE (1983) Radioprotection by WR-2721 in vitro at low oxygen tensions: implications for its mechanism of action. Br J Cancer 47: 387-392

Durkovsky J, Siracka-Vesela E (1958) Klinische Applikation von Cysteamin bei der Strahlungskrankheit. Neoplasma 5: 417-423

Ebel JP, Beck G, Keith G, Langendorff H, Langendorff M (1969) Study of the therapeutic effect on irradiated mice of substances contained in RNA preparations. Int J Radiat Biol 16: 201-209

Echols FS (1973) Normal and malignant tissue response to fractionated radiation exposure and the radioprotective durg WR-2721. Doctoral Dissertation. University of Florida, Gainesville, Florida, 1973. Cited by: Yuhas JM (1980a)

Ehling UH, Doherty DG (1962) AET protection of the reproductive capacity of irradiated mice. Proc Soc Exp Biol Med 110: 493-494

Eidus LK, Korystov YN, Kublik LN, Vexler AM (1982) Dependence of radioprotective effect of chemical modifying agents on their intracellular concentrations. Int J Radiat Biol 41: 625-632

Eldjarn L (1954) The conversion of cystinamine to taurine in rat, rabbit, and man. J Biol Chem 206: 483-490

Eldjarn L, Pihl A (1956) On the mode of action of x-ray protective agents. I. The fixation in vivo of cystamine and cysteamine to proteins. J Biol Chem 223: 341-352

Eldjarn L, Pihl A (1960) Mechanisms of protective and sensitizing action. In: Errera M, Forssberg A (eds) Mechanisms in radiobiology, vol II. Multicellular organisms. Academic, New York, pp 231-296

Eldjarn L, Pihl A, Shapiro B (1956) Cysteamine-cystamine: on the mechanism for the protective action against ionizing radiation. Proc 1st Intern Conf Peaceful Uses Atomic Energy, Geneva, 1955. 11: 335-342

Eltgen D, Koch R, Langendorff H (1961) Untersuchungen über einen biologischen Strahlenschutz 38. Mitt Der Einfluß von Cysteamin und Serotonin auf die Retikulozyten und den Eisenstoffwechsel bestrahlter Tiere. Strahlentherapie 114: 118-127

Ershoff BH, Brat V (1960) Failure of AET to protect against testes injury in the x-irradiated rat. Am J Physiol 198: 655-656

Faršatov MN (1986) Vlijanie radioprotektorov na effektivnost' lečenija kombinirovannych radiacionnych poraženij (Influence of radioprotectors on the therapeutic efficacy against combined radiation injuries). Voenno-medicinskij žurnal 7: 22-25

Finney DJ (1964) Probit analysis. A statistical treatment of the sigmoid response curve. University Press, Cambridge

Firket H, Lelièvre P (1966) Effet de la cystamine sur la respiration, la phosphorylation oxydative et l'ultrastructure des mitochondries du rat. Int J Radiat Biol 10: 403-415

Flemming K (1977) Some ideas concerning the mode of action of radioprotective agents. In: Locker A, Flemming K (eds) Radioprotection: chemical compounds-biological means. Birkhäuser, Basel, pp 79-86

Flemming K, Flemming C (1966) Die Wirkung von Pyrexal (Lipopolysaccharid aus Salm. abortus equi) auf die Überlebensrate nach Röntgenbestrahlung. Strahlentherapie 131: 150-159

Flemming K, Langendorff M (1965) Das Pro-Östrogen Chlortrianisen (Tace) als Strahlenschutzsubstanz. Strahlentherapie 128: 109-118

Floersheim GL, Floersheim P (1986) Protection against ionizing radiation and synergism with thiols by zinc aspartate. Br J Radiol 59: 597-602

Fowler JF, Kragt K, Ellis RE, Lindop PJ, Berry RJ (1965) The effect of divided doses of 15 MeV electrons on the skin response of mice. Int J Radiat Biol 9: 241-252

Francois J, Beheyt J (1955) Cataracte par rayons X et cystéamine. Ophthalmologica 130: 397-402

Gaugas JM (1982) Possible association of radioprotective and chemoprotective aminophosphorothioate drug activity with polyamine oxidase susceptibility. JNCI 69: 329-332

Gensicke F, Spode E, Venker P (1962) Die S^{35}-Verteilung und -Ausscheidung nach Injektion S^{35}-markierten Cystamins bei der Maus. Strahlentherapie 118: 561-569

Gerebtzoff MA, Bacq ZM (1954) Examen histopathologique de souris irradiées après injection de cystéamine. Experientia 10: 341-343

Gessner PK, Khairallah PA, Melsaac WM, Page IH (1960) The relationship between the metabolic fate and pharmacologic actions of serotonin, bufotenine and psilocybin. J Pharmacol 130: 126-133

Glick JH, Glover DJ, Weiler C, Blumberg A, Nelson D, Yuhas JM, Kligerman M (1982) Phase I clinical trials of WR-2721 with alkylating agent chemotherapy. Int J Radiat Oncol Biol Phys 8: 575-580

Glover DJ, Yuhas JM, Glick JH (1981) Stability of WR-2721 in solution. North Atlantic Council. AC/225 (Panel VII/ GEC) R/14. XIV. Meeting of Group of Experts on Chemoprophylaxis, Washington, DC, USA, Appendix II, p 9

Glover DJ, Glick JH, Weiler C, Fox K, Turrisi A, Kligerman MM (1986) Phase I/II trials of WR-2721 and cis-platinum. Int J Radiat Oncol Biol Phys 12: 1509-1512

Gray JL, Tew JT, Jensen H (1952) Protective effect of serotonin and paraaminopropiophenon against lethal doses of x-irradiation. Proc Soc Exp Biol Med 80: 604-607

Griem K, Weichselbaum RR, Umans RS, Gifford L, Little JB (1983) Work in progress: radioprotection of human breast cancer cells by elevation of intracellular cyclic AMP. Radiology 148: 289-290

Grigsby P, Maruyama Y (1981) Modification of the oral radiation death syndrome with combined WR-2721 and misonidazole. Br J Radiol 54: 969-972

Grigsby P, Maruyama Y (1982) Combined radiosensitization and radioprotection for oral cavity tumors: study with an oral cavity tumor model. Int J Radiat Oncol Biol Phys 8: 557-559

Gupta R, Uma Devi P (1986) Protection of mouse chromosomes against whole-body gamma irradiation by SH-compounds. Br J Radiol 59: 625-627

Gupta SM, Goyal PK, Dev PK (1981) Weight changes in mice after intrauterine treatment with MPG (2-mercaptopropionylglycine) against I^{131} irradiation. Experientia 37: 898-899

Haas E, Lorenz W (1960) Untersuchungen über die Wirkung der Strahlenschutzsubstanz β-Amino-äthylisothiuronium-chloridhydrochlorid (AET·Cl·HCl) auf die Strahlenempfindlichkeit des Yoshida-Sarkoms der Ratte. Strahlentherapie 112: 451-456

Hagen U, Ernst H, Langendorff H (1958) Untersuchungen über einen biologischen Strahlenschutz. 26. Mitt. Wirkung von Strahlenschutzsubstanzen auf die durch Röntgenstrahlen ausgelösten Stoffwechselveränderungen im Organismus. Strahlentherapie 107: 426-436

Hahn A, Lohmann W, Hillerbrand M, Deffner U (1975) Molecular mechanism of action of the radioprotective substance WR 2721. Radiat Environ Biophys 11: 265-269

Hajdukovic S, Hervé A, Vidovic V (1954) Diminution de radiosensibilité du rat adulte en hypothermie profonde. Experientia 10: 343-344

Hanna C, Colclough NV (1963) Toxicity and tolerance studies on AET. Arch Int Pharmacodyn Ther 142: 510-515

Hanna C, O'Brien JE (1963) Effect of AET on γ-ray radiation cataracts. Arch Int Pharmacodyn Ther 142: 198-205

Hanson WR, Ainsworth EJ (1985) 16,16-Dimethyl prostaglandin E_2 induces radioprotection in murine intestinal and hematopoietic stem cells. Radiat Res 103: 196-203

Harris JW, Meneses JJ (1978) Radioprotection of immunologically reactive T lymphocytes by WR-2721. Int J Radiat Oncol Biol Phys 4: 437-440

Harris JW, Phillips TL (1971) Radiobiological and biochemical studies of thiophosphate radioprotective compounds related to cysteamine. Radiat Res 46: 362-379

Harris JW, Power JA (1973) Diamide: a new radiosensitizer for anoxic cells. Radiat Res 56: 97-109

Hasegawa A, Landahl HD (1967) Studies on spleen oxygen tension and radioprotection in mice with hypoxia, serotonin, and p-aminopropiophenone. Radiat Res 31: 389-399

Healy JB (1960) A trial of cystamine in radiation sickness. Br J Radiol 33: 512-514

Heisler G (1969) Synthese und Untersuchung von N- und S-haltigen Polymeren als Strahlenschutzsubstanzen. Dissertation (Thesis), Naturw Fakultät der Phillips-Universität Marburg/Lahn

Hervé A (1957) L'irradiation des souris par or radioactif avec et sans protecteurs. Z Med Isotopenforsch 1: 128-131

Hervé A, Bacq ZM (1952a) Protection chimique contre le rayonnement X (essais thérapeutiques). J Radiol 33: 651-655

Hervé A, Bacq ZM (1952b) Modifications des effets du rayonnement x par certaines substances chimiques. Théorie, expériences. J Belge Radiol 35: 524-545

Heuwieser H (1954) Die Behandlung des Strahlenkaters mit Sulfhydrylkörpern und ihre Problematik. Strahlentherapie 95: 330-332

Hofmann D (1955) Über die lokale Wirksamkeit des Cysteins bei seiner Anwendung im aktiven Strahlenschutz. Strahlentherapie 96: 396-402

Hollcroft J, Lorenz E, Miller E, Congdon CC, Schweisthal R, Uphoff D (1957) Delayed effects in mice following acute total-body x irradiation: Modification by experimental treatment. JNCI 18: 615-640

Hope DB (1958) Radioprotective substances and hypothermia. Br J Radiol 31: 339

Hornsey S (1957) The effect of hypothermia on the radiosensitivity of mice to whole-body irradiation. Proc R Soc Lond [Biol] 147: 547-549

Hovestadt I, Ernst M, Mönig H, Fischer H (1983) The early effect of sublethal x-irradiation of phagocytic cells in mouse blood and the influence of cystamine as measured by chemiluminescence. Int J Radiat Biol 44: 563-573

Hunter N, Milas L (1983) Protection by S-2-(3-aminopropylamino)ethylphosphorothioic acid against radiation-induced leg contractures in mice. Cancer Res 43: 1630-1632

ICRU Report 30: Quantitative concepts and dosimetry in radiobiology. International Commission on Radiation Units and Measurements. Washington DC 1979

Ikebuchi M, Shinohara S, Kimura H, Morimoto K, Shima A, Aoyama T (1981) Effects of daily treatment with a radioprotector WR-2721 on Ehrlich's ascites tumors in mice: suppression of tumor cell growth and earlier death of tumor-bearing mice. J Radiat Res 22: 258-264

International Civil Defense Organization (1983) ABC self-protection kit in the Soviet Union. Bull Int Civil Defence Org 337: 8

Irie H, Yosihara H (1961) Influence of the radioprotective agents on the therapeutical effects of radiations for malignant tissues. Chemotherapia 3: 176-188

Ito H, Meistrich ML, Barkley HT, Thames HD, Milas L (1986) Protection of acute and late radiation damage of the gastrointestinal tract by WR-2721. Int J Radiat Oncol Biol Phys 12: 211-219

Jamieson D, van den Brenk HAS (1966) Studies of mechanisms of chemical radiation protection in vivo. III. Changes in fluorescence of intracellular pyridine nucleotides and modification by extracellular hypoxia. Int J Radiat Biol 10: 223-241

Jaskierowicz D, Genissel F, Roman V, Berleur F, Fatome M (1985) Oral administration of liposome-entrapped cysteamine and the distribution pattern in blood, liver and spleen. Int J Radiat Biol 47: 615-619

Jellum E (1965) Interaction of cystamine and cystamine derivatives with nucleic acids and nucleoproteins. Int J Radiat Biol 9: 185-200

Jones DP (1981) Hypoxia and drug metabolism. Biochem Pharmacol 30: 1019-1023

Kaplan WD, Lyon MF (1953) Failure of mercaptoethylamine to protect against the mutagenic effects of radiation. II. Experiments with mice. Science 118: 777-778

Kimura H, Yasui T, Aoyama T (1981) Modification of radiation sensitivity of cultured cells by pre- and postirradiation incubation with dibutyryl cyclic AMP. Radiat Res 85: 207-214

Kinnamon KE, Ketterling LL, Stampfli HF, Grenan MM (1980) Mouse endogenous spleen counts as a means of screening for anti-radiation drugs. Proc Soc Exp Biol Med 164: 370-373

Kligerman MM, Shaw MT, Slavik M, Yuhas JM (1980) Phase I clinical studies with WR-2721. In: Brady LW (ed) Radiation sensitizers. Masson, New York, pp 426-430

Kligerman MM, Glick JH, Turrisi AT, Glover D, Norfleet L, Gramkowski M (1984a) Clinical experiences with WR2721. In: Breccia A, Greenstock CL, Tamba M (eds) Advances on oxygen radicals and radioprotectors. Edizione Scientifiche-Lo Scarabeo, Bologna, pp 173-181

Kligerman MM, Glover DJ, Turrisi AT et al. (1984b) Toxicity of WR-2721 administered in single and multiple doses. Int J Radiat Oncol Biol Phys 10: 1773-1776

Knizner SA, Jacobs AJ, Lyon RC, Swenberg CE (1986) In vivo dephosphorylation of WR-2721 monitored by ^{31}P NMR spectroscopy. J Pharmacol Exp Ther 236: 37-40

Kobayashi J, Kitajima T, Katayama E, Kawamura F (1965) Biological protection mechanisms of radioprotectors on the mammalian cells. IV. Cytological effects of MEA on bone marrow cells and thymus cells. Tokushima J Exp Med 12: 35-39

Koch R (1962) Der Einfluß verschiedener Strahlenschutzstoffe auf die Strahlenempfindlichkeit von Tumoren. Nucl Med 2: 265-271

Koch R (1965) The radioprotective action of different radioprotectors for doses below 100 R. Prog Biochem Pharmacol 1: 427-431

Koch R (1967) Strahlenschutzsubstanzen und ihr Einfluß auf die Strahlenempfindlichkeit von Tumoren. Sonderbände zur Strahlentherapie (Strahlenbehandlung und Strahlenbiologie) 66: 338-344

Koch R, Schwarze W (1957) Toxikologische und chemische Untersuchungen an β-Aminoäthylisothiuronium-Verbindungen. Arzneimittelforschung 7: 576-579

Koch R, Seiter I (1964) Quantitative relations between doses of chemical protective agents and doses of x-irradiation. Nature 203: 984-985

Konermann G (1972) Die Dosis- und Phasenabhängigkeit von Cystamin im Verlaufe der Keimesentwicklung der Maus. Strahlentherapie 144: 96-116

Konermann G, Streffer C (1974) Histo-autoradiographic in-

vestigations on the distribution of injected 5-hydroxytryptamine (serotonin) in the brain and the pituitary of mice. Naunyn Schmiedebergs Arch Pharmacol 282: 349-365

Kovacs P, Hernadi F, Dezsi Z (1972) Effect of cysteine on haematopoietic tissues. Acta Biochim Biophys Acad Sci Hung 7: 67-74

Kulinskii VI, Lobyntsev KS, Krivenko ED (1977) Radioprotective effect of physiological concentrations of catecholamines and dibutyryl-cAMP on marrow stem cells. Dokl Akad Nauk SSSR 237: 1502-1505

Kumar A, Uma Devi P (1983) Chemical radiation protection of ovarian follicles of mice by MPG (2-mercaptopropionylglycine). J Nucl Med Allied Sci 27: 9-12

Kuna P (1979) Möglichkeiten des chemischen Strahlenschutzes im Falle einer Neutronenbestrahlung des Organismus (tschechisch). Vojenské Zdravotnické Listy 47: 237-244

Kuna P (1983) Duration and degree of radioprotection of WR-2721 in mice following its intraperitoneal, intramuscular and subcutaneous administration. Radiobiol Radiother 24: 357-364

Kuna P, Vodička I, Dostál M (1983a) Radioprotective and hemodynamic effects of 5-methoxytryptamine following different route of its administration in rats. Radiobiol Radiother 24: 365-376

Kuna P, Volenec K, Vodička I, Dostál M (1983b) Radioprotective and hemodynamic effects of WR-2721 and cystamine in rats: time course studies. Neoplasma 30: 349-357

Künkel H-A, Heckmann U (1958) Die Strahlenschutzwirkung von Cystein und Cysteamin bei reduziertem Stoffwechsel. Strahlentherapie 106: 256-259

Ladner H-A, Mitchell JS, King EA, Weisselberg R (1980) Das Verhalten einiger B-Vitamine beim Karzinompatienten während der Strahlentherapie. Strahlentherapie 156: 856-860

Langendorff H (1965) Grundlagen und Möglichkeiten eines biologisch-chemischen Strahlenschutzes bei äußerer Strahleneinwirkung. Arzneimittelforschung 15: 463-472

Langendorff H (1970) Zum Wirkungsmechanismus strahlenresistenzerhöhender Substanzen. Strahlentherapie 140: 428-432

Langendorff H, Catsch A (1956) Über die Schutzwirksamkeit des Cysteamins bei fraktionierter Ganzkörperbestrahlung von Mäusen. Strahlentherapie 101: 536-541

Langendorff H, Koch R (1955) Strahlenschaden und Narkose. Arzneimittelforschung 5: 677-680

Langendorff H, Koch R (1956) β-Aminoäthylisothiuronium als peroral wirksame Strahlenschutzsubstanz. Naturwissenschaften 43: 524-525

Langendorff H, Langendorff M (1971) Der Einfluß von adrenergen Rezeptoren-Blockern auf die Wirksamkeit von Strahlenschutzsubstanzen. Experientia 27: 1303-1304

Langendorff H, Langendorff M, Koch R (1958) Überprüfung des spezifischen Strahlenschutzeffektes der Cystein-Cysteamin-Gruppe unter Berücksichtigung eines Mercaptopyridinderivates. Strahlentherapie 107: 121-126

Langendorff H, Melching H-J, Rösler H (1960) Über den Anteil des Adenylsäuresystems und des Pyridoxal-5-phosphats am Strahlenschutzeffekt des Serotonins. Strahlentherapie 113: 603-609

Langendorff H, Melching H-J, Streffer C (1962) Zum Anteil des Adenylsäuresystems am Strahlenschaden des Säugetiers. Strahlentherapie 118: 341-347

Langendorff H, Langendorff M, Metzner R, Mönig H, Steinbach K-H, Temme W, Tumbrägel G (1971a) Radiobiological investigations with fast neutrons. II. The radioprotective action of different substances on male mice. Atom-kernenergie 18: 83-88

Langendorff H, Langendorff M, Steinbach KH, Weckesser J (1971b) Vergleichende Untersuchungen zur strahlenresistenzerhöhenden Wirkung von verschiedenen bakteriellen Lipopolysacchariden. Strahlentherapie 141: 214-220

Langendorff H, Langendorff M, Weckesser J, Steinbach KH (1971c) Die Wirkung einzelner Strukturkomponenten von bakteriellen Lipopolysacchariden auf die Strahlenempfindlichkeit der Maus. Strahlentherapie 141: 457-463

Langendorff H, Langendorff M, Mönig H (1974) Zum Problem der Induzierbarkeit einer erhöhten biologischen Strahlenresistenz durch chemische Stoffe bei einer Einwirkung schneller Neutronen. Strahlentherapie 147: 69-76

Langendorff M, Melching H-J, Langendorff H, Koch R, Jacques R (1957) Weitere Untersuchungen zur Wirkung zentralerregender und -dämpfender Pharmaka auf die Strahlenempfindlichkeit des Tieres. Strahlentherapie 104: 338-344

Latarjet R, Ephrati E (1948) Influence protectrice de certaines substances contre l'inactivation d'un bactériophage par les rayons X. C R Soc Biol (Paris) 142: 497-499

Lehnert S (1975) Modification of postirradiation survival of mammalian cells by intracellular cyclic AMP. Radiat Res 62: 107-116

Lehnert S (1979a) Modification of radiation response of CHO cells by methyl-isobutyl xanthine. 1. Reduction of D_0. Radiat Res 78: 1-12

Lehnert S (1979b) Modification of radiation response of CHO cells by methyl-isobutyl xanthine. 2. Increase in extrapolation number. Radiat Res 78: 13-24

Leonard A, Maisin JR (1964) Effect of 2-β-amino-ethylisothiourea (AET) against genetic damages induced by x-irradiation of male mice. Radiat Res 23: 53-62

Leonard A, Maisin JR, Mattelin G (1969) Effect of a mixture of chemical protectors against x-irradiation induced testis injury in mice. Strahlentherapie 138: 614-618

Liébecq-Hutter S, Bacq ZM (1958) Température interne de la souris après injection de radioprotecteurs. Arch Int Physiol Biochim 66: 469-471

Lipecka K, Domanski T, Daniszewska K et al. (1982) Lethal doses of ionizing radiation versus endogenous level of superoxide dismutase. Studia Biophys 89: 57-64

Liquier J, Fort L, Dai DN, Cao A, Taillandier E (1983) DNA protection by aminothiols: study of the cysteamine-DNA interaction by vibrational spectroscopy. Int J Biol Macromol 5: 89-93

Lohmann W, Momeni M, Nette P (1967) On the possible involvement of charge transfer complexes (redox systems) in radioprotection. Strahlentherapie 134: 590-594

Lowy RO, Baker DG (1973) Effect of radioprotective drugs on the therapeutic ratio for a mouse tumor system. Acta Radiol Ther Phys Biol 12: 425-433

Ludwig KH (1955) Sulfhydrylkörper in der gynäkologischen Strahlentherapie. Münch Med Wochenschr 1955: 823-824

Lüning KG, Frölén H, Nelson A (1961) The protective effect of cysteamine against genetic damages by x-rays in spermatozoa from mice. Radiat Res 14: 813-818

Maisin J, Moutschen J (1960) Chemical protection of the alimentary tract of whole-body X-irradiated mice. II. Chromosome breaks and mitotic activity. Exp Cell Res 21: 347-352

Maisin J, Maisin H, Dunjic A, Maldague P (1955a) Radiolésions cellulaires et tissulaires leurs conséquences et leurs réparations. J Belge Radiol 38: 394-429

Maisin J, Maisin H, Dunjic A, Maldague P (1955b) La radiobiologie comme méthode de travail en physiopathologie et en cancérologie expérimentale. Bull Schweiz Akad Med Wiss 10: 247-273

Maisin J, Dunjic A, Couvreur P (1964) Protective effects of cysteamine and cystamine on irradiated rats. J Belg Radiol 47: 755-771

Maisin JR (1969) Reduction of long term radiation lethality by mixtures of chemical protectors. Atomkernenergie 14: 226-228

Maisin JR, Lambiet-Collier M (1967) Influence of a mixture of radioprotectors on the cell renewal in the duodenum of X-irradiated mice. Int J Radiat Biol 13: 35-43

Maisin JR, Mattelin G (1967) Reduction in radiation lethality by mixtures of chemical protectors. Nature 214: 207-208

Maisin JR, Declève A, Gerber GB, Mattelin G, Lambiet-Collier M (1978) Chemical protection against the long-term effects of a single whole body exposure of mice to ionizing radiation. Radiat Res 74: 415-435

Maisin JR, Gerber GB, Lambiet-Collier M, Mattelin G (1980) Chemical protection against long-term effects of whole-body exposure of mice to ionizing radiation. III. The effects of fractionated exposure to C57Bl mice. Radiat Res 82: 487-497

Mandl AM (1959a) The effect of cysteamine on the survival on spermatogonia after x-irradiation. Int J Radiat Biol 1: 131-142

Mandl AM (1959b) The effect of β-mercaptoethylamine on the sensitivity of oocytes to x-irradiation. Proc R Soc Med 150: 72-77

Mayer SH, Patt HM (1953) Potentiation of cysteine protection against x-radiation by dinitrophenol or hypoxia. Fed Proc 12: 94-95

Mazur L (1985) Effects of AET, MEA, or 5-HT treatment before x-irradiation of pregnant C57B mice. Strahlentherapie 161: 433-435

Melching H-J (1965) The influence of serotonin on radiation effects in mammals. Curr Top Radiat Res 1: 93-137

Melching H-J, Streffer C (1966) Zur Beeinflussung der Strahlenempfindlichkeit von Säugetieren durch chemische Substanzen. Fortschr Arzneimittelforsch 10: 11-128

Melching H-J, Streffer C, Sauer H (1964) Untersuchungen über einen biologischen Strahlenschutz. 51. Der Einfluß der Splenektomie auf strahlenbedingte Veränderungen des Aminosäurestoffwechsels der Maus. Strahlentherapie 123: 571-599

Melville GS, Leffingwell TP (1962) Toxic and protective effects of AET upon normal and irradiated female rats. Br J Radiol 35: 563-571

Menard TW, Osgood TB, Clark JI, Spence AM, Steele JE, Krohn KA, Livesey JC (1986) Radioprotection against cataract formation by WR-77913 in gamma-irradiated rats. Int J Radiat Oncol Biol Phys 12: 1483-1486

Mendiondo OA, Connor AM, Grigsby P (1982) Toxicity and radiation protective effect of WR-77913 in BALB/c mice. Acta Radiol Oncol 21: 319-323

Mendiondo OA, Grigsby PW, Beach JL (1983) Radioprotection combined with hypoxic sensitization during radiotherapy of a solid murine tumor. Radiology 148: 291-293

Messerschmidt O (1977) Kombinationsschäden als Folge nuklearer Explosionen. In: Bundesamt für Zivilschutz (ed) Zivilschutz-Forschung, vol 5. Osang, Bad Honnef-Erpel, p 26

Messerschmidt O, Metzger E, Stevenson AFG (1978) Strahlenbiologische Untersuchungen mit schnellen Neutronen. Forschungsbericht aus der Wehrmedizin T/R750/R7500/41052

Mewissen DJ (1957) Action de la cystamine per os sur la survie des souris irradiées par le radiocobalt 60. Acta Radiol (Stockh) 48: 141-150

Mewissen DJ, Brucer M (1957) Late effects of gamma radiation on mice protected with cysteamine or cystamine. Nature 179: 201-202

Michel C, Fritz-Niggli H (1973) Die Beeinflussung der Fertilität der weißen Maus durch kleine Strahlenmengen und ein Hefepräparat (Bio-Strath). Radiol Clin Biol (Basel) 42: 222-233

Milas L (1984) Need for studies on factors that influence radioprotection of solid tumors by WR-2721. Int J Radiat Oncol Biol Phys 10: 163-165

Milas L, Hunter N, Reid BO (1982a) Protective effects of WR-2721 against radiation-induced injury of murine gut, testis, lung and lung tumor nodules. Int J Radiat Oncol Biol Phys 8: 535-538

Milas L, Hunter N, Reid BO, Thames HD (1982b) Protective effects of S-2-(3-aminopropylamino)ethylphosphorothioic acid against radiation damage of normal tissues and a fibrosarcoma in mice. Cancer Res 42: 1888-1897

Milas L, Hunter N, Ito H, Peters LJ (1984) Effect of tumor type, size, and endpoint on tumor radioprotection by WR-2721. Int J Radiat Oncol Biol Phys 10: 41-48

Millar JL, McElwain TJ, Clutterbuck RD, Wist EA (1982) The modification of melphalan toxicity in tumor bearing mice by S-2-(3-aminopropylamino)-ethylphosphorothioic acid (WR 2721). Am J Clin Oncol 5: 321-328

Mitznegg P (1973a) Die Wirkung der Strahlenschutzsubstanz auf die Leber weißer Mäuse in Abhängigkeit vom Lebensalter. Act Geront 3: 753-756

Mitznegg P (1973b) On the mechanism of radioprotection by cysteamine. II. The significance of cyclic 3',5'-AMP for the cysteamine-induced radioprotective effects in white mice. Int J Radiat Biol 24: 339-344

Modig HG (1973) Interaction of cysteamine with the thiol and disulphide groups in deoxyribonucleoproteins. Biochem Pharmacol 22: 1623-1631

Modlin RK, Morris JMcL (1961) The effect of certain chemical and physical agents on the radiation sensitivity of mouse tumors. Cancer 14: 117-125

Mondovi B, Tentori L, Marco C de, Cavallini D (1962) Distribution of cysteamine-^{35}S in the subcellular particles of the organs of the rat. Int J Radiat Biol 4: 371-378

Mönig H, Seiter I, Kofler E (1975) Untersuchungen über den Einfluß von Aminopropyl-aminoäthylthiophosphat auf die Radioeisenutilisation nach Ganzkörperbestrahlung von Mäusen im subletalen Dosisbereich. Strahlentherapie 150: 44-50

Mönig H, Sedlmeier H, Messerschmidt O (1986) The influence of the solvent on the toxicity and radioprotective effect of thiophosphate WR 2721 in mice. Strahlenth Onkol 162: 134-137

Nakken KF (1965) Radical scavengers and radioprotection. In: Ebert M, Howard A (eds) Current topics in radiation research, vol I. North Holland, Amsterdam, pp 49-92

Nikolov I, Rogozkin VD, Pantev T, Chertkov KS, Dikovenko EA, Davidova SA (1986) Protection of monkeys against prolonged gamma-irradiation. Strahlenth Onkol 162: 200-204

Oldfield DG, Doull J, Plzak V (1965a) Chemical protection against 440-MeV protons in mice pretreated with mercaptoethylamine (MEA) or p-aminopropiophenone (PAPP). Radiat Res 26: 12-24

Oldfield DG, Doull J, Plzak V (1965b) Chemical protection against absorber-moderated protons. Radiat Res 26: 25-31

Paget GE (1965) Toxicity tests: a guide for clinicians. In: Herrick AD, Cattell M (eds) Clinical testing of new drugs. Revere, New York, pp 31-39

Pant RD, Ghose A (1981) Effect of MPG and AET on erythrocytes in peripheral blood after gamma irradiation. Int J Radiat Biol 40: 227-228

Patt HM, Tyree EB, Straube RL, Smith DE (1949) Cysteine protection against X-irradiation. Science 110: 213-214

Patt HM, Smith DE, Tyree EB, Straube RL (1950) Further studies on modification of sensitivity to X-rays by cysteine. Proc Soc Exp Biol Med 73: 18-21

Patt HM, Clark JW, Vogel HH (1953) Comparative protective effect of cysteine against fast neutron and gamma irradiation in mice. Proc Soc Exp Biol Med 84: 189-193

Petkau A (1978) Radiation protection by superoxide dismutase. Photochem Photobiol 28: 765-774

Petkau A, Chelack WS, Pleskach SD, Meeker BE, Brady CM (1975) Radioprotection of mice by superoxide dismutase. Biochem Biophys Res Commun 65: 886-893

Petkau A, Chelack WS, Pleskach SD (1976) Protection of post-irradiated mice by superoxide dismutase. Int J Radiat Biol 29: 297-299

Petkau A, Chelack WS, Pleskach SD (1978) Protection by superoxide dismutase of white blood cells in x-irradiated mice. Life Sci 22: 867-882

Phillips TL (1977) Chemical modification of radiation effects. Cancer 39: 987-999

Phillips TL (1980) Rationale for initial clinical trials and future development of radioprotectors. In: Brady LW (ed) Radiation sensitizers. Masson, New York, pp 321-329

Phillips TL, Kane L, Utley JF (1973) Radioprotection of tumor and normal tissues by thiophosphate compounds. Cancer 32: 528-535

Pirie A, Lajtha LG (1959) Possible mechanism of cysteine protection against radiation cataract. Nature 184: 1125-1127

Poulsen SS, Szabo S (1977) Mucosal surface morphology and histological changes in the duodenum of the rat following administration of cysteamine. Br J Exp Pathol 58: 1-8

Power JA, Goldstein LS, Harris JW (1974) A test of the 'mixed-disulphide' hypothesis of cysteamine radioprotection. Int J Radiat Biol 26: 91-96

Prasad KN (1972) Radioprotective effect of prostaglandin and an inhibitor of cyclic nucleotide phosphodiesterase on mammalian cells in culture. Int J Radiat Biol 22: 187-189

Pratt NE, Sodicoff M, Liss J, Davis M, Sinesi M (1980) Radioprotection of the rat parotid gland by WR-2721: morphology at 60 days postirradiation. Int J Radiat Oncol Biol Phys 6: 431-435

Prütz WA, Mönig H (1987) On the effect of oxygen or copper (II) in radiation-induced degradation of DNA in the presence of thiols. Int J Radiat Biol 52: 677-682

Prütz WA, Siebert F, Butler J, Land EJ, Menez A, Montenay-Garestier T (1982) Charge transfer in peptides. Intramolecular radical transformations involving methionine, tryptophan and tyrosine. Biochim Biophys Acta 705: 139-149

Purdie JW, Inhaber ER, Schneider H, Labelle JL (1983) Interaction of cultured mammalian cells with WR-2721 and its thiol, WR-1065: implications for mechanisms of radioprotection. Int J Radiat Biol 43: 517-527

Rajewsky B (1956) Strahlendosis und Strahlenwirkung, 2nd edn. Thieme, Stuttgart

Rasey JS, Nelson NJ, Mahler P, Anderson K, Krohn KA, Menard T (1984) Radioprotection of normal tissues against gamma rays and cyclotron neutrons with WR-2721: LD$_{50}$

studies and ^{35}S-WR-2721 biodistribution. Radiat Res 97: 598-607

Rasey JS, Krohn KA, Menard TW, Spence AM (1986) Comparative biodistribution and radioprotection studies with three radioprotective drugs in mouse tumors. Int J Radiat Oncol Biol Phys 12: 1487-1490

Révész L (1985) The role of endogenous thiols in intrinsic radioprotection. Int J Radiat Biol 47: 361-368

Révész L, Modig H (1965) Cysteamine-induced increase of cellular glutathione-level: a new hypothesis of the radioprotective mechanism. Nature 207: 430-431

Révész L, Bergstrand H, Modig H (1963) Intrinsic non-protein sulfhydryl levels and cellular radiosensitivity. Nature 198: 1275-1277

Riklis E (1983) DNA repair as a probe of radiosensitivity and radioprotection. In: Nygaard OF, Simic MG (eds) Radioprotectors and anticarcinogens. Academic, New York, pp 363-380

Riklis E, Green M, Prager A (1983) Cell survival and DNA repair capacity affected by the radioprotective compound WR-2721. In: Broerse JJ, Barendsen GW, Kal HB, van der Kogel AJ (eds) Proceedings of the Seventh International Congress of Radiation Research. Martinus Nijhoff, Amsterdam, pp B6-25

Ringsdorf H (1967) Makromolekulare Verbindungen als potentielle Schutzstoffe gegen ionisierende Strahlen. Strahlentherapie 132: 627-635

Ringsdorf H (1978) Synthetic polymer drugs. Midl Macromol Monogr 5 (Polymer Delivery Syst) pp 197-225

Rix-Montel MA, Mallet G, Costa A, Vasilescu D (1982) Influence of ionizing radiations on DNA in the presence of sulfur containing radioprotectors. 2. Cysteamine protection against γ-radiations. Studia Biophysica 89: 205-211

Roberts JM (1970) Cysteamine protection against lethal and growth-inhibiting effects of prenatal X and gamma irradiation. Teratology 3: 319-324

Rojas A, Stewart FA, Denekamp J (1982a) Experimental radiotherapy with WR-2721 and misonidazole. Int J Radiat Oncol Biol Phys 8: 527-530

Rojas A, Stewart FA, Denekamp J (1982b) Interaction of radiosensitizers and WR-2721. 1. Modification of skin radioprotection. Br J Cancer 45: 684-693

Rojas A, Stewart FA, Denekamp J (1983) Interaction of misonidazole and WR-2721. 2. Modification of tumour radiosensitization. Br J Cancer 47: 65-72

Roman V, Bocquier F, Leterrier F, Fatome M (1982) Action radioprotectrice de la cystéamine incorporée dans des liposomes administrés par voie orale à la souris. C R Acad Sci [III] 295: 191-193

Ross WM, Peeke J (1986) Radioprotection conferred by dextran sulfate given before irradiation in mice. Exp Hematol 14: 147-155

Rostock RA, Stryker JA, Abt AB (1980) Evaluation of high vitamin E as a radioprotective agent. Radiology 136: 763-766

Rousanov AM, Novoselova GS (1962) The protective effects of aminoethylisothiorea against ionizing radiations and its toxic effects according to the age of the white mice. Ann Radiol 5: 225-241

Rowley DA, Halliwell B (1982) Superoxide-dependent formation of hydroxyl radicals in the presence of thiol compounds. FEBS Letters 138: 33-36

Rugh R, Clugston H (1955) Protection of the fetus against X-radiation death. Radiat Res 3: 342

Rugh R, Clugston H (1956) Protection of mouse fetus against x-irradiation death. Science 123: 28-29

Rugh R, Grupp E (1960a) Protection of the embryo against the congenital and lethal effects of x-irradiation. Part 1. Atompraxis 6: 143-148

Rugh R, Grupp E (1960b) Protection of the embryo against the congenital and lethal effects of x-irradiation. Part 2. Atompraxis 6: 209-217

Rugh R, Wolff J (1957) Evidence of some chemical protection of the mouse ovary against x-irradiation sterilization. Radiat Res 7: 184-189

Rupkey AK, Gold R, Rugh R, Wang SC (1963) Effects of drugs alone, or combined with refrigeration, in protecting female mice against x-irradiation-induced sterility. Radiat Res 19: 88-103

Saini MR, Uma Devi P (1980) Radiation protection of pronormoblasts and normoblasts by 2-mercaptopropionylglycine (MPG). Experientia 36: 448-449

Saltzstein EC, Rimm AA, Bortin MM (1970) Skin allograft survival in mice after radioprotection with sodium S-(2-aminoethyl) phosphorothioate and supralethal total body X-irradiation. Transplant 10: 451-453

Sasaki MS, Matsubara S (1977) Free radical scavenging in protection of human lymphocytes against chromosome aberration formation by gamma-ray irradiation. Int J Radiat Biol 32: 439-445

Savkovic NV, Radivojević DV, Hajduković SI, Radotić MM, Popović SH, Karanović J (1961) Histological analysis of testes in locally and whole body irradiated infantile rats. Bull Inst Nuclear Sci "Boris Kidrich" 12: 145-148

Sawyer DT, Valentine JS (1981) How super is superoxide? Acc Chem Res 14: 393-400

Schaeffer J, Schellenberg KA, Seymore CH, Schultheiss TE, El-Mahdi AM (1986) Radioprotective effect of polyethylene glycol. Radiat Res 107: 125-135

Schuman VL, Clement JJ, Levitt SH, Song CW (1983) Skin radioprotection by 5-thio-D-glucose. Radiat Res 93: 326-331

Schwartz EE, Shapiro B (1961) Radiation-induced changes in the gastrointestinal function of mice and their prevention by chemical means. Radiology 77: 83-90

Schwartz EE, Shapiro B, Kollmann G (1964) Selective chemical protection against radiation in tumor-bearing mice. Cancer Res 24: 90-96

Schweikert C-H, Ladner H-A (1964) Die Operationsmöglichkeit nach Röntgenganzkörperbestrahlung und Strahlenschutzgaben. Langenbecks Arch Klin Chir 308: 65-69

Sedlmeier H, Messerschmidt O (1980) Schutzeffekt von WR 2721 bei strahlen- und kombinationsgeschädigten Mäusen. Strahlentherapie 156: 572-578

Sedlmeier H, Metzger E, Jentzsch U, Weitzenegger E (1981) Schutzeffekt von Aminopropylaminoäthylthiophosphat (WR 2721) bei Neutronen-, Gamma- oder Röntgenbestrahlung von Mäusen. Strahlentherapie 157: 685-691

Shapira R, Doherty DG, Burnett WT (1957) Chemical protection against ionizing radiation. III. Mercaptoalkylguanidines and related isothiuronium compounds with protective activity. Radiat Res 7: 22-34

Shapiro B, Schwartz EE, Kollmann G (1963) The mechanism of action of AET. IV. The distribution and chemical forms of 2-mercaptoethyl-guanidine and bis(2-guanidoethyl) disulfide in protected mice. Radiat Res 18: 17-30

Shapiro NI, Tolkacheva EN, Spasskaya IG, Fedoseeva VM (1960) An experimental study of the possibility of using defense substances in radiation therapy of malignant neoplasm. Vopr Onkol 6: 71-79

Shaw LM, Turrisi AT, Glover DJ, Bonner HS, Norfleet AL,

Weiler C, Kligerman MM (1986) Human pharmacokinetics of WR-2721. Int J Radiat Oncol Biol Phys 12: 1501-1504

Shirai M (1941) Studies on the treatment of radiation sickness by glutathione (in Japanese). Nagoya Igakkai Zasshi 54: 183-192

Sigdestad CP, Connor AM, Scott RM (1975a) The effect of S-2-(3-aminopropylamino)ethylphosphorothioic acid (WR-2721) on intestinal crypt survival. I. 4 MeV X-rays. Radiat Res 62: 267-275

Sigdestad CP, Connor AM, Scott RM (1975b) Chemical radiation protection of the intestinal epithelium by MEA and its thiophosphate derivative. Int J Radiat Oncol 1: 53-60

Sigdestad CP, Connor AM, Scott RM (1976) The effect of S-2-(3-aminopropylamino)ethylphosphorothioic acid (WR-2721) on intestinal crypt survival. II. Fission neutrons. Radiat Res 65: 430-439

Sigdestad CP, Grdina DJ, Connor AM, Hanson WR (1986) A comparison of radioprotection from three neutron sources and ^{60}Co by WR-2721 and WR-151327. Radiat Res 106: 224-233

Smith DE (1959) Protection of the irradiated ground squirrel by cysteine. Radiat Res 10: 335-338

Smith DE (1960) Failure of cysteine given postirradiation to protect the hibernating ground squirrel. Radiat Res 12: 79-80

Smith WW, Alderman IM, Gillespie RE (1958) Hematopoietic recovery induced by bacterial endotoxin in irradiated mice. Am J Physiol 192: 549-556

Smith WW, Cornfield J, Luskus C, Miller C (1965) β-Mercaptoethylamine effect on radiation dose-response characteristics of granulocyte and lymphocyte counts. Radiat Res 26: 146-158

Smith WW, Budd RA, Cornfield J (1966) Estimation of radiation dose-reduction factor for β-mercaptoethylamine by endogenous spleen colony counts. Radiat Res 27: 363-368

Snyder SL, Walden TL, Patchen ML, MacVittie TJ, Fuchs P (1986) Radioprotective properties of detoxified lipid A from Salmonella minnesota R595. Radiat Res 107: 107-114

Sodicoff M, Conger AD (1983) Radioprotection of the rat parotid gland by WR-2721 and isoproterenol and its modification by propranolol. Radiat Res 94: 97-104

Sodicoff M, Conger AD, Trepper P, Pratt NE (1978a) Short-term radioprotective effects of WR-2721 on the rat parotid glands. Radiat Res 75: 317-326

Sodicoff M, Conger AD, Pratt NE, Trepper P (1978b) Radioprotection by WR-2721 against long-term chronic damage to the rat parotid gland. Radiat Res 76: 172-179

Sodicoff M, Ziskin MC, Conger AD (1986) Effect of WR-2721 on fetal development in the rat. Radiat Res 107: 49-57

Soltysiak-Pawluczuk D, Bitny-Szlachto S (1976) Effects of ionizing radiation and cysteamine (MEA) on activity of mouse spleen adenyl cyclase. Int J Radiat Biol 29: 549-553

Song CW, Clement JJ, Levitt SH (1977) Cytotoxic and radiosensitizing effects of 5-thio-D-glucose against hypoxic tumor cells. Radiology 123: 201-205

Spence AM, Krohn KA, Edmondson SW, Steele JE, Rasey JS (1986) Radioprotection in rat spinal cord with WR-2721 following cerebral lateral intraventricular injection. Int J Radiat Oncol Biol Phys 12: 1479-1482

Starkie CM (1961) The effect of cysteamine on the survival of foetal germ cells after irradiation. Int J Radiat Biol 3: 609-617

Stevenson AFG, Mönig H, Weckesser J (1981) Radioprotective and haemopoietic effects of some lipopolysaccharides from Rhodospirillaceae species in mice. Experientia 37: 1331-1332

Stewart FA, Rojas A (1982) Radioprotection of mouse skin by WR-2721 in single and fractionated treatments. Br J Radiol 55: 42-47

Stewart FA, Rojas A, Denekamp J (1983) Radioprotection of two mouse tumors by WR-2721 in single and fractionated treatments. Int J Radiat Oncol Biol Phys 9: 507-513

Storaasli JP, Rosenberg SA, Friedell HL (1953) The effect of cysteine on the radiosensitivity of rat lymphosarcoma. Cancer 6: 1244-1247

Straube RL, Patt HM, Smith DE, Tyree EB (1950) Influence of cysteine on the radiosensitivity of Walker rat carcinoma 256. Cancer Res 10: 243-244

Streffer C (1971) Biochemical post-irradiation changes and radiation indicators. In: Biochemical indicators of radiation injury in man. Int Atomic Energy Agency, Vienna, pp 11-32

Streffer C (1974) DNA-synthesis in the spleen of mice after whole-body X-irradiation and its modification by 5-hydroxytryptamine and NADH. Int J Radiat Biol 25: 425-435

Streffer C (1976) Biologische Grundlagen der Strahlentherapie. In: Scherer E (ed) Strahlentherapie. Radiologische Onkologie. Springer, Berlin Heidelberg New York, pp 172-230

Streffer C (1977) Studies on the mechanism of 5-hydroxytryptamine in radioprotection of mammals. In: Locker A, Flemming K (eds) Radioprotection. Chemical compounds - biological means. Birkhäuser, Basel, pp 71-77

Streffer C, Flügel M (1972) Die Steigerung der Strahlenresistenz von Mäusen nach der intracerebralen Injektion von 5-Hydroxytryptamin. Biophysik 8: 342-351

Streffer C, Flügel M (1973) Inhibitors of monoamine oxidase and the radioprotective effect of 5-hydroxytryptamine. Int J Radiat Biol 23: 323-335

Sugahara T, Horikawa M, Hikita M, Nagata H (1977) Studies on a sulfhydryl radioprotector of low toxicity. In: Locker A, Flemming K (eds) Radioprotection. Chemical compounds - biological means. Birkhäuser, Basel, pp 53-61

Sullivan MF (1964) Measurement of intestinal damage from neutrons, x-rays, or nitrogen mustard treatment. Radiat Res 22: 241-242

Sullivan MF, Thompson RC, Crosby AC (1964) The influence of fractionated X-irradiation on the intestine of rats protected by cysteine and partial-body shielding. Radiat Res 23: 551-563

Supek Z, Randic M, Lovasen Z (1961) Radioprotective action of some indolealkylamines. Int J Radiat Biol 4: 111-112

Sverdlov AG (1974) Biological effect of neutrons and chemical protection (in Russian). Leningrad, 1974. (cited by Kuna 1979)

Sverdlov AG, Mozzhukhin AS, Pavlova LM, Nikanorova NG (1969) Chemical protection from injury by neutrons. Radiobiologiia Translat Ser 9: 706-710

Sverdlov AG, Mosjuchin AS, Pavlova LM, Nikanorova NG, Postnikov LN (1973) Chemical protection against neutron irradiation. Adv Radiat Res 2: 611-617

Swynnerton NF, Huelle BK, Mangold DJ (1986) A method for the combined measurement of ethiofos and WR-1065 in plasma: application to pharmacokinetic experiments with ethiofos and its metabolites. Int J Radiat Oncol Biol Phys 12: 1495-1499

Szumiel I (1981) Intrinsic radiosensitivity of proliferating mammalian cells. Adv Radiat Biol 9: 281-321

Tabachnik F, Blackburn P, Peterson CM, Cerami A (1982) Protein binding of N-2-mercaptoethyl-1,3-diaminopropane via mixed disulfide formation after oral administration of WR 2721. J Pharmacol Exp Ther 220: 243-246

Takahashi I, Mitsuhashi N, Itoh J et al. (1984) Clinical use of radioprotector Amifostine (YM-08310). Nippon Act Radiol 44: 1396-1404

Takahashi I, Nagai T, Miyaishi K, Maehara Y, Niibe H (1986) Clinical study of the radioprotective effects of amifostine (YM-08310, WR-2721) on chronic radiation injury. Int J Radiat Oncol Biol Phys 12: 935-938

Takeda A, Yonezawa M, Katoh N (1981) Restoration of radiation injury by ginseng. I. Responses of x-irradiated mice to ginseng extract. J Radiat Res 22: 323-335

Tanaka Y, Sugahara T (1970) Chemical radiation protection in man as revealed by chromosome aberrations in peripheral lymphocytes. J Radiat Res 11: 166-168

Tanaka Y, Sugahara T (1980) Clinical experiences of chemical radiation protection in tumor radiotherapy in Japan. In: Brady LW (ed) Radiation sensitizers. Masson, New York, pp 421-425

Tarbell NJ, Rosenblatt M, Amato DA, Hellman S (1986) The effect of N-acetylcysteine inhalation on the tolerance to thoracic radiation in mice. Radiother Oncol 7: 77-80

Teličenas A, Karosene E (1973) Materialien über die klinische Erprobung des Präparates Cystafos (Mononatriumsalz der Beta-Aminoäthylthiophosphorsäure). Radiobiol Radiother (Berl) 6: 671-676

Tempel K, Wulfius-Kock M, Winkle J, Schmerold I (1982) Semikonservative und reparaturbedingte Desoxyribunukleinsäuresynthese in Rattenthymozyten unter dem Einfluß einiger radioprotektiver und radiosensibilisierender Verbindungen. Strahlentherapie 158: 112-122

Tikhomirova MV, Rogozkin VD (1979) Effect of adenosin diphosphate against radiation delivered at high and low dose rates. Radiobiologiia 19: 241-245. Translat Ser Radiobiol (Washington) 19: 94-99

Till JE, McCulloch EA (1961) A direct measurement of the radiation sensitivity of normal mouse bone marrow cells. Radiat Res 14: 213-222

Travis E, Luca AM de, Fowler JF, Padikal TM (1982) The time course of radioprotection by WR 2721 in mouse skin. Int J Radiat Oncol Biol Phys 8: 843-850

Travis EL (1984) The oxygen dependence of protection by aminothiols: implications for normal tissues and solid tumors. Int J Radiat Oncol Biol Phys 10: 1495-1501

Trevan JW (1927) The error of determination of toxicity. Proc R Soc Lond [Biol] 101: 483-514

Trocha PJ, Catravas GN (1983) Effect of radioprotectant WR 2721 on cyclic nucleotides, prostaglandins, and lysosomes. Radiat Res 94: 239-251

Trush MA, Mimnaugh EG, Gram TE (1982) Activation of pharmacologic agents to radical intermediates. Implications for the role of free radicals in drug action and toxicity. Biochem Pharmacol 31: 3335-3346

Twentyman PR (1983) Modification by WR 2721 of the response to chemotherapy of tumours and normal tissues in the mouse. Br J Cancer 47: 57-63

Uma Devi P, Jagetia GC (1981) Effect of 2 mercaptopropionylglycine (MPG) on thyroid function in sublethally irradiated mice. Experientia 37: 312-313

Uma Devi P, Kumar S (1981) Radioresponse of peripheral blood and its modification by MPG (2-mercaptopropionyl-

glycine) in mice. I. Erythrocytes. Strahlentherapie 157: 63-65

Uma Devi P, Saini MR, Saharan BR, Bhartiya HC (1979) Radioprotective effect of 2-mercaptopropionylglycine on the intestinal crypt of Swiss albino mice after cobalt-60 irradiation. Radiat Res 80: 214-220

Upton A, Doherty DG, Melville GS (1959) Chemical protection of the mouse against leukemia induction by roentgen rays. Acta Radiol [Ther] (Stockh) 51: 379-384

Urso P, Congdon CC, Doherty DG, Shapira R (1958) Effect of chemical protection and bone marrow treatment on radiation injury in mice. Blood 13: 665-676

Utley JF, Phillips TL, Kane LJ, Wharam MD, Wara WM (1974) Differential radioprotection of euoxic and hypoxic mouse mammary tumors by a thiophosphate compound. Radiology 110: 213-216

Utley JF, Marlowe C, Waddell WJ (1976a) Distribution of ^{35}S-labeled WR-2721 in normal and malignant tissues of the mouse. Radiat Res 68: 284-291

Utley JF, Phillips TL, Kane LJ (1976b) Protection of normal tissues by WR2721 during fractionated irradiation. Int J Radiat Oncol Biol Phys 1: 699-703

Utley JF, King R, Giansanti JS (1978) Radioprotection of oral cavity structures by WR-2721. Int J Radiat Oncol Biol Phys 4: 643-647

Utley JF, Quinn CA, White FC, Seaver NA, Bloor CM (1981) Protection of normal tissue against late radiation injury by WR-2721. Radiat Res 85: 408-415

Vachtel VS, Sinenko LF (1963) In: Arndt D, Ritter M (eds) Handbuch für medizinische Fragen des Strahlenschutzes. Militärverlag der Deutschen Demokratischen Republik, Berlin, 1979, p 91 f.

van den Brenk HAS (1969) The oxygen effect in radiation therapy. Curr Top Radiat Res 5: 197-251

van den Brenk HAS, Haas M (1961) Studies of mechanisms of chemical radiation protection in vivo. I. 5-hydroxytryptamine in relation to effect of antimetabolites, antagonista and releasing agents. Int J Radiat Biol 3: 73-94

van der Meer C, Bekkum DW van (1961) A study on the mechanism of radiation protection by 5-hydroxytryptamine and tryptamine. Int J Radiat Biol 4: 105-110

van Lancker J, Maisin J (1953) Le rôle de la β-mercaptoéthylamine dans la protection et la régénération des tissus chez les animaux irradiés in toto. C R Soc Biol (Paris) 147: 2057-2059

Vasilescu D, Rix-Montel MA (1980) Interaction of sulfur-containing radioprotectors with DNA: a spectrophotometric study. Physiol Chem Phys 12: 51-55

Vasilescu D, Broch H, Rix-Montel MA (1986) Mechanism of aminothiol radioprotectors action at the molecular level. J Mol Struct (Theochem) 134: 367-380

Virsik RP, Harder D (1982) Effect of L-cysteine on the dose-effect relationship for chromosome aberrations in irradiated human lymphocytes. Int J Radiat Biol 42: 211-214

Vladimirov VG, Golubentsev DA, Gusev IV, Libikova NI (1979) Effects of different radioprotective agents on cyclic adenosine-3'-5'-monophosphate and DNA synthesis in the spleen. Radiobiologiia 19: 114-116

Vogel HH, Hasegawa AT, Wang RI (1969) Comparative protection by a combination treatment in mice irradiated with fission neutrons or x-rays. Radiat Res 39: 57-67

von Sallmann L, Munoz CM (1952) Further efforts to influence x-ray cataract by chemical agents. Arch Ophthalmol 47: 21-22, 48: 276-291

Vos O (1980) Chemical protection against x-irradiation by an orally administered compound. Radiat Environ Biophys 17: 329

Vos O, Budke L, Vergroesen AJ (1962) Protection of tissue-culture cells against ionizing radiation. I. The effect of biological amines, disulfide compounds and thiols. Int J Radiat Biol 5: 543–557

Wang SC, Kuskin S, Rugh R (1959) Protective action of cysteinamine (β-mercaptoethylamine) against x-irradiation-induced sterility in CF₁ male mice. Proc Soc Exp Biol Med 101: 218–221

Washburn LC, Rafter JJ, Hayes RL, Yuhas JM (1976) Prediction of the effective radioprotective dose of WR-2721 in humans through an interspecies tissue distribution study. Radiat Res 66: 100–105

Wenz W (1956) Die Wirkung von Becaptan (β-Merkaptoaethylamin) auf bestrahltes und unbestrahltes Ehrlich-Karzinom der weißen Maus. Oncologia 9: 310–315

Westland RD, Holmes JL, Mouk ML, Marsh DD, Cooley RA, Dice JR (1968) N-substituted S-2-aminoethyl thiosulfates as antiradiation agents. J Med Chem 11: 1190–1201

Withers HR, Elkind MM (1970) Microcolony survival assay for cells of mouse intestinal mucosa exposed to radiation. Int J Radiat Biol 17: 261–267

Withers HR, Hunter N, Barkley HT, Reid BO (1974) Radiation survival and regeneration characteristics of spermatogenic stem cells of mouse testis. Radiat Res 57: 88–103

Woollam DHM, Millen JW (1958) The influence of cysteamine on the teratogenic action of X-radiation. Nature 182: 1801

Wright EA, Shewell J (1965) Modification of radiation "cerebral death" by hypoxia. Nature 208: 904–905

Yarmonenko SP, Avrunina GA, Shashkov VS, Govorun RD (1962) A study of biological protection against irradiation with high energy protons. Radiobiologiia Translat Ser 2: 188–192

Yuhas JM (1970) Biological factors affecting the radioprotective efficiency of S-2-[3-aminopropylamino]ethylphosphorothioic acid (WR-2721). LD$_{50(30)}$ doses. Radiat Res 44: 621–628

Yuhas JM (1972a) Radioprotective and toxic effects of S-2-(3-aminopropylamino)ethylphosphorothioic acid (WR-2721) on the development of immunocompetent cells. Cell Immunol 4: 256–263

Yuhas JM (1972b) Improvement of lung tumor radiotherapy through differential chemoprotection of normal and tumor tissue. JNCI 48: 1255–1257

Yuhas JM (1973) Radiotherapy of experimental lung tumors

in the presence and absence of a radioprotective drug, S-2-(3-aminopropylamino)-ethylphosphorothioic acid (WR-2721). JNCI 50: 69–78

Yuhas JM (1979) Differential protection of normal and malignant tissues against the cytotoxic effects of mechlorethamine. Cancer Treat Rep 63: 971–976

Yuhas JM (1980a) On the potential application of radioprotective drugs in solid tumor radiotherapy. In: Sokol GH, Maickel RP (eds) Radiation–drug interactions in the treatment of cancer. Wiley, New York, pp 113–135

Yuhas JM (1980b) Active versus passive absorption kinetics as the basis for selective protection of normal tissues by S-2-(3-aminopropylamino)-ethylphosphorothioic acid. Cancer Res 40: 1519–1524

Yuhas JM (1980c) A more general role for WR-2721 in cancer therapy. Br J Cancer 41: 832–834

Yuhas JM (1983) Efficacy testing of WR-2721 in Great Britain or everything is black and white at the Gray lab. Int J Radiat Oncol Biol Phys 9: 595–598

Yuhas JM, Storer JB (1969a) Chemoprotection against three modes of radiation death in the mouse. Int J Radiat Biol 15: 233–237

Yuhas JM, Storer JB (1969b) Differential chemoprotection of normal and malignant tissues. JNCI 42: 331–335

Yuhas JM, Proctor JO, Smith LH (1973) Some pharmacologic effects of WR-2721: their role in toxicity and radioprotection. Radiat Res 54: 222–233

Yuhas JM, Yurconic M, Kligerman MM, West G, Peterson DF (1977) Combined use of radioprotective and radiosensitizing drugs in experimental radiotherapy. Radiat Res 70: 433–443

Yuhas JM, Spellman JM, Culo F (1980a) The role of WR-2721 in radiotherapy and/or chemotherapy. Cancer Clin Trials 3: 211–216

Yuhas JM, Spellman JM, Jordan SW, Pardini MC, Afzal SM, Culo F (1980b) Treatment of tumours with the combination of WR-2721 and cisplatinum(II)dichlorodiamine or cyclophosphamide. Br J Cancer 42: 574–585

Yuhas JM, Davis ME, Glover D, Brown DQ, Ritter M (1982) Circumvention of the tumor membrane barrier to WR-2721 absorption by reduction of drug hydrophilicity. Int J Radiat Oncol Biol Phys 8: 519–522

Znamenskii VV, Zherebchenko PG, Terekhov AV, Dikovenko EA, Zeitunyan KA, Semenov LF (1975) Radioprotective effectiveness of cystaphos in the case of intragastric administration in experiments on monkeys. Radiobiologiia 15: 79–82. Translat Ser Radiobiology (Washington) 15: 102–107

Subject Index

Medical Radiology

Diagnostic Imaging and Radiation Oncology

Edited by:
L. W. Brady, Philadelphia;
M. W. Donner, Baltimore;
H.-P. Heilmann, Hamburg;
F. Heuck, Stuttgart

J. H. Anderson, Johns Hopkins University, Baltimore, MD (Ed.)

Innovations in Diagnostic Radiology

With contributions by numerous experts
Foreword by M. W. Donner, W. R. Brody, F. Heuck
1989. XIII, 212 pp. 144 figs., some in colour. Hardcover
ISBN 3-540-19093-7

The current status and future applications of various diagnostic imaging procedures are discussed in this volume. The main focus is on research and development, in particular the potential they hold for further development and clinical application in various specialties. Those concerned with the academic growth of radiology as a discipline, such as basic scientists, young clinicians in training, and chairmen responsible for department planning, will be especially interested in the trends presented here.

The subject matter includes magnetic resonance imaging, interventional radiology, ultrasound, image analysis and management, positron emission tomography, and research training programs.

The volume emphasizes the multidisciplinary direction of medical imaging; both basic scientists and clinicians in a wide variety of medical research areas should be aware of this tendency and the opportunities which will be available in the future.

R. R. Dobelbower, Jr., Medical College of Ohio, Toledo (Ed.)

Gastrointestinal Cancer

Radiation Therapy

Foreword by L. W. Brady and H.-P. Heilmann
1989. XIV, 298 pp. 81 figs. 77 tabs. Hardcover ISBN 3-540-50505-9

The intricate role of radiation therapy in the management of GI cancer is comprehensively reviewed in this timely volume. It offers unique coverage of the entire gastrointestinal tract, from the esophagus to the anus. Although the work focuses clearly on the radiotherapeutic management of tumors of the GI tract, reviews of anatomy, epidemiology, and other pertinent topics are given. The volume is generally organized by disease site, but extensive special sections amplify important aspects related to GI cancer. These include the radiographic evaluation of GI malignancy and tumor markers, an up-to-date review of chemotherapeutic treatment of gut cancers, and patient follow-up. Edited by an internationally known leader in the field of radiotherapy, this volume represents a compilation of years of research and experience utilizing today's technology and state-of-the-art techniques.

Springer-Verlag
Berlin Heidelberg New York
London Paris Tokyo Hong Kong